Disaster Communication and Early Warning Systems

Disaster Communication and Early Warning Systems

Preeti Banzal

Anil K Gupta

2018

Daya Publishing House®

A Division of

Astral International Pvt. Ltd.

New Delhi – 110 002

ISBN: 9789388173605 (Int. Edn.)

Published by : **Daya Publishing House®**
A Division of
Astral International Pvt. Ltd.
– ISO 9001:2015 Certified Company –
4736/23, Ansari Road, Darya Ganj
New Delhi-110 002
Ph. 011-43549197, 23278134
E-mail: info@astralint.com
Website: www.astralint.com

Digitally Printed at : **Replika Press Pvt. Ltd.**

ACKNOWLEDGEMENTS

This book is an outcome of appreciations and encouragements from many people who rated the research treatise by the first author (PB) on 'Disaster Communication and Early Warning Systems' as a noble work, worth extending to larger audience specifically to the scholars and practitioners. Authors are deeply grateful, for all the support and inspiration in mentoring the research work by the first author, to Prof. Pardeep Sahni, Head of the Faculty of Public Administration at School of Social Sciences in the Indira Gandhi Open University, New Delhi. Authors mention special record of thanks to Shri B. H. Anil Kumar, Executive Director of NIDM, for all positive spirits, motivation and encouragements.

Constant encouragements from exemplary leaders in the field of disaster management in India, Dr. P K Mishra, Addl. Principal Secretary to Hon'ble Prime Minister of India; Mr. R K Jain and Mr. Kamal Kishore, Member(s) National Disaster Management Authority of India are duly acknowledged for being a source of energy and motivation tothe Author (AKG). The Author is also grateful to Mr. Sanjeev Jindal, Joint Secretary (Disaster Management) in Government of India Ministry of Home Affairs, Prof. Santosh Kumar and Prof. Chandan Ghosh Professors of NIDM; for their cooperation and facilitation. All time support of present Chairman of University Grants Commission and formerly Vice-chancellor of BHU, HSGU and DAU, Prof. D. P. Singh is worth special mention.

Dr. Shirish Ravan of United Nations Office for Outer Space Affairs, Vienna; Dr. Sanjay Srivastava of United Nations Economic and Social Commission for Asia Pacific, Bangkok; Dr. Muralee Thummarukuddy of the United Nations Environment Programme (Disaster Management Branch) Geneva, Florian Bemmerlein-Lux, Senior Advisor-GIZ at Ifanos C&P Germany; Prof. Rajib Shaw (Keio University, Japan, and former Executive Director UN-IRDR, Beijing), Dr. Akhilesh Gupta, Advisor and Head – Large and Strategic Programmes at DST-Govt. of India, Dr. K J Ramesh,

Director General of the India Meteorology Department; as professional colleagues and personal friends have always been kind and supportive during the endeavor. Ms. Sreeja Nair, Dr. Sushma Guleria, Mr. Ashish Panda, Mr. Shashikant Chopde and Prof. Shiraz A Wajih, deserve special thanks. Critical insights from Dr. P G Dhar Chakrabarti, former ED NIDM/Secretary NDMA, and Former Distinguished Fellow TERI, and Dr. R K Bhandari, Member of NDMA Advisory Committee, former Director CBRI, and Prof. J P Gupta, Former Professor IIT Kanpur, Director Gujarat Institute of Energy, Director Rajiv Gandhi Petroleum University, have been specially important.

Authors are deeply obliged to their family members for their patience, cooperation and encouragements especially to Mr. Sanjeev Banzal, Ms. Srishti Banzal, Mr. Siddharth Banzal, Mr. Sanat Jain, Mrs. Sushila Jain, Mrs. Sushila Banzal (Husband, Daughter, Son, Father, Mother and Mother-in-law, respectively, of PB), and Dr. Alka Gupta, Ms. Vanya Badal Gupta, Mr. Vibhu Badal Gupta, Dr. D.L. Gupta and Mrs. Urmila Devi (Wife, Daughter, Son, Father and Mother, respectively, of AKG); without their support this work would not have been possible.

Last but not the least, authors are thankful for the cooperation and support from Mr. Anil Mittal, and the editorial team of Astral International Publishing for timely production and release of the book.

Preeti Banzal
Anil Kumar Gupta

FOREWORD

Since the first International Early Warning Conference held in Potsdam, Germany in 1998, the world has seen significant improvements in early warning systems for all types of disasters. The Potsdam conference recognized that forecasting and warning of hazardous events by itself does not constitute an early warning "system". An early warning system is not merely a scientific and technical process of issuing alerts. The working group on early warning systems established under the International Decade for Natural Disaster Reduction recognized that an effective early warning system must be made up of four inter-linked sub-systems: a warning sub-system for hazards monitoring and forecasting; a risk information sub-system, which generates impact scenarios for impending hazardous events; a preparedness sub-system, which delineates the actions required by different actors in response to the warning; and a communication sub-system, which ensures the timely communication of early warning.

This systems view of early warning remains as relevant today as it was two decades ago when it was first proposed. The Sendai Framework for Disaster Risk Reduction adopted in 2015 also includes a specific global target on access to early warning. This target cannot be achieved unless we place adequate emphasis on all the four sub-systems mentioned above.

As is evident from the title this book deals more specifically with the fourth sub-system dealing with communication and dissemination of early warning. The two authors bring complementary strengths and expertise to this topic. They have provided a welcome addition to the growing body of literature in this area of work.

I hope that both researchers and practitioners working on disaster risk management issues will find the book useful.

Kamal Kishore
Member
National Disaster Management Authority
New Delhi, India

FOREWORD

Rightly communicated early warnings prevent natural disasters from becoming human tragedies!

In their quest towards faster development and economic progress, Asian countries are adding complexity to the hazard risk scenarios and the need of risk projections and early warning of disaster incidence. The new set of international frameworks like Sendai Framework of Disaster Risk Reduction, Paris Climate Agreement, and 2030 Agenda for Sustainable Development are showing a way and seeking commitment to build resilient societies. When it comes to implementing the targets set up by these international frameworks, the role of disaster related communication and early warning attains central importance.

The World is witnessing a reform in the way disasters are managed after adopting Hyogo Framework for Action in 2005 and renewed focus on reducing risks after adoption of the Sendai Framework for Disaster Risk Reduction in 2015. In parallel, information and communication technologies (ICT) are revolutionizing the way disaster related information is gathered, processed and communicated, reaching the beneficiary on the last mile.

The paradigm shift in disaster management is leading to more proactive, organized, integrated approach to reducing losses and damages from hazard events. Effective risk management depends on prudent understanding of anticipatory risk scenarios and advance communication to all those who matter in containing or responding to a disaster situation. A critical challenge is in integrating disaster risk information from various sources, combining it with precise early warning and its dissemination to ensure safety and sustainability of humankind and its resources. ICTs are a powerful tool to build resilience of individuals, communities and societies at large.

The thematic context in this book, titled 'Disaster communication and early warning systems', is not just limited to emergency response. It reveals a much broader scope of ICT such as dissemination of science-based methodologies and tools to record disaster risk and share relevant data and statistics, as well as to strengthen disaster risk modelling, assessment, mapping, monitoring and multi-hazard early warning systems. The book provides in-depth discussion for the benefit of the disaster management community on broadband access technologies, broadband-enabled services and ICT applications that offer new opportunities for transparent e-governance, interaction among people, sharing knowledge and expertise and empowering people and transforming lives. These technologies have a crucial role to play in disaster risk reduction by highlighting risk areas, vulnerabilities and potentially affected populations by producing geographically referenced information. A number of ICT tools namely, radio, television, broadband, satellite radio and phones, fixed and wireless phones, emails, short message service (SMS), computer network hardware and software, websites, portals, earth observation systems, geographic information systems, global navigation systems, disaster and climate change early warning systems, Machine to machine (M2M) communication, Internet of Things (IOT), and Big Data Analysis System Integration can play a significant role in all phases of disasters.

The book also encompasses a broad spectrum of disaster management paradigm shift, interlinkages of climate change adaptation and disaster management, prediction and early warnings, communication technologies and emergency communication vis-à-vis ICT as enabler of disaster risk management. It provides insight on disaster knowledge management, Public Protection and Disaster Relief (PPDR), M2M communication and common alert protocol for early warnings. The book navigates the reader through case studies on information crowd sourcing with free and open source software (FOSS) based systems viz "Sahana", Emergency.lu, Disaster Inventory System, DesInventar (Sistema de Inventario de Desastres) etc. and highlights the role of these technological innovations in all phases of disaster management.

The authors are well recognized professionals in their fields and their integrated effort has resulted in providing a truly comprehensive account of disaster management and the role of ICT by keeping the focus on disaster communication and early warning. I am sure the book will be a good reference book for the disaster management community as well as the technologists involved in telecommunication and ICT.

Dr. Shirish Ravan

United Nations Office for Outer Space Affairs

Dr. Shirish Ravan leads the United Nations Platform for Space based Information for Disaster Management and Emergency Response (UN-SPIDER), the programme administered by the United Nations Office for Outer Space Affairs.

PREFACE

Preparing for extreme events and managing disaster risk effectively requires technical skills of experts and agencies as well as new technologies, however, it is even more important that people on the Last Mile are getting informed and prepared. Consequently, reducing the vulnerability of people and communities exposed to extreme events and climate change is essential. Improving disaster communication and early warning systems is key in this context.

The importance of reducing human vulnerability through effective disaster communication and preparedness including effective early warning systems that reach people on the Last Mile can be underscored by looking at past events. While an earthquake in Haiti (2010) with a moment magnitude of 7 caused approximately 230,000 fatalities, a nearly similar event – an earthquake with a moment magnitude of 8.8 struck Chile and caused approximately 500 fatalities. These significant differences cannot sufficiently be explained by the physical event (earthquake). Rather the different levels of vulnerability are key factors that explain the different levels of impact seen in both countries. The differential vulnerability depends on specific human, social, economic, institutional and environmental factors and conditions that make exposed communities vulnerable to natural hazards and extreme events (see IPCC 2014). While it is difficult in some cases to reduce the hazard frequency (e.g. storm frequency), it is possible to reduce human vulnerability for example through the improvement of risk and disaster communication and education as well as early warning systems that must go beyond just technical details, but also need to actively involve people on the Last Mile in order to ensure anticipated responses.

In this context, I am very pleased to support the work of Dr. Preeti Banzal from the Ministry of Communications and Information Technology, and Dr. Anil K. Gupta from the National Institute of Disaster Management. The book "Disaster Communication and Early Warning Systems" provides new insights on factors that enable effective disaster risk management particularly in the context of India. In addition, the book encompasses a systematic overview of different components of disaster risk communication – including future trends in communication. In the first part the volume also explores risk and disaster trends in India and explains the paradigm shift in disaster management in the past decades and synergies between disaster risk reduction and sustainable development. Consequently the topic of disaster risk management is also linked to the Sustainable Development Goals as outlined by the United Nations.

Overall, the book is an important read for the scientific community, but also for practitioners interested in new development trends and different concepts of communication and disaster early warning tools.

Prof. Dr.-Ing. habil. Jörn Birkmann
Director of the IREUS
(Institute of Spatial and Regional Planning),
University of Stuttgart, Germany,
Member of the Scientific Committee of IRDR
(Integrated Research on Disaster Risk) and
Coordinating Lead Author of the Sixth Assessment Report
of the IPCC (Intergovernmental Panel on Climate Change)

CONTENTS

ABBREVIATIONS

AWS: Automatic Weather Stations
BPR: Bottom Pressure Recorders
BS: Base Station
BSC: Base Station Controller
BSNL: Bharat Sanchar Nigam Limited
BTS: Base Station Terminal
CAG: The Comptroller and Auditor General of India
CBRN: Chemical, Biological, Radiological and Nuclear
CDMA: Code Division Multiple Access
CMC: Crisis Management Committee
CMG: Crisis Management Group
CMTS: Cellular Mobile Telephone Services
CRF: Calamity Relief Fund
CRPF: Central Reserve Police Force
CSC: Common Service Centers
CWC: Central Water Commission
DAMA: Demand Assigned Multiple Access
DAT: Distress Alert Transmitter
DCRB: District Crime Records Bureau
DDMA: District Disaster Management Authority
DIT: Department of Information Technology
DM: Disaster Management
DM Act: Disaster Management Act

DMS: Disaster Management Support
DRM: Disaster Risk Management
DRDO: Defence Research and Development Organisation
DRR: Disaster Risk Reduction
DOT: Department of Telecommunications
DOS: Department of Space
DSS: Decision Support System
DTR: Diurnal Temperature Range
DWDM: Dense Wave Division Multiplex Technology
EOC: Emergency Operations Centre
EPON: Ethernet Passive Optical Network
ETSI: European Telecommunication Standardization Institute
FC: Finance Commission
FCC: Federal Communications Commission
FDMA: Frequency-Division Multiple Access
FEMA: Federal Emergency Management Agency
FTS: Fixed Telephone Services
GEO: Geostationary Earth Orbit
GIS: Geographical Information Service
GMPCS: Global Mobile Personal Communication by Satellite
GMDSS: Global Maritime Distress and Safety System
GPON: Gigabit Passive Optical Networks
GSI: Geological Survey of India
GSM: Global System for Mobile Communication
GTS: Global Telecommunication System
HPC: High Powered Committee
HFA: Hyogo Framework for Action
HLC: High Level Committee
ICDM: Inter-Governmental Committee on Disaster Management
ICT: Information and Communication Technologies
IDNDR: International Decade for Natural Disaster Reduction
IEC: Information, Education and Communication
IMG: Inter-Ministerial Group
IMD: India Meteorological Department
INCOIS: Indian National Centre for Ocean Information Services
IRIS: Consortium of Incorporated Research Institutions in Seismology

ISDN: Integrated Services Digital Network
ISDR: International Strategy for Disaster Reduction
ISP: Internet Service Provider
ISRO: Indian Space Research Organisation
ITU: International Telecommunication Union
LAN: Local Area Network
LBSNAA: Lal Bahadur Shastri National Academy of Administration
LEO: Low Earth Orbit
MHz: Mega Hertz
MHA: Ministry of Home Affairs
MPLS: Multiprotocol Label Switching
MSC: Mobile Switching Centre
NADAMS: National Agricultural Drought Assessment and Management System
NCCF: National Calamity Contingency Fund
NCMC: National Crisis Management Committee
NDMO: National Disaster Management Office
NDMF: National Disaster Mitigation Fund
NDPMC: National Disaster Prevention and Mitigation Committee
NDRF: National Disaster Response Force/National Disaster Response Fund
NCRB: National Crime Records Bureau
NCRMP: National Cyclone Risk Management Project
NDMA: National Disaster Management Authority
NEC: National Executive Committee
NEMA: National Emergency Management Authority
NEOC: National Emergency Operations Centre
NGO: Non-Governmental Organisation
NIDM: National Institute of Disaster Management
NKN: National Knowledge Network
NMEICT: National Mission on Education through ICT
NMS: Network Management Systems
NNP: National Numbering Plan
NRSA: National Remote Sensing Agency
NRSC: National Remote Sensing Centre
NSDC: National Seismological Database Centre

NSDG: National e-Governance Service Delivery Gateway
NSN: National Seismological Network
NTP: National Telecom Policy
OFC: Optical Fiber Communication
OFDA: Office of Foreign Disaster Assistance
PMRTS: Public Mobile Radio Terminal Service
POP: Point of Presence
PON: Passive Optical Networks
POLNET: Police Telecommunication Network
PPDR: Public Protection and Disaster Relief
PSTN: Public Switched Telephone Network
RTSMN: Real Time Seismic Monitoring Network
RTSP: Regional Tsunami Advisory Service Providers
SASE: Snow Avalanche Study Establishment
SCPC: Single Channel Per Carrier
SCMC: State Crisis Management Committee
SCRB: State Crime Records Bureau
SDMA: State Disaster Management Authority
SDMF: State Disaster Mitigation Fund
SDRF: State Disaster Response Fund
SEC: State Executive Committee
SFDR: State Fund for Disaster Response
SSDG: State e-Governance Service Delivery Gateway
SWAN: State Wide Area Network
TDMA: Time Division Multiple Access
TRAI: Telecom Regulatory Authority of India
UHF: Ultra High Frequency
UN: United Nations
UNDP: United Nations Development Programme
UNFCCC: United Nations Framework Convention On Climate Change
UNHCR: United Nations High Commission for Refugees
UNISDR: United Nations International Strategy for Disaster Risk Reduction
USA: United States of America
USAID: United States Agency for International Development
USOF: Universal Service Obligation Fund
UTs: Union Territories

VHF: Very High Frequency
VOIP: Voice Over Internet Protocol
VSAT: Very Small Aperture Terminals
WLL: Wireless in Local Loop
WMO: World Meteorological Organization

Chapter 1

INTRODUCTION AND OVERVIEW OF DISASTER MANAGMENT

Disaster Definitions

United Nations Development Programme (UNDP, 2004), defines disaster as *"a social crisis situation occurring when a physical phenomenon of natural, socio-natural or anthropogenic origin negatively impacts vulnerable populations, causing intense, serious and widespread disruption of the normal functioning of the affected social unit".* United Nations International Strategy for Disaster Reduction (UNISDR) defines disaster as *"a serious disruption of the functioning of a community or a society at any scale due to hazardous events interacting with conditions of exposure, vulnerability and capacity, leading to one or more of the following: human, material, economic and environmental losses and impacts"*[1]. As per *the* Disaster Management Act of India (2005) "disaster" *means a catastrophe, mishap, calamity or grave occurrence in any area, arising from natural or manmade causes, or by accident or negligence which results in substantial loss of life or human suffering or damage to, and destruction of, property, or damage to, or degradation of, environment, and is of such a nature or magnitude as to be beyond the coping capacity of the community of the affected area".*

UNISDR defines Hazard as *"a process, phenomenon or human activity that may cause loss of life, injury or other health impacts, property damage, social and economic disruption or environmental degradation".*[2]

UNISDR defines Vulnerability as *"the conditions determined by physical, social, economic and environmental factors or processes which increase the susceptibility of an individual, a community, assets or systems to the impacts of hazards".*[3] The Disaster risk as defined by UNISDR is " *The potential loss of life, injury, or destroyed or damaged assets which could occur to a system, society or a community in a specific period of time, determined probabilistically as a function of hazard, exposure, vulnerability and capacity".*[4]

International federation of Red Cross and Red Crescent Society defines Disaster Management as " *the organization and management of resources and responsibilities for dealing with all humanitarian aspects of emergencies"*[5]. Therefore Disaster management may be considered as the holistic process of risk reduction, vulnerability reduction and reduction of losses due to a disaster. Therefore it encompasses Risk and Vulnerability Assessment, Risk Reduction, Preparedness, Mitigation, Response, Relief and Recovery.

Disaster Trends: Global Scenario

Globally, "*natural disasters have caused damages of nearly $4 trillion over the past 30 years. In smaller developing countries, such disasters can wash away decades of development gains and impede social and economic progress for years to come. And when population growth is combined with rapid urbanization, and climate change, the frequency and intensity of these events will continue to rise, further outpacing humanitarian assistance and derailing sustainable development goals."*[6] These figures are eye-opener for all and provides sufficient trigger to create awareness about impact of disasters on economy of a country and specifically to its impact on developing countries. Normally damages are measured on the basis of direct losses incurred due to disaster. However they normally do not point to long-term micro and macro socio-economic losses suffered by the community and the region.

In developing world climate stresses are compounded by rapid urbanization, underinvestment in basic infrastructure, and dysfunctional urban land markets pushing the poor to settle in the most dangerous areas. These factors can turn cities into hotbeds of exposure to natural hazards. Developing countries are therefore exposed to large fiscal impacts on public expenditure, as governments shoulder an increasing financial responsibility for post-disaster recovery and reconstruction. For example in the Solomon Islands, the earthquake and tsunami of 2007 caused losses estimated at 95 per cent of the national budget. The urbanization of disasters, with frequent flooding, rising complexity, and greater cross-regional impacts, calls for urgent actions.[7]

Total economic loss from natural calamities and man-made disasters across the world, including Nepal earthquake and heat wave in India, was at $ 37 billion in the first half of 2015, Swiss Re[8] said in a report. Total economic losses from all disasters, including both natural and man-made events, were USD 92 billion in 2015 (vs USD 113 billion in 2014). Around USD 80 billion were due to natural catastrophes, with the earthquake in Nepal causing the most damage. Global economic losses were well below the previous 10-year annual average of USD 192 billion. Of the USD 37 billion global insured losses, USD 28 billion were attributed to natural catastrophes, about the same as in 2014. The biggest insured loss of the year an estimated property loss of between USD 2.5 billion and USD 3.5 billion was caused by two large explosions at the Port of Tianjin in China in August. The Report further wrote that economic losses from all events in Asia were close to USD 38 billion. The earthquake in Nepal was the biggest disaster of the year globally, killing close to 9000 people, the largest loss of life in a single event[9,10]

Disaster Trends: Regional Scenario

Asia is the most disaster-prone region in the world. The region has borne brunt of the physical and economic damages of natural disasters. Statistics provide hunting reminder that no country is without risk of being hit by natural disasters. Floods, earthquakes and cyclones have wreaked havoc across Asia, killing people, wiping out homes, livelihoods and leaving economies in distress. The region occupies 30 per cent of the world's land mass, but 40 per cent of the world's disasters occurred in the region in the past decade, resulting in a disproportionate 80 per cent of the world's disaster deaths. During the last 10 years, according to the Centre for Research on the Epidemiology of Disasters, majority of deaths in Asia were caused by earthquakes. Floods have killed 17,442 people while as cyclones were responsible for 13,368 deaths in the region. Earthquakes have led to massive loss of human lives in China, Pakistan, Indonesia and India. Floods and cyclones have killed more people in Thailand and the Philippines.[11] From floods to earthquake to other natural disasters, the past century has seen devastating disasters. Each disaster has left behind death and destruction and huge economic loses.

Past century has seen many devastating disasters. Each disaster leaves behind a very big trail of economic loses apart from physical and material losses. And according to the Emergency Events Database (EM-DAT) on international disasters from the Centre for Research on the Epidemiology of Disasters, the occurrence of all natural disasters has risen about 600 per cent in the last 60 years.[12]

To put the financial losses for one year in perspective, in 2012 alone Asia lost US$15 billion due to natural disasters, according to data released by the Center for Research on the Epidemiology of Disasters (CRED).[13] This is a drop from 2011, when the region recorded a staggering $300 billion loss mostly due to Japan's tsunami and Thailand's floods. Disasters have taken a heavy toll on the region's economy in the last 100 years. Particularly floods have caused more financial damage than any other natural disasters.

An estimation of economic losses due to various disasters in Asia is shown in Figure 1.1.

As per facts documented in report "The rise of natural disasters in Asia and the Pacific: Learning from ADB's experience" [14] Asia and the Pacific are exceedingly vulnerable to natural disasters. The Asia Pacific are facing natural disasters which are four times more likely to affect people in the region than those in Africa, and 25 times more likely than those in Europe. One climate change vulnerability index indicates that all seven cities globally classified as "extremely risky" are in Asia: Dhaka, Manila, Bangkok, Yangon, Jakarta, Ho Chi Minh, and Kolkata. As per ibid report Asia accounts for almost half of the estimated global economic cost close to

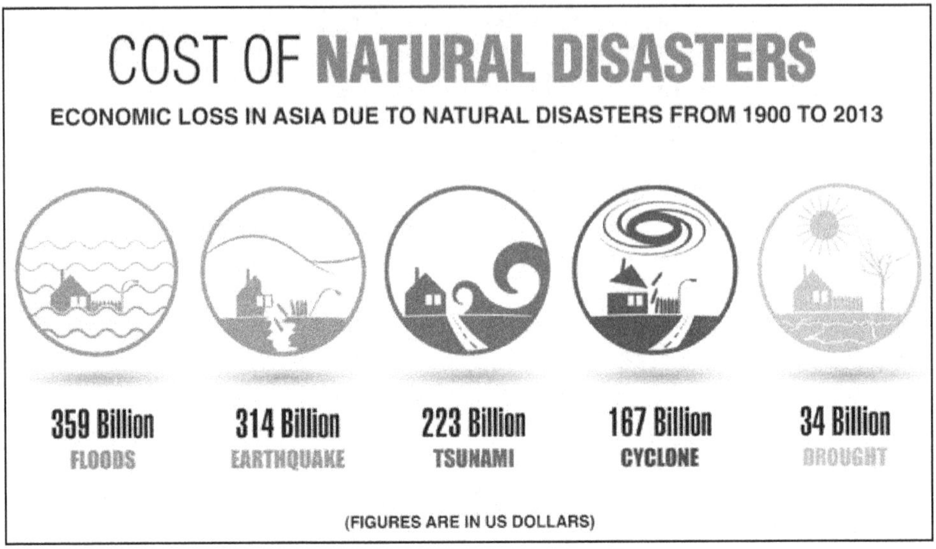

Figure 1.1: Cost of Natural Disasters.
(*Source*: CRED-EMDAT. Retrieved from http://ejap.org/environmental-issues-in-asia/natural-disasters-asia.html).

$1 trillion caused by natural disasters since the early 1990s. Increasing unplanned urbanisation due to sharply rising populations in some of Asian countries are forcing people to live in high risk and vulnerable areas. The only ways to restrain this trend is better disaster preparation and disaster risk reduction.

An earthquake of magnitude 9.0 occurred on 11th March 2011 in the Pacific Ocean off the coast of Japan's Tohoku region. The quake lasted for several minutes and shook the ground as far away as western Japan. A half hour later, a tsunami of unprecedented force broke over 650 kilometers of coastline, toppling sea walls and other defenses, flooding more than 500 km of land, and washing away entire towns and villages. The devastation left around 20,000 people dead or missing, with most of the deaths caused by drowning. The tsunami leveled 130,000 houses and severely damaged 260,000 more. About 270 railway lines ceased operation immediately following the disaster, and 15 expressways, 69 national highways, and 638 prefectural and municipal roads were closed. The areas worst hit were the Fukushima, Iwate, and Miyagi prefectures.[15]

When disasters engulf industrial stock, such as a power plant, it prompts a chain reaction of consequences beyond the initial disaster zone. This occurred in Japan in 2011 when the Japan earthquake and tsunami triggered a nuclear accident at Fukushima Daiichi Power Plant, spreading alarm throughout the region. "Compound disaster is much harder to respond to," says ADB's Groff. "Increasing regional integration has been a key driver of economic growth in the region, but it has also diffused the impact of disasters across multiple provincial and international boundaries." [16]

The incidence of 2011 Japan earthquake indicates that a developed and well prepared country like Japan had to bow down before a devastating calamity. This clearly indicates that natural disasters cannot be completely prevented and, therefore, communities have to live with certain level of risk. Though it is not feasible to have complete control over nature and put an end to its devastating activities, efforts need to be made to minimize effect of disasters on human lives, livestock, infrastructure and environment by managing disaster risk and emergencies in an planned and effective way. There are methods and guidelines described and developed by various organisations, tirelessly working on the subject, to make disasters easier to combat with. If proper measures are taken with right intentions in pre-disaster phase, substantive losses can be minimised in the event of a disaster. If preventive measures are already drafted in a systematic way; disaster management strategies are developed in a participative manner; community is made empowered to respond; number and sufferings of the victims can be reduced. Consequences of effective pre-disaster preparedness and efficient disaster response pave ways for effective post-disaster recovery and rehabilitation measures.

Disaster Trends: Indian Scenario

Within Asia, 24 per cent of deaths due to disasters occur in India because of its Size, population and vulnerability (Planning Commission, 2008). Each year India suffers disaster losses of US$1 billion, according to World Bank studies. According to National Policy on Disaster Management in India, India is vulnerable, in varying degrees, to a large number of natural as well as man-made disasters. 58.6 per cent of the landmass is prone to earthquakes of moderate to very high intensity; over 40 million hectares (12 per cent of land) is prone to floods and river erosion; of the 7,516 km long coastline, close to 5,700 km is prone to cyclones and tsunamis; 68 per cent of the cultivable area is vulnerable to drought and hilly areas are at risk from landslides and avalanches.[17] India's topography, geo-climatic conditions as well as the prevalence of socio-economic vulnerability among the marginalised and weaker sections of the population make it one of the most disaster prone countries in the world. Further, during the last almost five decades, vulnerability to disasters, both natural as well as man-made, have been increasing due to rapid and unplanned urbanisation, high population density poor compliance to industrial and environmental safety laws, India is also vulnerable to Chemical, Biological, Radiological and Nuclear (CBRN) disasters. Besides, natural disasters, anthropogenic hazards or human-made hazards can result in the form of a human-made disaster. It results in huge loss of life and property. It further affects a person's mental, physical and social well-being. AMRI hospital Fire incidence is the example of what can be called a Human induced negligence converting to a disastrous situation.[18] In view of India's high vulnerability profile, the recurrent phenomena of a range of Geophysical as well as hydro-meteorological hazards impact millions across the country leaving behind a trail of heavy loss of lives, property and livelihoods. India's topography, geo-climatic conditions as well as the prevalence of socio-economic vulnerability among the marginalised and weaker sections of the population make it one of the most disaster prone countries in the world.

India is highly vulnerable to natural disasters on account of its unique geo-climatic conditions, coupled with ecological and socio-economic settings. Due to India's high vulnerability profile, the recurrent phenomena of a range of Geophysical as well as hydro-meteorological hazards impact millions across the country leaving behind a trail of heavy loss of lives, property and livelihoods. India's topography, geo-climatic conditions as well as the prevalence of socio-economic vulnerability among the marginalized and weaker sections of the population make it one of the most disaster prone countries in the world. Further, during the last almost five decades, vulnerability to disasters, both natural as well as man-made, have been increasing due to rising population, haphazard urbanization, structural development in high risk zones, environmental degradation, climate change *etc.* In many areas of the country, disaster losses tend to outweigh the development gains. Floods, drought, cyclone, wild-fire, earthquake, landslides, pest epidemics, are common disasters related with natural or anthropogenic environmental origin whereas technological ones like fire, explosion, toxicity related accidents are examples of technological or operational mishaps sometimes also called as man-made or anthropogenic disasters.[19,20]

Even in ancient times, vulnerability for various disasters of Indian region was well known. In this regard, Kautilya's prescription for disaster management is worth noting, particularly the anticipation of disasters and prior preparation for preventing them (Basu, 2011). Kautilya has mentioned in the rules that there are 8 great calamities, namely: fire, floods, disease, famines, rats, wild animals, serpents and evil spirits, from which the king should protect the country. Rules have been prescribed for the City Superintendent and the people, regarding do's and don'ts for fire hazards. The prevention measures[21] suggested by Kautilya for flood hazards are highly remarkable. He suggests that during rainy seasons, villagers situated near water should live away from the levels of floods and they should keep a collection of wooden planks, bamboos and boats. He also emphasizes on mass participation in rescue works, which is done by moral persuasion and legal measures (Ibid). The role of the state in famine management, as prescribed by Kautilya, is highly noteworthy. The ruler should have prior preparation for anticipated famines so that he is not caught napping. When the disaster actually occurs, he should take quick measures so as to minimize the harmful effects on the masses. The most interesting aspect in these prescriptions is that the ruler should relinquish, if he fails to handle the famine situation (Ibid).[22] Help of foreign governments in this regard can also be sought, if it is not possible to manage disaster by the efforts of the government of the affected country alone[23].

It can be stated that even though disasters are a common phenomenon to India, the past experience shows that the Indian society is not appropriately prepared to contain the destructive incidences of such hazards. This is evident from the devastations caused by Orissa Super Cyclone of 1999, Bhuj earthquake of 2001, Tsunami on 2004, *etc.*, to name a few. As per the World Disaster Report, 2007, the number of people killed in the disasters in India during 1987-1996 and 1997-2006 were 43085 and 83,252 respectively. The Report makes it apparent that disasters are not something new to India. Though the report referred here dates back to 2007, it is still valid as situation on front of disaster management is not changed yet. Even latest incidences of Leh Cloudburst (August 6: 2010), Sikkim Earthquake (September 18:2011), AMRI hospital fire incident (December 9: 2011), Uttarakhand cloud burst (June 16-17: 2013), Jammu and Kashmir flood (September: 2014) have again shown that India is not adequately prepared. Despite knowing about vulnerability of the country, and consequences of disasters, the preventive measures have not been put in place. Consequently, decade after decade there has been considerable increase in the number of deaths due to disasters in India. Had the preventive measures been taken in an effective way, so many lives would not have lost in the major disasters during the last decade.

In "Disaster Management in India - A Status Report" (2004) the Government of India had enumberated the broad features of the draft national policy on disaster management. These are: A holistic and pro-active approach towards prevention, mitigation and preparedness to be adopted for disaster management; Each Ministry/Department of the Central/State Government to set apart an appropriate quantum of funds under the Plan for specific Schemes/projects addressing vulnerability reduction and preparedness; Each project in a hazard prone area will have mitigation as an essential term of reference; Community involvement and awareness generation particularly that of the vulnerable segments of population and women have been emphasised as necessary for sustainable disaster risk reduction. This is to be a critical component of the policy since communities are the first responders to disasters and, therefore, unless they are empowered and made capable of managing disasters, any amount of external support cannot lead to optimal results; States are to formulate State Disaster Management Policies with the broad objective to minimise the loss of lives and social, private and community assets and contribute to sustainable development.

Accordingly the objectives of National Policy on Disaster Management are:

- ☆ Promoting a culture of prevention, preparedness and resilience at all levels through knowledge, innovation and education;
- ☆ Encouraging mitigation measures based on technology, traditional wisdom and environmental sustainability;
- ☆ Mainstreaming disaster management into the developmental planning process;
- ☆ Establishing institutional and techno-legal frameworks to create an enabling regulatory environment and a compliance regime;

- ☆ Ensuring efficient mechanism for identification, assessment and monitoring of disaster risks;
- ☆ Developing contemporary forecasting and early warning systems backed by responsive and failsafe communication with information technology support;
- ☆ Promoting a productive partnership with the media to create awareness and contributing towards capacity development;
- ☆ Ensuring efficient response and relief with a caring approach towards the needs of the vulnerable sections of the society;
- ☆ Undertaking reconstruction as an opportunity to build disaster resilient structures and habitat for ensuring safer living; and
- ☆ Promoting productive and proactive partnership with media in disaster management.

In development circles today, disaster management is often treated holistically rather than as a single issue. It is an essential component of any development framework. Proper disaster management had been recognised as a key requirement towards achieving the Millennium Development Goals (MDGs) by the specified target of 2015[24]. The link between disasters and development began to be seriously examined from various perspectives in the 1990s. What the different perspectives share is recognition that disaster impacts depend to a large extent on historical and current development activities[25,26]. New Sustainable Development Goal (SDG) have also recognized that ending poverty must go hand-in-hand with strategies that build economic growth and addresses a range of social needs including education, health, social protection, and job opportunities, while tackling climate change and environmental protection.[27]

The paradigm shift in disaster management worldwide influencing national policies, as well as the climate-change adaptation regime highlights the focused need to address the pre-disaster phase that is actually "hazard risk reduction" broadly understood as Disaster Risk Reduction (DRR). The Sendai Framework for DRR identify development agenda with seven targets and four priority actions for "*The substantial reduction of disaster risk and losses in lives, livelihoods and health and in the economic, physical, social, cultural and environmental assets of persons, businesses, communities and countries.*"[28] The DRR strategy for an administrative unit need to incorporate following components: (a) hazard identification and assessment; (b) assessment of vulnerabilities; (c) identification of hazard reduction and impact mitigation options; (d) analysis of risk; and (e) building resilience of the community.

It is only in the recent times disaster management has started gaining momentum in India. Now India's main focus is on mainstreaming DRR in all developmental plans of the government. Efforts are being made to link Disaster Risk Reduction with the key areas of poverty alleviation, education, health, infrastructure and environment. Strengthening the legislative frameworks, promoting multi-stakeholder coordination and educating and creating awareness among communities had become a matter of concern in recent years in India. Today, after considerable and meticulous planning and concerted efforts, a paradigm shift in the approach of the Government departments and agencies as well as of other stakeholders including the community, the corporate sector and other institutions is being brought about for building holistic capabilities for disaster management. Institutions like National Disaster Management Authority (NDMA), National Institute of Disaster Management (NIDM), National Disaster Response Force (NDRF), State Disaster Management Authorities (SDMA) and various central/State Government Ministries and Departments, local Governance, academia, corporate sectors, NGOs, *etc.*, are trying to emphasize Disaster Risk Reduction. Now focus is being shifted to a balanced approach for disaster management which assigns equal importance to pre-disaster aspects such as disaster prevention, mitigation and preparedness besides getting prepared for response and recovery. It is understood that appropriate mitigation measures can substantially, if not wholly, reduce the heavy toll of lives and property, the dissipation of developmental, industrial and infrastructural gains and the hard-earned socio-economic infrastructure.

Disasters in the contemporary society, whether of natural or technological origin, are complex phenomena that demand comprehensive and multi-disciplinary assessments and actions. For example, earthquakes are considered today not simply as a technical issue of maintaining structural robustness, but as an interrelated area of study that involve and require contributions from a wide spectrum of expertise. Understanding of natural and social systems and processes, as well as their interactions is an essential step in the study of disasters. Furthermore, there is the need to develop a capacity to manage such systems, be it in the context of emergency/crisis environments, or alternatively in the avoidance and management of risks prior to disasters.[29]

Effective disaster management strategy demands a multi-dimensional approach involving scientific, engineering, financial and social processes for developing risk reduction strategies. There are multi-disciplinary functions like mitigation, preparedness and response which have to be initiated since the very beginning to manage a disaster. This becomes possible only if Institutional mechanisms are put in place and they work towards sustainable development. Institutions should function with orientation towards such sustainable development by paying proper attention to disasters with required focus on undertaking timely regulatory measures and initiatives for development in the long run.

Every day in cities and towns, emergency response personnel respond to incidents of varying scope and magnitude. Their ability to communicate in real time is critical to establishing command and control at the scene of an emergency, to maintaining event situational awareness, and to operating overall within a broad range of incidents. However, as numerous post-action reports and national assessments have revealed, there are still communications deficiencies that affect the ability of responders to manage routine incidents and support responses to natural disasters, acts of terrorism, and other incidents.

Disasters always need quick response to reduce the quantum of loss to human lives and property. Efficient management of disasters, rather than mere response to their occurrence has, in recent times, received increased attention. Holistic management of disaster calls for assigning priority in the prevention, mitigation and preparedness activities in the pre-disaster scenario of the Disaster Management (DM) continuum, while strengthening the efforts for faster and more efficient response through properly trained personnel equipped with advanced equipment/instruments and building better recovery during rehabilitation and reconstruction period. Past experiences has shown that the countries in which the government, the people and trained personnel joined hands together to face the calamities, recovered faster than the countries that were not so prepared and organised. Preparedness enhances resilience of a community which builds on response and coping mechanism of the system. Early warning systems supported by an efficient emergency communication system plays prominent role in quick and effective response, in addition to various preparedness, mitigation, and capacity building measures.

In India, despite having multiple organisations making individual efforts for creation and dissemination of early warnings on various disasters in India and south Asian region, there is lack of fail-safe disaster communication systems. In India, though, National Disaster Management Authority Guideline on National Disaster Management Information and Communication System, by NDMA, establishes the need of dedicated disaster communication network, no concerted effort, so far, has been initiated by stakeholder agencies. Besides there has not been any concrete effort to study status of emergency communication and early warning systems in India in the light of impending need for integration of existing systems. Therefore, benefits of Information Communication Technologies (ICT) and Science and Technology (S&T) could not so far been available till the last person who is affected by the disaster.

Climate Change and Disaster Management

All over the world, the frequency and the severity of natural disasters, caused by climate related functions and events have been increasing. Further it has been recognized that climate-change is going to pose impacts on global and regional scales. Rapid changes in climate have already resulted in glacial retreat, global sea level rise, changes in temperature and rainfall patterns, affecting the natural resources productivity and quality. There is also an observed increase in the frequency and intensity of hydro-meteorological disasters like drought, floods, heat and cold waves, desertification and coastal hazards like cyclone, coastal and sea erosion, storm surges and flooding in several parts of the sub-continent.

Climate change is a significant and lasting change in the statistical distribution of weather patterns over periods ranging from decades to millions of years. It may be a change in average weather conditions, or in the distribution of weather around the average conditions (*i.e.*, more or fewer extreme weather events). Climate change is caused by factors that include oceanic processes (such as oceanic circulation), biotic processes, variations in solar radiation received by Earth, plate tectonics and volcanic eruptions, and human-induced alterations of the natural world; these latter effects are currently causing global warming, and "climate change" is often used to describe human-specific impacts.

In the context of climate variation, anthropogenic factors are human activities which affect the climate. The scientific consensus on climate change is "that climate is changing and that these changes are in large part caused by human activities," and it "is largely irreversible. Of most concern in these anthropogenic factors is the increase in CO_2 levels due to emissions from fossil fuel combustion, followed by aerosols (particulate matter in the atmosphere) and cement manufacture. Other factors, including land use, ozone depletion, animal agriculture and deforestation, are also of concern in the roles they play - both separately and in conjunction with other factors - in affecting climate, microclimate, and measures of climate variables.

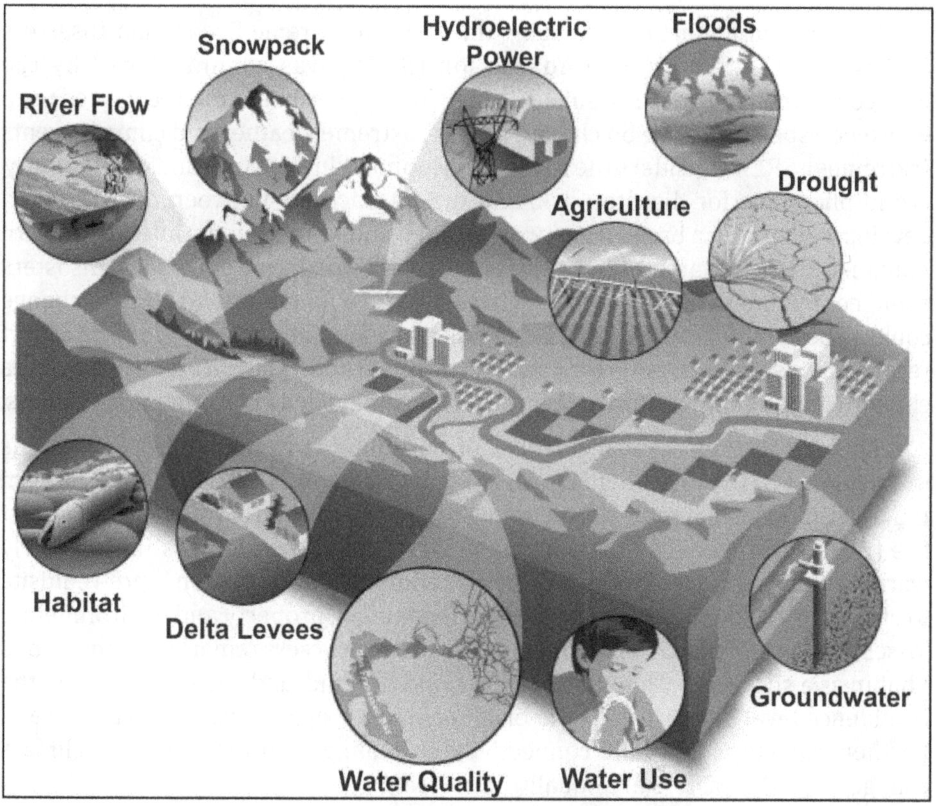

Figure 1.2: Impact of Climate Change.
(*Source*: http://www.christiansforthemountains.org).

Intergovernmental Panel on Climate Change (IPCC) is the leading international body for the assessment of climate change. It was established by the United Nations Environment Programme (UNEP) and the World Meteorological Organization (WMO) in 1988 to provide the world with a clear scientific view on the current state of knowledge in climate change and its potential environmental and socio-economic impacts. In the same year, the UN General Assembly endorsed the action by WMO and UNEP in jointly establishing the IPCC.

The United Nations Framework Convention on Climate Change: The UNFCCC entered into force on 21 March 1994. Today, it has near-universal membership. The 195 countries that have ratified the Convention are called Parties to the Convention. IPCC's 4th Assessment Report (2007) have revealed its observations and has categorized them as high, medium and low confidence impacts on ecosystem structure and functions, bioproductivity, food security, livelihood, health, water and especially on the occurrence – time, intensity and frequency of climatic events like rainfall, winds, sea level rise, *etc.* thus aggravating natural hazards. At the same time climate change impacts are aggravating the factors causing socio-economic vulnerability by reducing ecosystem services and affecting people's resources and capacities and making them occupy hazardous locations and situations.

The Special Report on Managing the Risks of Extreme Events and Disasters to Advance Climate Change Adaptation (SREX) was commissioned by the Intergovernmental Panel on Climate Change (IPCC) in response to a recognized need to provide specific advice on climate change, extreme weather and climate events (extremes). SREX considers the role of development in exposure and vulnerability, the implications for disaster risk, and the interactions between disasters and development under a broad framework of environmental systems interaction with human-system. It examines how humen response to extreme events and disasters could contribute to adaptation objectives, and how adaptation to climate change could be better integrated with disaster risk management practice. The SREX report represents a significant step forward for the integration and harmonization of the climate change adaptation, disaster risk management and environmental sciences.

Looking to the recent incidences of flashflood (Uttarakhand 2013), floods in Kashmir (2014) and Chennai (2015), fatal heatwaves (2015, 2016 Southern India), El-nino led hazard risks including drought (Latur 2016), devastating forest fire (2016 Uttarakhand) following many other incidences, forecasting and early warning along with a foolproof disaster risk communication system is pre-requisite to efficiently tackle the challenge of reducing human, property and environmental losses. Cyclone forecasting and tracking has proven success remarked with cyclone Phailin as a success story followed by cyclone Hudhud, and has strengthened the confidence level in addressing the brunt of climatic events risks in coastal areas. Further, aspects of last mile connectivity in dealing with the impact of climate change related disasters are equally important.

With a major shift in global policy environment for dealing with disaster risk and climate change, especially in the broad framework of sustainability and inclusive - green growth, we have following new set of international protocols (2015-30) dealing with these:

(a) Paris Climate Agreement

(b) Sendai Framework for Disaster Risk Reduction, and

(c) Sustainable Development Goals.

Details of these policy protocols for these 15 years are given in the chapter 2. Interestingly, when it comes to integrated mechanism for implementing these diverse policy protocols at national, sub-national and local levels where most stakeholders – actors and beneficiaries are almost common and many objectives and outputs directly linked to each other, an easily understood roadmap is critical. Indian Prime Minister's Agenda 10 on Disaster Risk Management offered great coherence into these by way of ten thematic areas of interventions announced while inaugurating the 7th Asian Ministerial Conference on Disaster Risk Reduction on 3rd November 2016 at New Delhi. The ten point agenda enumerated as following:

1) All development sectors must imbibe the principles of disaster risk management
2) Work towards risk coverage for all
3) Encourage greater participation and leadership of women in disaster risk management
4) Invest in risk mapping globally
5) Leverage technology to enhance the efficiency of disaster risk management efforts
6) Develop a network of Universities to work on disaster issues
7) Utilize the opportunities provided by social media and mobile technologies
8) Built local capacities and initiatives
9) Ensure the opportunity to learn from a disaster is not wasted
10) Bring about greater cohesion in international disaster response

Information Communication Technology in Disaster Management

Even though the global average of number of disasters had been rising rapidly, improved early warning systems and mitigation programmes have significantly reduced the number of human lives lost in the technologically advanced developed nations. Advancements in Information and Communication Technology (ICT) have made it possible not only to forecast some of the disasters but also have made available means for quick and effective rescue and relief operation, thereby minimising the deadly impacts of some of the worst disasters. For instance, compared to more than 10,000 killed during the cyclone which hit Andhra Pradesh in 1979, improved communication techniques limited the loss to less than 1000 during the May 1990 cyclone in the same state. Similarly when cyclone Phailin crossed the Bay of Bengal on the eastern coast provinces (states) of Andhra Pradesh and Odisha on Saturday the October 12, 2013; Red Message, the highest alert from the Indian Meteorology Department in Delhi, was usable to a large extent by the local district officials and could drastically reduced loss of human lives. The early warning messages for this cyclone prompted India's biggest evacuation in 23 years with more than 550,000 people moved up from the coastline in Odisha and Andhra Pradesh to safer places.[30],[31]

By definition, an Emergency is a situation which requires immediate response. An emergency situation might develop into a disaster, either due to its nature, or as a consequence of insufficient response to the initial event. Almost every year/ alternate year, India is facing a natural hazard converted into a disaster, incidence of Leh Cloudburst (August 6: 2010); Sikkim Earthquake (September 18:2011); AMRI hospital fire incident (December 9: 2011); Uttarakhand cloud burst (June 16-17:2013); Cyclone Phailin in Odisha and Andhra Pradesh (October 12 : 2013); and now flood in Jammu and Kashmir (September 2014) are very recent examples. The breakdown of crucial communications is one of the most widely shared characteristics of all disasters. India has multiple agencies issuing early warning messages for different disasters and there are many public and private pan India telecommunication networks. However convergence of early warning systems and a dedicated disaster communication system is non-existent. NDMA has prepared a guideline on disaster communication: NDMA National Disaster Management Guidelines on Disaster Management Information and Communication System (NDMICS) 2011, but so far nothing has been initiated on ground by either the Ministry of Home affairs or by Ministry of Communication and Information Technology. If existing emergency communication systems in India are not worked upon, it may further lead to conversion of many impending hazard situations to big disasters. The time has come for building a robust, infallible integrated emergency communication and early warning system in India.

A Review of Relevant Literature

Today, almost every part of the world is being affected by disasters. There are various reasons for a disaster to strike. Some disasters are caused by nature whereas some are human induced. Irrespective of the causes of disasters, the repercussions of disasters are always painful, devastating and distressful. In recent times, importance of handling disasters in a holistic way is getting momentum. In the succeeding text an attempt has been made to review some of the important works related to disaster management in general and importance of disaster communication in particular.

When the needs of disaster relief are unknown, the estimate for the same has to be done with much care as poor estimate of the same would result in heavy losses (May, 1982). This article gives a picture of the relief needs that arose out of the Mount St. Helens volcanic eruptions in May 1980 which was greatly overestimated.

It points out that technical, bureaucratic, and political considerations contribute to the upward bias in such cases. It suggests how abstract errors in early estimates of relief needs can be reduced by systematic means. It further infers that more important than obtaining good early estimates is maintaining an effective control over actual disaster relief expenditures. As it turns out, the more effective systems of control usually go hand in hand with poorer early estimates, creating a dilemma for the management of disaster relief. However this thesis deals with how ICT can be leveraged for Post Disaster Need Analysis (PDNA) to remove anomaly in disaster relief estimation.

Wallace and Balogh (1985) have made an analysis of Decision Support System (DSS) and its differences from Management Information System (MIS) in the public sector. It further analyses the stages in the life cycle of a disaster and the types of events occurring within stages. It has created a framework for conceptualizing decision making in disaster management at various levels-operational, tactical, and strategic. It also makes a study of the information and modeling needs of planners and decision makers within each stage and selects examples of DSS applications and prototypes, which have been developed on microcomputers which will serve to illustrate the role of DSS in the disaster management. This important work on DSS is still valid, as most of developing countries are yet to deploy DSS for disaster management. Even in India we do not have integrated DSS for DM activities.

Public administration, as a discipline, has generally neglected to consider emergency management within the mainstream of its activities. Historically, emergency management was considered only a function of law enforcement and fire departments, with support in the event of a major catastrophe from public health and civil defense organisations. Specifically, both public policy and public institutions have been slow to respond and meet the challenges posed by a complex technological society, and society has in many respects become a victim of our technological genius. Thus, Petak (1985) explores emergency management in the backdrop of public administration and brings to light how it becomes the responsibility of public institutions to handle the emergency situation.

India Disaster Management Training Country Workshop, Disaster Assessment", the Report on Disaster Assessment by IIPA (1993) presents an analysis of disaster management and assessment on the basis of the UNDP/UNDRO Disaster management manual and its principles, procedures, and terminology. Part one of the module clarifies the assessment process and charts the changing objectives of assessment through the various phases of a disaster and relates the assessment activities to the decision making process. Part two provides practical guidelines for the collection of accurate and usable data during the various phases of a disaster and gives specific insights to the differing assessment needs acquired by disaster types. Part three of the module addresses the role of the UN in disaster assessment. Part four discusses the preparedness planning measures, which must be in place prior to a disaster facilitating rapid and accurate assessment when required.

The scope of disaster management is wide as it embraces decisions made at all levels, from central government to the grassroots. Plans for disaster preparedness, however, tend not to be made on the ground but they are made in capital cities by government ministries (Myers, 1994), In many disaster-prone countries of the South, a national disaster plan is often inherited, or copied directly, from the former colonial power. Though, governments are improving and updating their plans, often this is being done in the light of lessons painfully learned in the aftermath of disasters of horrendous proportions. Gender is an issue which tends to have a low profile in national disaster plans. In contrast to development, where gender considerations have become almost obligatory for planners, in the area of disasters they are still seen as somewhat of a 'luxury'. Particularly in rapid-onset disasters, the received idea is that gender issues must be put `on hold' for the duration of an emergency. Planners at central government level, most of whom are men, perhaps consider women's needs and position after a disaster strikes, but certainly not 'in the heat of the moment'. In effect, when it comes to disaster preparedness, the old maxim 'Women and Children First' turns out in practice to be the exact opposite. There emerges skepticism on how to introduce the idea of gender into disaster preparedness strategy and how to insert gender issues into a country's national disaster plan. Such issues are dealt in detail in this article and throws light on the gaps in the disaster plans in meeting the needs of women and children. Further, it gives various suggestions on how to mainstream the activities of disaster mitigation and relief for the target group of women and children. Recently NDMA in association with UNFPA developed an emergency kit meant for circulation amongst women.

Singh (1995) not only analyses the drought problems of India and its related impacts on social life of the country but also brings to the fore the point that the behaviour of the human society is equally responsible along with nature for frequent visits of the natural calamity. A set of interaction models based upon physical, political, agro climatic and socio-economic perspectives provide great inspiration for their application in understanding and planning drought disaster reduction measures in India and this has been analysed by the author by exploring the core details. The author through his painstaking study dealt elaborately with the subjects like Water resource development, Drought Impact Assessment, Crop weather relation, Nature of Drought episodes in different parts of India.

Misra and Mathur (1995) explore the ways to reduce natural disasters in the context of mitigation, preparedness and management of natural disasters. This work makes deep insight into the aspects of dealing with disasters and analyses the issue not only in technical and scientific lines but also on humanitarian, social and economic lines concerning prevention and mitigation of natural disasters, thus, giving a holistic way to cope with natural disasters effectively. It also throws light on various aspects of disaster mitigation including the technological advances that have been accomplished and gives meaningful suggestions to reduce the ill effects of disasters.

Pearson and Clair (1998) attempted to integrate and build upon current knowledge to create a multidisciplinary approach to crisis management research, using psychological, social-political, and technological-structural research perspectives which were missed out in the earlier research on crisis management. It further makes definitions of organisational crisis and crisis management, as well as gives a framework that depicts the crisis management process and researchable propositions for the integration of these perspectives. It also suggests implications for research and practice.

Tampere 1998, At the Intergovernmental Conference on Emergency Telecommunications (ICET- 98) at, Finland, held on 18 June, 1998, a legally binding international treaty on deployment of Telecommunication Resources was unanimously adopted by 75 countries and it was recognised that the effective, timely deployment of telecommunication resources for rapid, efficient, accurate and truthful information flows are essential for reducing loss of life, human suffering and damage to property and the environment caused by natural disasters. India became signatory to this Tampere Convention on 29th November, 1999.The Tampere Convention calls on States to facilitate the provision of prompt telecommunication assistance to mitigate the impact of a disaster, and covers both the installation and operation of reliable, flexible telecommunication services. Regulatory barriers that impede the use of telecommunication resources for disasters are waived. These barriers include the licensing requirements to use allocated frequencies, restrictions on the import of telecommunication equipment, as well as limitations on the movement of humanitarian teams. The Convention describes the procedures for request and provision of telecommunication assistance, recognising the right of a State to direct control and coordinate assistance provided under the Convention within its territory. It calls on States to provide prompt telecommunication assistance to mitigate the impact of a disaster, and to ensure the installation and operation of reliable telecommunication resources. Regulatory obstacles to the use of telecommunication services to mitigate the impact of disasters are removed, including those relating to the allocation of frequencies and payment for using them and protection is provided for technicians using equipment. The treaty sets out procedures for requesting and providing telecommunication assistance and recognises the right of a State Party to direct, coordinate and supervise telecommunication assistance provided within its territory under the terms of this treaty. It defines specific elements and aspects of the provision of telecommunication assistance, such as termination of assistance. It requires States to make an inventory of the resources both human and materials available for disaster mitigation and relief, and to develop a telecommunication action plan that identifies the steps necessary to deploy those resources.[32]

A Report on Chamoli Earthquake of 29th March 1999" (12 WCEE 2000)[33] consists of the damage survey of the earthquake by the team and its interpretation in terms of isoseismic and performance of buildings and bridges. It also contains the seismological aspects of the data acquired, the causes of damage, construction quality, remedial measures and a report on the scientific issues raised with the comments and suggestions on various reports published in India. The recommendation on further course of action as a result of lesson learnt from the damages in this earthquake are also present and gives a deep insight on how to handle disasters when it actually strikes.

Developing countries are more vulnerable to the debilitating consequences of disasters and especially the poor of the developing economies are the worst hit as disasters cause economic and social disruptions of their lives. Disaster Mitigation becomes a major component in managing disasters in this backdrop and it entails measures to reduce the physical, economic and social vulnerability of a community of disasters. The edited work by Sahni *et al.* (2001) makes an excellent analysis of different mitigation strategies and offers deep insight into the different dimensions of disaster preparedness and mitigation. It also illuminates the pertinence of the mitigation efforts that would prepare everyone related with disaster management to comprehend and approach the problem more holistically.

Sharma (2001) gives a holistic picture of disaster management in general and gives much useful information on natural disasters, such as disaster history report, progress in International Strategy for Disaster Reduction (ISDR), concepts and guidelines of Hazards and Vulnerability analysis for administrators and academics. Dinham and Sarangi (2002) describes the inadequacies in the response of the Union Carbide Corporation to the accidental release of the highly toxic gas, methylisocyanate, from its plant in Bhopal, India in 1984. Over 20,000 people are estimated to have died from exposure to this gas since 1984, with some 1,20,000 chronically ill survivors. Union Carbide fought to avoid compensation or to keep it very low. The long, much delayed process of distributing compensation focused on minimizing payouts to victims. The corporation tried to blame the accident on a disgruntled employee, whereas key parts of the safety equipment designed to stop the escape of the gas were not functioning or were turned off. The corporation has always sought to underplay the health effects and has refused to release its research on the health impacts of the gas (which could have helped develop more effective treatment). In addition, the medical services in Bhopal have failed to develop a health care service that offers sustained relief and treatment to the communities most affected. This paper makes a study of all the factors listed above and also describes the work of the Sambhavna Trust, a charitable body set up to work with the survivors, and its programme to develop simple, more effective, ethical and participatory ways of carrying out research, monitoring and treatment. It also throws deep insight on the work of government, non-governmental organisations and other private players and community in general that engaged itself in mitigating the suffering faced by the people in terms of relief measures and health services and identifies the critical gaps involved in it.

"Design Challenges for an Integrated Disaster Management Communication and Information System"[34] for the natural and man-made disasters, such as earthquakes, floods, plane crashes, high-rise building collapses, or major nuclear facility malfunctions, aim at all time readiness of the public emergency services to meet such ever-present challenges. In order to cope with such disasters in a fast and highly coordinated manner, the optimal provision of information concerning the situation is an essential pre-requisite. Police, fire departments, public health, civil defense and other organisations have to react not only efficiently and individually, but also in a coordinated manner. This results in the need for both intra and inter organization coordination at several hierarchy levels[35]. Since coordination requires current information, and such information must be communicated upstream and downstream within and between organisations in real-time, the need arises for an integrated communication and information system for disaster management that provides efficient, reliable and secure exchange and processing of relevant information.

An Emergency communication system (ECS) is any system (typically, computer-based) that is organised for the primary purpose of supporting one-way and two-way communication of emergency messages between both individuals and groups of individuals. These systems are commonly designed to integrate the cross-communication of messages between varieties of communication technologies, forming a unified communication system intended to optimise communications during emergencies.

Sahni and Ariyabandu (2003) make a study of the disasters and its consequences in general and analyses the efforts of South Asia towards disaster risk reduction by adhering to effective participatory strategies. Further the edited work explores how the risk of disasters can be reduced by structural and non-structural measures and makes an analysis of challenges that are inherent in disasters. It also gives various suggestions in meeting the challenges of disasters so as to foster a culture of preparedness among the community, society, government and non-governmental actors involved in risk reduction. It further throws light on the factors of risk reduction from the view point of development efforts and gives a thorough insight on how to manage disasters without impeding developmental initiatives and reflects a serious and meaningful point of concern to be put to practice for reducing disaster risk.

As a measure for disaster management, relief measures in terms of medical facilities and provision of rescue teams has always been a great challenge as the non availability of these timely service results in huge loss of lives. Dave *et al.* (2003) advocated a practical approach with holistic view of the Emergency Medical Services (EMS) that can be made available at the time of disaster. It covers all aspects of EMS including medical transportation, train accidents, ambulance services and telemedicine.

Indian Government's report (MHA, 2003) on "Relief and Rehabilitation Measures in Natural Calamities" tells about the potential risk of natural calamities in India and the administrative response to the natural calamities in the form of setting up of various funds - (Calamity Relief Fund (CRF), National Fund for Calamity Relief (NFCR), National Calamity Contingency Fund (NCCF). It also mentions the inadequacy of such funding and about misappropriation of funds, the need for immediate assistance in the wake of calamities and long term assistance and the setting up of organisational structure for monitoring the process. However now these funds with change in budget head and operation and release mechanism have been converted and merged as National Disaster Response Fund (NDRF) and State Disaster Response Fund (SDRF).

"Application of Information Technology in Disaster Risk Reduction " as argued by Prof. N Vinod Chandra Menon, then Member Indian National Disaster Management Authority, brought forth a comprehensive understanding of disaster management as the process by which the uncertainties that exist in potentially hazardous situation can be minimized and public safety is maximized. The objective is to limit the human material and economic cost. This is achieved through a range of strategies from hazard management and prevention to speedy restoration of affected community (Granger, 1993).[36]

According to Disaster Emergency Communications Division of US-FEMA[37], "Disaster Emergency Communications is a specialized field within the broader field of emergency communications. Emergency communications covers all technical means and modes for public safety agencies at all levels of government (*e.g.* law enforcement, fire services, emergency medical services) to perform their routine, daily communications. Disaster emergency communications applies to those technical means and modes required to provide and maintain operable and interoperable communication before, during, and after presidentially declared emergencies, disasters, or planned National Special Security Events".

DFID's study on "Learning Lessons from the Cyclone" (2004) draws out practical lessons from the events of and after the 1999 Orissa cyclone, using the livelihood framework as an analytical tool and contributing to its development as an approach. It reviews and compares the situation in Orissa before and over the period since the cyclone, where the elements of sustainable livelihoods in relation with the position of the poor are considered for the three time phases of the cyclones.

The working paper by Gurenko and Lester (2004) provides a conceptual framework for designing a comprehensive risk management strategy for rapid onset natural disasters at the country level, with a particular emphasis on the role of catastrophe loss funding. The paper discusses the key policy and technical issues involved in building financially sustainable catastrophe risk transfer and funding programmes in disaster prone countries, and their links to risk mitigation. The paper also deals with the cognitive and political economy issues that are likely to arise and ways to accommodate them.

India's National Disaster Management Division in Union Government (2004) recommended the broad features of the draft national policy on disaster management. These are: a holistic and pro-active approach towards prevention, mitigation and preparedness to be adopted for disaster management; each Ministry/Department of the Central/State Government to set apart an appropriate quantum of funds under the Plan for specific Schemes/projects addressing vulnerability reduction and preparedness; each project in a hazard prone area will have mitigation as an essential term of reference; community involvement and awareness generation particularly that of the vulnerable segments of population and women have been emphasized as necessary for sustainable disaster risk reduction. This is to be a critical component of the policy since communities are the first responders to disasters and, therefore, unless they are empowered and made capable of managing disasters, any amount of external support cannot lead to optimal results; states are to formulate State Disaster Management Policies with the broad objective to minimize the loss of lives and social, private and community assets and contribute to sustainable development.

As per Joint Commission Report (2005), Disaster Management mechanism can be effective only if pre-disaster preparations are executed in a proper manner. In order to concentrate on preparedness, Emergency Management Planning Process has to be taken care of; so that the small rural and sub-urban communities are able to create proper health care facilities through rescue teams and prepare themselves towards relief and rehabilitation measures. According to UNISDR (2005) "Risk information communication/early warning is one of the key priorities for risk reduction. Effective risk communicating and early warning system needs collaborative and participatory approaches within the different levels (especially local level) and actors of Disaster Risk Management (DRM) during planning and decision making related to Disaster Risk Reduction (DRR)".

"The Hyogo Framework for Action 2005-2015: Building the Resilience of Nations and Communities to Disasters" (2005) recommended by the World Conference on Disaster Reduction, Kobe, Hyogo, Japan, states that "disaster loss is on the rise with grave consequences for the survival, dignity and livelihood of individuals, particularly the poor and hard-won development gains. The disasters in the past two decades, on average more than 200 million people have been affected every year by disasters. Disaster risk arises when hazards interact with physical, social, economic and environmental vulnerabilities. Events of hydro meteorological origin constitute the large majority of disasters. The scope of this Framework for Action encompasses disasters caused by hazards of natural origin and related environmental and technological hazards and risks. Specific gaps and challenges have been identified in five main areas: Governance – organisational, legal and policy frameworks; Risk identification, assessment, monitoring and early warning; Knowledge management and education; Reducing underlying risk factors; and Preparedness for effective response and recovery. These are the key areas for developing a relevant framework for action for the decade 2005–2015. To cope with disaster challenges effectively, the major strategic goals adopted include: more effective integration of disaster risk considerations into sustainable development policies, planning and programming at all levels, with a special emphasis on disaster prevention, mitigation, preparedness and vulnerability reduction; development and strengthening of institutions, mechanisms and capacities at all levels, in particular at the community level, that can systematically contribute to building resilience to hazards; systematic incorporation of risk reduction approaches into the design and implementation of emergency preparedness, response and recovery programmes in the reconstruction of affected communities".

Ertur and Yum (XXX) (http://www.preventionweb.net/files/2312_toc.pdf) emphasized that need of central and local authorities to reach people is much greater during and immediately after an emergency. The media then play a very significant role as an important tool for authorities in meeting their need to reach, inform, mobilise and calm people. In most instances, during and in the wake of an emergency, authorities should manage information dissemination and updates through the media. This is important in order to correct false or incorrect news and information. It is also important to increase the effectiveness of the recovery efforts, the transparency of actions and the participation of parties.

The International Telecommunication Union, ITU (2005) defines "emergency is simply a situation requiring urgent response. Depending on the circumstances initial response will be provided by whosoever is present, using whatever means are available on site. Any other additional intervention deemed necessary can best be mobilised mainly through telecommunication. An emergency situation might develop in to a disaster either due to its very nature or as a consequence of insufficient response to initial event. The magnitude of the event will require resource mobilization at regional level or even at international scale; communication related to a disaster will however include activities well beyond an alert requesting emergency response, made through normally available means of telecommunications".

ITU's Radiocommunication Sector (ITU–R) is continuously working for Public Protection and Disaster Relief (PPDR). When disaster strikes agencies involved in response and relief functions need to communicate. To cater to such needs of emergency communication and also to ensure access to communication to victims, it is necessary that some spectrum be identified as dedicated to emergency communication. ITU–R has defined the term "public protection radiocommunication" as referring to radiocommunication used by agencies and organizations dealing with the maintenance of law and order, the protection of life and property and with emergencies. An important agreement concerning Public Protection and Disaster Relief (PPDR) was reached at the World Radiocommunication Conference in 2003 (WRC-03) in its Resolution 646. It paves the way for the deployment of new technologies for enhanced applications involving higher data rates, real-time full motion video and multimedia services that should facilitate the work of PPDR agencies around the world. ITU-R is also focusing on other programmes related to disaster relief operations. The humanitarian activities of amateur radio and amateur satellite services are highlighted in Recommendation ITU–R M.1042, which encourages the development of networks within these services that are capable of providing communications during disasters and relief operations. [38].

ITU has recognized national/international use of radio communication in case of disaster and administrations have been urged to take all practical steps to facilitate the rapid deployment and effective use of telecommunication resources for disaster mitigation and disaster relief operations by reducing and, where possible, removing regulatory barriers and strengthening transborder cooperation between States as reflected in Articles 40 and 46 of its Constitution and in resolution 209 (Mob-87). ITU Radiocommunication Sector is to continue to study, on aspects of radio communication that are relevant to disaster mitigation and relief operations, such as decentralised means of communications that are appropriate and generally available, including amateur radio facilities and mobile portable satellite terminals. ITU Recommendation ITU-R F.1105 on "Transportable Fixed Radiocommunication Equipment for Relief Operation" deals with various types of transportable wireless equipment and their broad characteristics. ITU-D has also issued a Handbook on Disaster Communication against Question No. 16/2. This document deals with this subject in a comprehensive manner with specific needs of developing countries.

EWC III (2006) provided the opportunity to present new and innovative early warning projects and to discuss natural hazards and risks around the world and how their impacts can be minimised through the implementation of people-centered early warning. "Developing Early Warning Systems: A Checklist" was developed as a conference outcome, to both inform and draw upon the discussions and practical examples raised during the conference, and to support the implementation of the early warning components of the Hyogo Framework for Action. The Checklist, was structured around the four key elements of effective early warning: risk knowledge; technical monitoring and warning service; communication and dissemination of warnings; and community response capability is included to emphasize the major components that comprise an effective people-centred early warning system, and why each is important. In addition to the four elements, a number of cross-cutting issues that are critical to the development and sustainability of effective early warning systems were also outlined. These include effective governance and institutional arrangements, a multi-hazard approach to early warning, involvement of local communities and consideration of gender perspective and cultural diversity.

Grasso, Ashbindu and Pathak (2012) further ratifies that development plays a key role and has a significant impact on disaster risk. Almost 85 per cent of the people exposed to the deadliest hazards, earthquakes, floods, cyclones and droughts live in the developing world. According to this report, "impact of disasters is mostly influenced by development choices made previously. By integrating disaster mitigation strategies into planning and policies, effects of disasters can be sensibly reduced and managed". "Disaster risk is not inevitable, but on the contrary can be managed and reduced through appropriate development actions" (United Nations Development Programme-UNDP, 2004). "It is through risk-sensitive development planning that country can help reduce disaster risks". The report suggests that key targets for capacity building are:

- ☆ Development of national research, monitoring and assessment capacity, including training in assessment and early warning;
- ☆ Support to national and regional institutions in data collection, analysis and monitoring of natural and man-made hazards;
- ☆ Access to scientific and technological information, including information on state-of-the-art technologies;
- ☆ Education and awareness-raising, including networking among universities with programmes of excellence in the field of the emergency management;
- ☆ Organization of training courses for local decision makers and communities;
- ☆ Bridge the gap between emergency relief and long-term development;
- ☆ Bridge the gaps between science and decision making and Strengthen coordination and communication links;
- ☆ Provide simple, effective and timely warnings and promote the use of standard protocols for warning dissemination and information exchange;

Scientific and technological advances in modeling, monitoring and predicting capabilities would bring incredible benefits to early warning if science would be translated into effective disaster management actions. The first important steps towards reducing disaster impact are to correctly analyse the potential risk and identify measures that can prevent, mitigate or prepare for emergencies (Wattegama, 2007). ICT can play a significant role in highlighting risk areas, vulnerabilities and potentially affected populations by producing geographically referenced analysis through, for example, a geographic information system (GIS). The importance of timely disaster warning in mitigating negative impacts can never be underestimated.

Tang (2007) specified on the intensive use of science and technology for earthquake disaster preparedness in China. It gives a fare idea about how a determined government intends to manage an unpredictable disaster like earthquake in a planned manner. The paper has cited that considering a high seismic risk in China, the Chinese government has paid great attention to the preparedness and mitigation of seismic disaster in the country. Paper provides information that new trends of seismic disaster management in China include contents: laws and policies on protecting against and prevention of seismic disaster, systems of seismic monitoring and quick prediction, systems of seismic disaster prevention based on the modern developed technology, and systems of emergency response and rescuing. Many modern high-new technologies and theories can be found in seismic disaster prevention system, including a seismic design and seismic control of all kinds of structures, education and popularization of the earthquake knowledge. As per Vanderford (2007) disasters are large scale events that demand multiagency coordination. The coordination of multiple agencies becomes a pivotal communication concern.

Fakhruddin (2007) elaborates that before disasters strike; telecommunications can be used as a conduit for disseminating information about the impending danger thus, making it possible for people to take the necessary precautions to mitigate the impact of these hazards. Other telecommunication applications, including remote sensing and global positioning system (GPS), have critical roles to play in tracking approaching hazards, alerting authorities, warning affected populations, coordinating relief operations, assessing damages and mobilising support for reconstruction. It is clear therefore that telecommunications play a pivotal role in disaster prevention, mitigation, and management.

Sahni (2008) gives a unique interpretation that Disaster Management is nothing, but in essence it is everything. When it is said that disaster management is nothing, it is because that various facets given attention in disaster management are from other disciplines are professions. For example, when policies and programmes for managing disasters are referred to, it is found as having taken from political science; geographical considerations, regional planning, spatial planning, *etc.*, stem from geography; reference to community, society, individuals, groups, and the like are from the field of sociology; statistical resources, decision making, computation models are from the area of statistics and mathematics; plan formulation, micro finance, macro finance emanate from economics; ecological foundations are from the field of biology; and past practices as well as treatment to earlier disasters are from the field of history. It makes one to feel that disaster management has nothing of its own and so it is nothing. The other side of it reveals that disasters affect the polity, society, economics, environment, development planning, administration, management, *etc.*, and without appropriate management of disasters each of the professions and disciplines that contribute to the management of disasters is out of place unless there are appropriate mechanisms for disaster management. Thus, disaster management is everything. The scope of disaster management is quite vast and could be understood through two sub-divisions. These are: Theoretical Foundations and Practical Concerns. Theoretical foundations include elements and facets like planning, organising, staffing, directing, coordinating, reporting, and budgeting. The Practical concerns of the scope of disaster management include subject matter or issue specific namely administration, agriculture, education, food, health, livelihood, livestock, reconstruction, recovery, rehabilitation, relief, rescue, shelter, *etc.* It also comprises human elements touching upon psychological, social, and economic dimensions.

Barrantes (2009) suggests that information is the most valuable commodity during emergencies or disasters. It is what everyone needs to make decisions. It is an essential aspect in an organization's ability to gain (or lose) visibility and credibility. Above all, it is necessary for rapid and effective assistance for those affected by a disaster. Information is the main element in the damage and needs assessment process and is the basis for coordination and decision making in emergency situations. Right information communicated/gathered in appropriate time has very powerful impact on how national and international resources are mobilised. It is essential for post disaster analysis, evaluation, and lessons learned. Moreover, public and social communication and media relations have become key elements in efficient emergency management. Technical operations in highly charged political and social situations must be accompanied by good public communication and information strategies that take all stakeholders into account.

Legarda (2009) comments that "Timely, preventive response to disaster risk requires effective early warning systems that are technically sound, politically viable and communally acceptable. To curb increasing disaster risks and climate change impacts, as highlighted in this year's World Disasters Report, it is high time to unite and to take concrete concerted actions, for securing human life and livelihoods and protecting socio-economic gains and opportunities." The Report further describes that while natural hazards cannot be prevented, they only become disasters because affected communities are vulnerable and unprepared. Early warning systems have been proved effective beyond doubt to save lives and reduce economic losses at all levels, but they are still not an integral part of disaster management and risk reduction globally.

World Bank Handbook (2010) describes that "strategic communication builds trust, consensus, and active participation which are key factors for positive outcomes in development programs. It promotes credibility, transparency, legitimacy, and ownership for the project and ensures that the right messages are reaching all relevant stakeholders".[39] Particularly in post-disaster situation, good communication is the foundation for acceptance, sustainability, and mutual understanding when rebuilding people's lives. It further emphasises that communication should be initiated as early as possible in reconstruction projects and provides project managers, partner organisations, and governments with tools and guidelines for development and carrying out a successful communications strategy.

UNISDR Handbook (2010) explains that the "objective of people-centred early warning systems is to empower individuals and communities exposed to hazards to act in sufficient time and in an appropriate manner to reduce the possibility of personal injury, loss of life and damage to property and the environment. A complete and effective early warning system comprises four inter-related elements, characterised by knowledge of hazards and vulnerabilities, education and training for people at risk which are well harmonised with monitoring tools, warning systems, dissemination strategies as well as preparedness and capacity to respond". Good practice in early warning systems also includes in having strong inter-linkages, coordination and effective communication channels between all disaster management stakeholders. Responsibilities need to be clearly defined and people need to be well-informed and feel a degree of ownership of the implementation process. Much depends on establishing institutional capacities to ensure that early warning systems are well integrated into governmental policy and decision-making process.

World Bank Handbook (2010) mentions about utility of Communications Action Plan (CAP). The report guides the implementation of the communications strategy. The CAP covers institutional coordination, media relations, grassroots communication, capacity building, external relations, and coordination with program implementation units (environmental, resettlement unit, *etc.*). The CAP details the production, training, hiring, budgeting, and timing of all initiatives. The nature of a post-disaster intervention will lead to overlaps and doubling of messages. Lessons learned show that too much communication is better than too little. However, retracting or correcting information can be difficult. "Silence" promotes rumors that can be exploited for political or economic reasons, which can lower community participation levels.

WCNIS (2010) identifies that quickly establishing a temporary communication system to support emergency management is one of the most urgent tasks in the disaster relief mission. To facilitate the rescue teams and victim people to communicate inside and outside the disaster site, this paper proposes an integrated communication system by composing heterogeneous wireless networks. Firstly, temporary Wireless Ad hoc Network are to be immediately deployed on the disaster site for local communication and information collection. To communicate with the remote disaster safe areas, satellite gateway may be used for the local networks to interconnect with the satellite mobile network. Furthermore, cellular gateway is used as an alternative remote communication means when the expansion of local networks reaches a working cellular base station.

Sarun (2011) mentions that "risk communication and early warning system needs detailed information about hazard characteristics and vulnerability for effective prognoses and preparedness. However, this kind of information is often lacking in many countries that limits the capabilities for effective DRM. During risk information communication, it is important to ensure that significant required information is disseminated to the citizens. To the same effect information received from the public about the hazards can play an important role in disaster management and risk reduction". It is widely recognised that information and communications technology (ICT), plays an important role in establishing effective linkage between various actors that enables risk reduction. "Participatory Geographic information System (PGIS) is one of the well-known methods that were developed from participatory approaches combined with 4different ICT tools to gather Local (Spatial) Knowledge (LSK) for effective communication linkages between communities at risk and higher level government in the various stages of disaster management (mitigation, preparedness, response and recovery)".

YAP (2011) mentions that "climate change presents two types of disaster threats in developing countries. One is the potentially devastating impacts on vulnerable communities of more frequent and more intense extreme weather events. This contributes to the second threat, the compounding of what are already complex development problems, potentially leading to a downward development spiral for the world's poor". The IFRC reminds us, "Whether the changing climate with all its uncertainties, contributes to more disasters that affect more people, does not depend merely on what happens to the weather. Climate change and the accompanying risks will be superimposed on an unequal world, where vulnerability to disasters is directly linked to poverty".

Aizu (2011) citing to Japan earthquake has written " It is perhaps one of the first massive disasters that hit a well-developed country equipped with broadband and 3G mobile networks and other ICT infrastructure and services. Many citizens were using internet and smart phones in addition to the conventional mass media such as TV and radio broadcasting to find information or call for rescue. However, most telephone lines were inaccessible. Given the massive call demand from people immediately after the quake, the telephone operators blocked 90 per cent of calls in most devastated areas a standard practice to insure critical connections, such as those used by emergency services, could be made. However this also meant that many citizens could not talk to their families and friends for hours, and even days in some areas. In coastal areas, the Tsunami waves destroyed most physical infrastructure roads and railways, telephone and power lines and radio towers. These areas became "information black holes" and that continued for a week to a month or even longer. This article emphasizes that a developed country like Japan which is considered to be a well prepared country also could not utilize ICT to its full potential due to multiple reasons.

Identifying importance of communication during various phases of DM, Department of Telecommunications, Ministry of Communication and IT, Government of India in its draft New Telecom Policy (NTP)-2011 has emphasized the need to built resilient telecom networks in India. References made in the draft NTP-11 are:

Para 13 of Preamble – "NTP-2011 recognizes the importance of creation of the robust and resilient telecom networks for adequately addressing the need for proactive support for mitigating disasters, natural and manmade."

Para 5.13 of Strategies - "To prescribe sectoral Standard Operating Procedures for effective and early mitigation during disasters and emergencies. To mandate Telecom Service Providers to provide alternative reliable means of communication at the time of disaster by creating appropriate regulatory framework."

Para 7.7 of Strategies – "To promote creation of robust, reliable and resilient communication networks"

TRAI (2012) refers to NTP 2011 and has cited that provisions of the draft telecom policy have far reaching implications for a country like India which is prone to natural disasters like floods, earth-quakes, coastal cyclones *etc.* on one hand and manmade disasters like accidents, terrorist attacks *etc.* on other. Hence, putting in place processes and structures to ensure reliable means of communications during disasters/emergencies in India will definitely help in better response and recovery.

NDMICS (2011) advocates to establish a dedicated National Disaster Communication Network (NDCN). It highlights "The aim for establishing NDCN is to create a digitised, converged, adequate, reliable, responsive, inter-operable, self healing, redundant and diverse technology based Communication and IT Network with the desired NDMIS capability in support of the complete disaster continuum. Essentially, being a network of networks, NDCN will be operationalised based on existing and planned National Terrestrial and Satellite Infrastructure, whether public or private. In accordance with organizational and functional structures of various DM stakeholders, Communications and IT support activities of NCDN could be viewed as a system involving number of existing interdependent but geographically separate sub-systems or basic networks connected together to provide ICT support to Disaster Management. In terms of additionalities, however, NDCN would be extended to all NDRF battalion stations in addition to Mini Mobile Communication Pack and transportable Mobile Emergency Operations Center (MEOC) with added bandwidth to be air flown by NDRF. This network of networks will be backed up by VSAT network (to be created in the proposed NDCN) to provide its fail-safe character under all phases of disaster scenarios."

NTP (2012) in its preamble recognises the importance of creation of the robust and resilient telecom networks for adequately addressing the need for proactive support for mitigating disasters, natural and manmade. It further reiterates:

"To prescribe sectoral Standard Operating Procedures for aiding effective and early mitigation during disasters and emergencies";

"To create appropriate regulatory framework for provision of reliable means of public communication by Telecom Service Providers during disasters";

"To encourage use of ICTs in prediction, monitoring and early warning of disasters and early dissemination of information"; and

"To facilitate an institutional framework to establish, a nationwide Unified Emergency Response Mechanism by providing nationwide single access number for emergency services".

TRAI (2013) explains that telecommunication has a significant social, cultural and economic impact on modern society. Many research studies have shown the macroeconomic link between good telecommunication infrastructure and economic growth. Telecommunication has played a significant role in social relationships by helping people staying connected to family and friends. Culturally, telecommunication has impacted the society by way of increased access to music and film. It has also transformed the way people receive their news. Apart from playing significant role in economic, social and cultural development, modern telecommunications infrastructure has also provided effective tools to enable human establishments to cope with crisis, and quickly restore disrupted social and economic activities. Role of telecommunication has been significant in 'response and recovery' during emergency situations.

Prashanta (2013) has specifically written about importance of micro insurance products for poor. He wrote that "opportunities for development are extremely restricted by different conditions like political, economic, living in disaster prone areas like coastal areas prone to cyclone, floods as was seen in Cyclone Phailin. Through different micro finance and insurance policies, the poor could attain a better standard of life through programmes prepared in the field of financial business services. Demand of micro insurance is very huge and constantly increasing. There are several factors contributing to this reality including frequency of disaster event due to high vulnerability, increasing population and awareness of insurance. The poor and vulnerable families really required micro insurance that cover non-life component especially shelter and livelihood. Single party cannot do this effectively. A combination of insurance companies and non-government organisations with focus on effective facilitation and time-to-time follow up with poor communities can produce desired result. The above-mentioned case is one such initiative. For micro insurance to be successful for the insured and for the risk bearers many elements are important; such as simple and affordable insurance products reaching large numbers of people; stream-lined administration, including premium payment; a simplified claims procedures and verifications; and rapid delivery of benefits. If most of these elements are present, it can be possible for micro insurance schemes to become sustainable, to perform well and to provide "real value" to the poor".

SIPA (2013), highlights when disaster strikes, humanitarian aid organisations often play a pivotal role in relief efforts. Communicating with affected populations is one of the many challenges these groups face during a response operation. How organisations contact local communities "depends on the context and also the stage of the disaster. In the immediate aftermath of a major emergency the aim is to reach as many people as possible, as fast as possible.

General Dynamics (2013) specifies that "public safety agencies know that when disaster strikes effective, uninterrupted communications is the key to containing any situation, protecting citizens, and saving lives. Continuous communications between first responders on site and command and control centers ensures that everyone at all levels has the most accurate information available to properly assess a situation, make operational decisions, and adjust response efforts by fire, police, and emergency medical teams. In these situations, a reliable, fully operational communications network is critical. It provides the vital link between personnel in the field and their colleagues, and with forward and central command centers. A disaster situation requires an immediate coordinated response by multiple public safety agencies. In these situations, communication and coordination among first response teams is critical for the success of the operation. To be effective, all teams arriving on a scene need a communications network that enables instant collaboration between in-field personnel, with central command, and with other agencies. That network must support the deployment of emergency communications network elements in remote areas. It must easily scale from a small, on-site, ad-hoc network to a larger multi-site network, if required. And it must easily interconnect with similar networks, as needed".

World Disasters Report 2013 explains that "technology enables affected communities to quickly transform themselves into first responders, send requests and messages, provide critical information, match assistance needs with providers or support rapid damage assessments. This is also the case in rural areas around the world, which are increasingly connected and have access to information and communication resources that are unprecedented. Local communities are now becoming more fully engaged in humanitarian action than ever. It is also true on the web, where individuals are mobilising in the aftermath of disasters to provide assistance to affected communities and humanitarian actors. Humanitarian technology refers to the use and new applications of technology to support efforts at improving access to and quality of prevention, mitigation, preparedness, response, recovery and rebuilding efforts. While much has been learned from a range of pilot projects and field implementation, the practice of humanitarian technology remains first and foremost defined by its potential contribution rather than by an actual integration in standard practices. The rise of humanitarian technology is also the result of technology fusion or "the integration of information network, mobile technology hardware and applications, and social media and mapping platforms into a readily available single mobile device such as a laptop, a mobile smartphone, or a tablet with access to unlimited amount of data from multiple sources and in multiple formats (big data)" (Pham and Vinck, 2012)".

Sakurai and Watson (2015) have described how in 2011, the biggest earthquake on record hit Japan resulting in suspension of ICT services. This made it impossible for officials to respond as per the existing disaster management plan. The plan turned out not to be useful in response to such an unexpected event. This paper shows how municipal governments responded to the disaster and leads us to propose that unpredictable, large-scale disasters need communication systems that are founded in familiar contexts rather than technological complexity. "They have suggested unique solution of Combining daily-use and disaster-only channels. For example, the Japanese mobile company, Softbank Mobile Co., launched a new satellite phone called 202TH in 2014, which can be used in conjunction with the iPhone. A solid, hand-held bracket firmly holds the iPhone in place and connects to it via Bluetooth using a special application called "SatSleeve." This enables the iPhone to be used as a satellite phone: making calls by phone number lookup, sending text messages, and transmitting data. Its initial and operating costs are almost half of a conventional satellite phone. This illustrates the possibility of daily-use channels as the foundation for multi-functional cover".

Sendai Framework for Disaster Risk Reduction (2015-2030) taking disaster management goals further emphasise on disaster risk reduction and have defined 4 priorities as;

- Priority 1: Understanding disaster risk.
- Priority 2: Strengthening disaster risk governance to manage disaster risk.
- Priority 3: Investing in disaster risk reduction for resilience.
- Priority 4: Enhancing disaster preparedness for effective response and to "Build Back Better" in recovery, rehabilitation and reconstruction

All above stated references touch upon multidimensional views and perspectives about various disasters with cause and effect analysis of such uncalled for events. However one common outcome of all studies suggests that prevention, mitigation and effective response are based on pre-disaster preparedness. It emphasise on disaster risk reduction and building resilience by leveraging science and technology. Use of science and technology at all stages of DM continuum is accepted as best strategy for effective disaster management in all recent protocols and strategies signed by international communities. Priority actions as identified in HFA, demands extensive use of Science and Technology for attaining desired level of preparedness, preventing, mitigating effects of hazards from turning in to a disasters. During disasters, information is as much a necessity as water, food and medicine [40.]The right kind of information communicated and used at the right time can save lives, livelihoods and resources[41.] Responding effectively to disasters demands rapid access to reliable and accurate data, mechanism to convert data in to useful ground information and the capacity to assimilate, analyse, and taking timely decisions.It requires speedy communication with administration, all stakeholders,key actors, to marshal resources (*e.g.,* food, water, medical supplies, and other key emergency relief supply materials); to mobilise, deploy and coordinate emergency personnel; and to coordinate response activities among agencies involved. Communication and information sharing is thus very critical.

All references cited in this chapter specifically do not delve into peculiarity of problems prevalent in India due to lack of integration of early warning systems and emergency communications facilities, nor do they offer India specific solutions. NDMA National Disaster Management Guidelines on Disaster Management Information and Communication System (NDMICS) 2011 has though summarised status of existing scenario of emergency communication and early warning systems in India, the analysis part is incomplete and the solution offered is cost intensive in nature. Besides, the guideline does not touch upon other very relevant issues such as requirement of single emergency number in India, Priority Call Routing, adoption of Common Alerting Protocol (CAP), adaptation of existing resource inventory database to web based multistakeholder platforms *etc.* Even latest important policy documents such as NTP 2011(draft) and 2012; Disaster Management Guidelines for Telecommunications by Telecom Engineering Centre and Standard Operating Procedure for Telecommunication Services for Responding to Disasters as issued by Department of Telecommunications (DOT) still require on ground execution. There is impending need for a study of existing emergency communication set up in India, with an aim to identify gaps and to offer pragmatic solutions based on long term and short term goals and global best practices. This book is meant to fill this void with a purpose to offer desired solutions of existing problems. The solutions offered in this book are not only applicable for India but for all developing countries.

Aim and Scope of the Book

The book "Disaster Communication and Early Warning Systems" is written with an aim to serve broad purpose comprising as professional and research reference and also a textbook for the students studying the subjects of disaster management and telecommunication technology in Universities or colleges. The book sets the conceptual background of disaster management and paradigm shift from response and relief centric to proactive preparedness centric approach with reference to international strategies on disaster management, more particularly the Sendai Framework for Disaster Risk Reduction, Sustainable Development Goals and Paris Climate Agreement.

The book discusses about holistic approach of disaster management, with ICT as an enabler in Disaster Management. It brings systematic knowledge base on the strategic role played by early warning systems, emergency communication - need analysis and existing setups as well as futuristic technological trends. A case study of India's disaster Communication and Early Warning System and is takes solutions widely applicable for many developing countries are discussed in detail.

Chapter 1 on Introduction and Overview, discusses the context and purpose of this book, abstract of the contents chapter wise, and provides a broad overview on the theme of the book title, *i.e.* Disaster Early Warning and Communication System. Chapter 2 on Paradigm Shift in Disaster Management, delves in the recent past when disaster management was an act of providing relief when disaster strikes. It documents how incidences like Bhuj Earthquake, Odisha super cyclone, and the Indian Ocean Tsunami shattered our belief system about disasters, which paved way of realization for enactment of the Disaster Management Act 2005 and formation of effective institutional framework. It also discusses the paradigm shift in disaster management at international level and the disaster management mechanisms in United States and SAARC nations. It enumerates on contextual references from Sendai Framework for DRR, Paris Climate Agreement and the new Sustainable Development Goals, besides other UN charters on the theme.

Role of ICT in Disaster Management is the theme of Chapter 3, which provides details on how Information and Communication Technology (ICT) is useful in all the phases of Disaster Management Continuum. In this chapter role of ICT in different phases of disaster management has been discussed systematically so as to familiarise the reader on basic conceptual contexts, needs, opportunity and scope of technological interventions and perspectives for their implementation. This elaborates upon development and suitability of various technologies to meet out specific needs of Pre –disaster, during- disaster, and post- disaster phases of disaster management continuum.

Chapter 4 focused on Early Warning Systems, and presented detailed account on the basic concept and scope of the Early Warning Systems for different disasters as a people centred approach of Disaster Management. It explains about four essential components of early warning systems, *viz.*,

a) Risk Knowledge
b) Monitoring, Simulations and Predictions
c) Disseminating relevant Information
d) Coordinated Response

In this chapter Early warning systems their components, institutional management and network framework has been discussed in a very systematic way. The chapter describes about early warning needs of all major hazards, with some case studies on implementation and application.

Components of Disaster Communication are enumerated in the Chapter 5. It provides a systematic coverage on various components of Communication technology in general and its connotation for disaster management, in particular. It established a prelude and context for the chapter on existing communication base in India for disaster management. This chapter has taken 360° approach in describing Communication. The chapter explains about communication definitions, concepts, components, equipments and technologies and tools of all verbal, non-verbal communication. In this chapter all contemporary technologies being used for Wireline and Wireless Communications have been detailed out with their basic fundamentals, technological background and application in disaster specific perspectives.

Chapter 6 give detailed account on ICT enabled Disaster Communication and Early Warning in India. There is urgent need to understand importance of integration of disaster communication and early warning systems. In order to provide a better and effective understanding of disaster communication and warning using ICT tools, the situation analysis of India has been presented as reference example here. This chapter provides a review and assessment of state of technological advances and application for disaster management, from assessments forecasting and early warning, to disaster response, relief communication and incident reporting, in the country. It brings in the systematic knowledge about existing communication facilities available through public networks in India. The outcome of situation analysis is summed up very nicely to provided intended understanding of emergency communication requirements.

Future Directions and Scoping in area of early warning and disaster communication are discussed in the last chapter, *i.e.* Chapter 7. Besides outlining the plausible solutions, it indicates on new options and best practices, which may be adopted in the field of Disaster Management and Emergency Communication in near future in a country, more particularly in developing countries. Chapter enumerates the essential actions as long term and short terms measures to address existing deficiencies of disaster communication facilities. Future directions detailed out in this chapter are not limited to the field of communication; it transcends to key cross cutting issues requiring immediate attention. This chapter may be seen as an overall guide for attaining intended paradigm shift in existing approach and methodology to be adopted for Disaster Management.

Notes

1. https://www.unisdr.org/we/inform/terminology
2. https://www.unisdr.org/we/inform/terminology
3. Ibid as 1,2 above
4. Ibid
5. http://www.ifrc.org/en/what-we-do/disaster-management/
6. http://www.worldbank.org/en/news/feature/2014/09/08/from-adversity-to-opportunity-how-the-aftermath-of-disaster-can-lead-to-safer-future
7. http://www.worldbank.org/en/news/press-release/2013/06/03/losses-from-disasters-in-east-asia-and-pacific-raise-concerns-for-poverty-reduction
8. Swiss Re is a leading wholesale provider of reinsurance, insurance and other insurance-based forms of risk transfer.
9. http://economictimes.indiatimes.com/news/international/world-news/
10. http://www.swissre.com/media/news_releases/Global_insured_losses_from_disasters_reach_USD_37_billion_in_2015.html
11. http://ejap.org/environmental-issues-in-asia/natural-disasters-asia.html
12. The rise of natural disasters in Asia and the Pacific: Learning from ADB's experience Mandaluyong City, Philippines: Asian Development Bank, 2013. As retrieved from https://www.adb.org/sites/default/files/evaluation-document/36114/files/rise-natural-disasters-asia-pacific.pdf
13. http://www.emdat.be/
14 The rise of natural disasters in Asia and the Pacific: Learning from ADB's experience Mandaluyong City, Philippines: Asian Development Bank, 2013. As retrieved from https://www.adb.org/sites/default/files/evaluation-document/36114/files/rise-natural-disasters-asia-pacific.pdf
15. http://wbi.worldbank.org/wbi/megadisasters
16. http://www.development.asia/PDF/issue16/devasia16-from-aceh-to-tacloban.pdf
17. http://ndmindia.nic.in/NPDM-101209.pdf

18. There were 160 people inside the AMRI hospital in south Kolkata, West Bengal India when the fire began in the hospital at about 3 am on December 9, 2011, when a fire that started in the basement of AMRI Hospital led to the death of 92 patients. Families of the victims and the government accused the privately-owned and posh hospital of ignoring basic fire safety laws; they also said that after the fire began, doctors and staff members abandoned patients, many of who were immobile.
19. http://ndmindia.nic.in/NPDM-101209.pdf
20. http://ndma.gov.in
21. Rules IV/3/8, IV/3/9, of Arthasastra
22. Rule IV/3/17: During a famine, the king should make a store of seeds and food-stuffs and show favour to the subjects with grant of food or share his provisions with them or entrust the country to another king.
23. Rule IV/3/18: Or, he should take shelter with allies or cause a reduction or shifting (of the population).
24. The eight Millennium Development Goals (MDGs) – which range from halving extreme poverty to halting the spread of HIV/AIDS and providing universal primary education, all by the target date of 2015 – form a blueprint agreed to by all the world's countries and all the world's leading development institutions. They have galvanised unprecedented efforts to meet the needs of the world's poorest
25. Allen 2006; Berke *et al.,* 1993; Drabek and Key 1976; Lewis.; Paton and Johnson 2001; Tobin 1999; Wisner *et al.,* 2003
26. Disaster management developing country communities and climate change The Role of ICT:NONOTA T. YAP 2011
27. http://www.un.org/sustainabledevelopment/development-agenda/
28. http://www.unisdr.org/we/coordinate/sendai-framework
29. http://eqs.metu.edu.tr/node/119
30. http://www.ndtv.com/article/india/cyclone-phailin-triggers-india-s-biggest-evacuation-operation-in-23-years-431466
31. https://en.wikipedia.org/wiki/Cyclone_Phailin
32. http://www.itu.int/ITU-D/emergencytelecoms/tampere.html
33. By Manish SHRIKHANDE, Durgesh C RAI, Jayprakash NARAYAN And Josodhir DAS Department of Earthquake Engineering, University of Roorkee, Roorkee, INDIA as retrieved from http://home.iitk.ac.in/dcrai details/ChamoliEQ.pdf
34. (Andreas Meissner, Thomas Luckenbach, Thomas Risse, Thomas Kirste, Holger Kirchner 1 {meissner, risse, hkirch}@ipsi.fhg.de, luckenbach@fokus.fhg.de, kirste@rostock.igd.fhg.de 1 Fraunhofer IPSI - Integrated Publication

and Information Systems Institute, Darmstadt, Germany 2 Fraunhofer FOKUS - Institute for Open Communication Systems, Berlin, Germany 3 Fraunhofer IGD - Institute for Computer Graphics, Rostock, Germany)

35. The First IEEE Workshop on Disaster Recovery Networks (DIREN 2002), June 24, 2002, New York City, co-located with IEEE INFOCOM 2002.
36. Disaster Risk Reduction in South Asia edited by Pardeep sahani and Madhavi Malalgoda Ariyabandhu
37. www.fema.gov/disaster-emergency-communications
38. Retrieved from https://www.itu.int/itunews
39. World Bank Independent Evaluation Group, 2006, Hazards of Nature, Risks to Development, (Washington, DC: World Bank), p. 116, http://www.worldbank.org/ieg/naturaldisasters/docs/natural_disasters_evaluation.pdf.
40. Disaster management developing country communities and climate change The Role of ICT:NONOTA T. YAP 2011.
41. Coyle and Meier 2009; Denning 2006; Gunawardene and Noronha 2007.

Chapter 2

PARADIGM SHIFT IN DISASTER MANAGEMENT

Paradigm Shift at Global Level

The concept of sustainable development dates back a long way but it was at the UN Conference on Human Environment (Stockholm, 1972) when the international community met for the first time to consider the global environment and highlighted the need to support people in this process[1]. The Conference indicated that industrialized environmental problems, such as habitat degradation, toxicity and acid rain, were not necessarily relevant issues for all countries and communities. However, it was the existing environmental problems that dominated the meeting and led to wider public environmental awareness.

In the 1980s, the UN set up the World Commission on Environment and Development, also called the Brundtland Commission. They produced Our Common Future, otherwise known as the Brundtland Report, which framed much of what would become Agenda 21 and the 27 principles of the Rio Declaration on Environment and Development. It defined sustainable development as development which "meets the needs of present generations without compromising the ability of future generations to meet their own needs."[2]

International Decade on Natural Disaster Reduction

In December 1987, it was decided to designate the 1990s as a decade in which the international community, under the auspices of the United Nations, would pay special attention to fostering international co-operation in the field of natural disaster reduction:[3]

1. Considering that natural disasters had adversely affected the lives of a great number of people and caused considerable damage to infrastructure and property worldwide, especially in developing countries;

2. Recognizing the necessity for the international community to demonstrate the strong political determination required to mobilize and use existing scientific and technical knowledge to mitigate natural disasters, bearing in mind in particular the needs of developing countries; and
3. Emphasizing that appropriate emergency planning for natural disasters and its integration in national development plans could also be very helpful in preventing other kinds of disasters, such as those of an industrial or technological nature.

The prime goals decided to be achieved during the International Decade on Natural Disaster Reduction were the following;

1. To improve the capacity of each country to mitigate the effects of natural disasters expeditiously and effectively, paying special attention to assisting developing countries in the assessment of disaster damage potential and in the establishment of early-warning systems and disaster-resistant structures when and where needed;
2. To devise appropriate guidelines and strategies for applying existing scientific and technical knowledge, taking into account the cultural and economic diversity among nations;
3. To foster scientific and engineering endeavours aimed at closing critical gaps in knowledge in order to reduce loss of life and property;
4. To disseminate existing and new technical information related to measures for the assessment, prediction and mitigation of natural disasters; and
5. To develop measures for the assessment, prediction, prevention and mitigation of natural disasters through programmes of technical assistance and technology transfer, demonstration projects and education and training, tailored to specific disasters and locations, and to evaluate the effectiveness of those programmes.

Expected Role of Governments[4]

For attainment of goals identified to be achieved during the decade, various stakeholders had to play their respective parts. Expected role of Governments of member countries included the following;

- ☆ To formulate national disaster-mitigation programmes, as well as economic, land use and insurance policies for disaster prevention, and particularly in developing countries, to integrate them fully into their national development programmes;
- ☆ To participate during the Decade, in concerted international action for the reduction of natural disasters and, as appropriate, establish national committees in co-operation with the relevant scientific and technological communities and other concerned sectors with a view to attain the objective and goals of the Decade;

- To encourage their local administrations to take appropriate steps to mobilize the necessary support from the public and private sectors and to contribute to the achievement of the purposes of the Decade;
- To take measures, as appropriate, to increase public awareness of risk probabilities and of the significance of preparedness, prevention, relief and short-term recovery activities with respect to natural disasters and to enhance community preparedness through education, training and other means, taking into account the specific role of the news media;
- To pay due attention to the impact of natural disasters on health care, particularly to activities to reduce the vulnerability of hospitals and health centres, as well as the impact on food storage facilities, human shelter and other social and economic infrastructures;
- To improve the timely availability of appropriate emergency supplies through the storage or earmarking of such supplies in disaster-prone areas.

Role of Scientific and Technological Institutions, was focused for attainment of goals identified to be achieved during the decade, various stakeholders had to play their respective parts. Scientific and technological institutions, financial institutions, including banks and insurance companies, and industrial enterprises, foundations and other related non-governmental organizations were also encouraged to support and participate fully in the programmes and activities of the Decade prepared and implemented by the international community, including governments, international organizations and non-governmental organizations.

Yokohama World Conference on Natural Disaster Reduction

The States Members of the United Nations and other States, met at the World Conference on Natural Disaster Reduction, in the city of Yokohama, Japan, during 23 - 27 May 1994, in partnership with non-governmental organizations, and with the participation of international organizations, the scientific community, business, industry and the media, for deliberation on the framework of the International Decade for natural Disaster Reduction and came up with Yokohama Strategy and Plan of Action for a Safer World. In this conference it was acknowledged that, "*Disaster prevention, mitigation and preparedness are better than disaster response in achieving the goals and objectives of the Decade. Disaster response alone is not sufficient, as it yields only temporary results at a very high cost, whereas prevention contributes to lasting improvement in safety and is essential to integrated disaster management*".[5]

The adopted Yokohama Strategy and related Plan of Action for the rest of the Decade and beyond are as under:

- Will note that each country has the sovereign responsibility to protect its citizens from natural disasters;

- Will give priority attention to the developing countries, in particular the least developed, land-locked countries and the small island developing States;
- Will develop and strengthen national capacities and capabilities and, where appropriate, national legislation for natural and other disaster prevention, mitigation and preparedness, including the mobilization of non-governmental organization and participation of local communities;
- Will promote an strengthen sub regional, regional and international cooperation in activities to prevent, reduce and mitigate natural and other disasters, with particular emphasis on:
  - Human and institutional capacity-building and strengthening;
  - Technology sharing, the collection, the dissemination and utilization of information; and
  - Mobilization of resources.

International Strategy for Disaster Reduction (ISDR)

Formulated by the United Nations in 2000, the DRR Framework propagated by the UN International Strategy on Disaster Reduction (ISDR) focused on risk awareness and assessment, hazard analysis and vulnerability/capacity analysis, knowledge development, public commitment, institutional frameworks, environmental management, land-use and urban planning, protection of critical facilities, application of science and technology, partnership and networking, financial instruments, early warning systems, research and development.[6]

The United Nations Framework Convention on Climate Change (UNFCCC): The Convention binds its member states (155 as on now) to:

- Promote and cooperate in the development, application, diffusion and transfer of those technologies, practices and processes that control, reduce or prevent anthropogenic (manmade) emissions of greenhouse gases;
- Encourage sustainable development;
- Cooperate in adapting to climate change;
- Promote scientific and technological research in information dissemination and management on climate change; and
- Impart/promote education, awareness and training related to climate change.

Kyoto Protocol

The Kyoto Protocol that came into existence in 2005 focused on good governance, development and poverty reduction, reduction in emissions of greenhouse gases during a five-year period (2008-2012) by involving all major sectors of economy. At Cancun 2010, the countries had struggled hard to reach to consensus on cuts on emissions, which shows how different countries are weighing their stakes in their bid at 'development Vs. environment' debate.[7] The Kyoto Protocol is now replaced by the new international accord on climate change, called as Paris Agreement (2015-30).

The Millennium Development Goals

The UN Millennium Summit in New York in September 2000, attended by 189 world leaders, adopted the UN Millennium Declaration (A/RES/55/2). Targets, known as the Millennium Development Goals were established, setting a new milestone and providing guiding principles for the international community, national governments and, in particular, the United Nations. Many of them touch on areas which are closely linked to vulnerability to natural hazards, such as eradicating extreme poverty and hunger, achieving universal primary education, promoting gender equality, ensuring environmental stability and partnerships for development. For example, the goal of improving the lives of thousands of slum dwellers around the world living in high-risk areas by 2020, involves poverty eradication, proper land use planning and the improved understanding of vulnerability to disasters in densely populated areas[8]

Millennium Declaration

"We recognize that, in addition to our separate responsibilities to our individual societies, we have a collective responsibility to uphold the principles of human dignity, equality and equity at the global level. As leaders, we have a duty therefore to al the world's people, especially the most vulnerable and, in particular, the children of the world, to whom the future belongs." Under "Protecting our common environment" the declaration resolves "to adopt in all our environmental action a new ethic of conservation and stewardship and, as first steps, resolves.to intensify cooperation to reduce the number and effects of natural and man-made disasters." Strategies linked to ISDR for moving ahead on this goal as described in the "Road map towards the implementation of the United Nations Millennium Declaration" (Secretary-General Report to GA A/56/326) include the following mandates:

- ☆ Developing early warning systems, vulnerability mapping, technological transfer and training;
- ☆ Supporting interdisciplinary and inter sectoral partnerships, improved scientific research on the causes of natural disasters and better international cooperation to reduce the impact of climate variables, such as El Niño and La Niña;
- ☆ Encouraging governments to address the problems created by megacities, the location of settlements in high-risk areas and other manmade determinants of disasters; and
- ☆ Encouraging governments to incorporate disaster risk reduction into national planning processes, including building codes.

Government of India has also been one of the signatories of MDG. India shows mixed progress report in achieving targets set under MDG. In some indicators like improvement in child and maternal health care it has fared well but in some indicators like hunger eradication and ensuring sanitation facilities its status is dismal.[9] The MDGs are now replaced by the more holistic framework known as Sustainable Development Goals (2015-30).

Hyogo Framework for Action (HFA) 2005-2015

Despite the rising global trend in the occurrence of disasters and associated economic losses, global loss of life associated with geophysical, meteorological, hydro or climate-related hazards in 2005, decreased to one-tenth of levels in the 1950's. This remarkable decline was a demonstration that mitigation, preparedness and capacity development combined with effective early warning systems and disaster management can significantly reduce the impacts of hazards on human life. This was acknowledged during the Second World Conference on Disaster Risk Reduction (Hyogo, Kobe, Japan 18-22 January 2005) when 4000 participants from 168 countries adopted the Hyogo Framework for Action (HFA) 2005- 2015 for Building the Resilience of Nations and Communities to Disasters. The broad objective of the conference was building the resilience of nations and communities to substantially reduce the losses in lives and social, economic and environmental assets of communities. [10]

The specific objectives were:

- ✰ Conclude the review of the Yokohama Strategy and Plan of Action with a view to updating the guiding framework on disaster reduction for the 21st century;
- ✰ Identify specific activities aimed at ensuring the implementation of relevant provisions of the Johannesburg Plan of Implementation (JPOI), adopted in 2002 at the World Summit on Sustainable Development (WSSD);
- ✰ Share best practices and lessons learned to support and facilitate disaster reduction within the context of attaining sustainable development, and identify gaps and challenges;
- ✰ Increase awareness of the importance of disaster reduction policies to facilitate and promote the implementation of those policies;
- ✰ Increase the reliability and availability of appropriate disaster-related information to the public and disaster management agencies in all regions, as set out in the relevant provisions of the JPOI.

Thus, HFA calls for a paradigm shift from emergency response to DRM in a more proactive and holistic approach where multiservice (of voice, data and video) knowledge-based Information Services would play a crucial role. The Hyogo Framework for Action (HFA) is the first plan to explain, describe and detail the work that is required from different sectors and actors to reduce disaster losses. It was developed and agreed on with many partners needed to reduce disaster risk governments, international agencies, disaster experts and many others bringing them into a common system of coordination. The HFA outlines five priorities for action, and offers guiding principles and practical means for achieving disaster resilience. Its goal is to substantially reduce disaster losses by 2015 by building the resilience of nations and communities to disasters. This means reducing loss of lives and social, economic, and environmental assets when hazards strike.HFA has identified 5 important Priority Actions, as given under:

Priority Action 1

Ensure that disaster risk reduction is a national and a local priority with a strong institutional basis for implementation.

Countries that develop policy, legislative and institutional frameworks for DRR and that are able to develop and track progress through specific and measurable indicators have greater capacity to manage risks and to achieve widespread consensus for, engagement in and compliance with disaster risk reduction measures across all sectors of society

Priority Action 2

Identify, assess and monitor disaster risks and enhance early warning.

The starting point for reducing disaster risk and for promoting a culture of disaster resilience lies in the knowledge of the hazards and the physical, social, economic and environmental vulnerabilities to disasters that most societies face, and of the ways in which hazards and vulnerabilities are changing in the short and long term, followed by action taken on the basis of that knowledge.

Priority Action 3

Use knowledge, innovation and education to build a culture of safety and resilience at all levels.

Disasters can be substantially reduced if people are well informed and motivated towards a culture of disaster prevention and resilience, which in turn requires the collection, compilation and dissemination of relevant knowledge and information on hazards, vulnerabilities and capacities.

Priority Action 4

Reduce the underlying risk factors.

Disaster risks related to changing social, economic, environmental conditions and land use, and the impact of hazards associated with geological events, weather, water, climate variability and change are addressed in sector development planning and programmes as well as in post-disaster situations.

Priority Action 5

Strengthen disaster preparedness for effective response at all levels.

At times of disaster, impacts and losses can be substantially reduced if authorities, individuals and communities in hazard-prone areas are well prepared and ready to act and are equipped with the knowledge and capacities for effective disaster management. Accordingly many countries, adopted HFA and prepared Disaster Management plans and strategies.

Sendai Framework for Disaster Risk Reduction (2015-2030)

The Sendai Framework for DRR (SFDRR) was adopted at the Third UN World Conference in Sendai, Japan, on March 18, 2015. It is the successor instrument of HFA 2005-2015 and is founded on elements which ensure continuity with the work done by States and other stakeholders under HFA. It is a non-binding voluntary agreement for 15 years which recognizes the responsibility of States for DRR and the shared responsibility with local government, private sector and other stakeholders. SFDRR improves on HFA by identifying the gaps, good lessons learned and future challenges. Key features of SFDRR are:

- ✰ Shifting focus from disaster management to disaster risk management by focusing on the underlying drivers of risk.
- ✰ For the first time one global goal and outcome is defined.
- ✰ Seven global targets are defined to support the assessment of global progress in achieving the outcome and goal of the present Framework.
- ✰ The Framework emphasizes the need of strengthening the disaster risk governance by placing governments at the centre of disaster risk reduction.
- ✰ A wider scope of DRR focusing on both natural and man-made hazards and related environmental, technological and biological hazards and risks.
- ✰ A set of guiding principles are provided for the implementation of Framework.
- ✰ Learning from the experience gained by the implementation of HFA and to achieve the expected outcome and goal, the Framework prioritizes the actions into four priority areas.
- ✰ Along with social vulnerability, great emphasis is given to environmental aspects by strongly recognizing that implementation of integrated environmental and natural resource management techniques are needed for disaster reduction.
- ✰ DRR is identified as a policy concern which cuts across many sectors, including health and education.

According to SFDRR, disaster risk reduction can be achieved by addressing the existing challenges and preparing for future challenges by putting emphasis

Scope of Sendai Framework

<u>Small scale, biological and man-made hazards are added to scope-</u> "The framework will apply to the risk of small-scale and large-scale, frequent and infrequent, sudden and slow-onset disasters, caused by natural or man-made hazards as well as related environmental, technological and biological hazards and risks."

<u>Scope of action in recovery, rehabilitation and reconstruction is increased-</u> "Enhancing disaster preparedness for effective response and to "Build Back Better in recovery, rehabilitation and reconstruction"

Global Outcome

GOAL	OUTCOME
"The substantial reduction of disaster risk and losses in lives, livelihoods and health and in the economic, physical, social, cultural and environmental assets of persons, businesses, communities and countries."	"Prevent new and reduce existing disaster risk through the implementation of integrated and inclusive economic, structural, legal, social, health, cultural, educational, environmental, technological, political and institutional measures that prevent and reduce hazard exposure and vulnerability to disaster, increase preparedness for response and recovery, and thus strengthen resilience."

13 Guiding Principles & 7 Global Targets

1. Substantially reduce global disaster mortality by 2030, aiming to lower the average per 100,000 global mortality rate in the decade 2020–2030 compared to the period 2005–2015
2. Substantially reduce the number of affected people globally by 2030, aiming to lower the average global figure per 100,000 in the decade 2020–2030 compared to the period 2005–2015
3. Reduce direct disaster economic loss in relation to global gross domestic product (GDP) by 2030
4. Substantially reduce disaster damage to critical infrastructure and disruption of basic services, among them health and educational facilities, including through developing their resilience by 2030
5. Substantially increase the number of countries with national and local disaster risk reduction strategies by 2020
6. Substantially enhance international cooperation to developing countries through adequate and sustainable support to complement their national actions for implementation of the present Framework by 2030
7. Substantially increase the availability of and access to multi-hazard earlywarning systems and disaster risk information and assessments to people by 2030

The Four Priorities for Action

- 1st: Understanding disaster risk
- 2nd: Strengthening disaster risk governance to manage disaster risk
- 3rd: Investing in disaster risk reduction for resilience
- 4th: Enhancing disaster preparedness for effective response and to "Build Back Better" in recovery, rehabilitation and reconstruction.

Figure 2.1: Summary of Sendai Framework for Disaster Risk Reduction.

on monitoring, assessing, and understanding disaster risk and sharing such information. The Framework highlights that it is "*urgent and critical to anticipate, plan for and reduce disaster risk*" to cope with disaster. It identifies the importance of strengthening disaster risk governance and coordination across various institutions and sectors. It recognizes the primary role of states and the shared responsibility of other stakeholders at different levels. The need for investment in economic, social, health, cultural and educational resilience at all levels is identified. For the enhancement of multi-hazard Early Warning Systems (EWS), preparedness, response, recovery, rehabilitation, and reconstruction, it emphasizes on the need of investments in research and appropriate use of technology.

Sustainable Development Goals (SDGs)

The UN Summit in September 2015 on Post-2015 Development Agenda formally adopted the 'Transforming Our World: The 2030 Agenda for Sustainable Development'. It includes 17 global sustainable development goals with 169 targets. 193 UN Member States endorsed the SDGs at the 70th Session of UN General Assembly. The three focus areas of SDGs are: i) Poverty eradication, ii) Safeguard the planet from degradation while making sure that economic, social and technological advancement occurs in harmony with nature; and (iii) Encouraging global peace and just and inclusive societies.

DRR, CCA and Resilience cut across different aspects and sectors of development. Twenty five targets of the new SDG Framework are directly or indirectly related to DRR in 10 of the 17 SDGs. The agenda identifies and asserts the immediate needs to reduce climate and disaster risk and emphasizes on resilience building of communities and nations to achieve the SDGs. Explicit references for DRR, CCA and resilience can be observed in goals and targets specially related to poverty, hunger, healthy lives, building resilient infrastructure, education, sustainable management of water, climate change, resilient cities and marine and terrestrial ecosystem. These references show that the new SDG framework is a further step towards DRR and it largely rectifies the drawback of previous development framework *i.e.* the MDG framework which overlooked the importance of CCA and DRR to reduce the impacts of disasters on societies and economies.

Paris Climate Agreement

The Paris Agreement on Climate Change was adopted at the UN Climate Change Conference held in December 2015. The agreement was adopted by 195 countries and called for a commitment to work together in order to safeguard the planet, promote sustainable human development and build more resilient and equitable world for all. In Paris, the member countries agreed to:

- Hold the increase in global average temperature to well below 2°C above pre- industrial levels and to pursue efforts to limit the increase to 1.5°C.
- The agreement is a first ever goal for global adaptation. Building on The Cancun Adaptation Framework 2010, which is based on DRR,

Table 2.1: DRR-CCA Targets in SDGs and Sendai Framework

Sustainable Development Goal	*Related DRR or CCA target*	*How Sendai Framework will Help to Achieve the Goal/Target*
Goal 1: End poverty in all its forms everywhere	Target 1.5	*"To achieve this goal and target, Sendai Framework proposes for the promotion and development of social safety nets linked with livelihood enhancement programmes in order to ensure resilience of household and communities to disasters."*
Goal 2: End hunger, achieve food security and improved nutrition and promote sustainable agriculture	Target 2.4	*"To achieve this goal and target in context of Sendai Framework, relevant actions including strengthening productive assets such as livestock, working animals, tools and seeds are required."*
Goal 3: Ensure healthy lives and promote well-being for all at all ages	Target 3.d	*"This target in particular is complemented by the outcome of Sendai Framework which has placed strong emphasis on the resilience of health systems and integration of disaster risk reduction into health care provision at all levels."*
Goal 4: Ensure inclusive and equitable quality education and promote lifelong learning opportunities for all	Target 4.7 and Target 4.a	*"In order to progress these target actions, implementation needs to consider promoting disaster risk knowledge at all levels including in professional education and training as recommended by the Sendai Framework."*
Goal 6: Ensure availability and sustainable management of Water and Sanitation for all.	Target 6.3, 6.4, 6.5, 6.6, 6.a and 6.b	*"Target 6.6 indirectly provides an opportunity to mainstream ecosystem-based approaches for disaster risk reduction and further highlight their value as a 'win-win' and 'no regrets' solution to the increasing disaster and climate risks underlined in the Sendai Framework."*
Goal 9: Build resilient infrastructure, promote inclusive and sustainable industrialization and foster innovation	Target 9.1 and Target 9.a	*"In order to progress these targets and goal, the Sendai Framework recommends strengthening disaster resilient public and private investments through structural, non-structural and functional disaster risk prevention and reduction measures in critical facilities, in particular schools and hospitals and other physical infrastructure."*
Goal 11: Make cities and human settlements inclusive, safe, resilient and sustainable	Target 11.1,11.3, 11.4, 11.5, 11.b, 11.c	*"Measures to achieve these targets and goal, as outlined in the Sendai Framework, require mainstreaming of disaster risk assessments into land-use policy development and implementation, including urban planning, land degradation assessments and informal and non-permanent housing, and the use of guidelines and follow-up tools informed by anticipated demographic and environmental changes."*
Goal 13. Take urgent action to combat climate change and its impacts	Target 13.1, 13.2, 13.3, 13.a and 13.b	*"In order to achieve these targets and the overall goal, the Sendai Framework recommends to strengthen disaster risk modeling, assessment, mapping, monitoring and multi-hazard early warning systems; promote the conduct of comprehensive surveys on multi-hazard disaster risks and the development of regional disaster risk assessments and maps, including climate change scenarios; and maintain and strengthen in situ and remotely sensed earth and climate observation."*

Sustainable Development Goal	*Related DRR or CCA target*	*How Sendai Framework will Help to Achieve the Goal/Target*
Goal 14. Conserve and sustainably use the oceans, seas and marine resources for sustainable development	Target 14.2	*"The Sendai Framework explicitly seeks to account for the environmental damages caused by disasters – in many cases damages are attributable to the removal of disaster waste and to impacts associated with recovery and reconstructions planning that have by-passed existing environmental legislation."*
Goal 15. Protect, restore and promote sustainable use of terrestrial ecosystems, sustainably manage forests, combat desertification, and halt and reverse land degradation and halt biodiversity loss	Target 15.1,15.2, 15.3, 15.4 and 15.9	*"These targets are also in line with the Sendai Framework's focus on building environmental resilience through the inclusion of ecosystems in risk analysis and planning.* *As per marine ecosystems, the Sendai Framework proposes similar priority actions for their terrestrial equivalents - mountains, rivers, coastal flood plain areas, dry lands and wetlands, among others."*

Sources: Transforming Our World: The 2030 Agenda for Sustainable Development, 2015 and, Disaster Risk Reduction and Resilience in the 2030 Agenda for Sustainable Development, 2015).

this agreement considers enhancing adaptive capacity, strengthening resilience and reducing vulnerability to climate change.

- *"Enhance understanding, action and support with respect to loss and damage associated with the adverse effects of climate change with focus on early warning systems and emergency preparedness, comprehensive risk assessment and management and risk insurance facilities, climate risk pooling and other insurance solutions"*[11]

Paris Agreement is legally binding to all the parties unless specifically excluded. It contains all greenhouse gas emissions from all sectors and human activities, sources and sinks. The agreement will bring about a process of maintaining a balance between mitigation and adaptation. In future approaches of climate adaptation, the loss and damage caused by disasters will be better incorporated to accomplish the goal of climate resilience. The preamble of agreement mentions about the adoption of Sendai Framework for Disaster Risk Reduction. There is a synergy in the language of agreement and goals of SFDRR to reduce risks and disaster losses. The role of sustainable development in order to reduce the loss and damage associated with climate change impacts, including extreme events and slow onset events is given great importance. National ratification by governments and implementation of national climate action plans are the steps towards implementing Paris Agreement.

Synergizing SDGs, Paris Climate Agreement and Sendai Framework

"Disaster Risk Reduction (DRR) is the concept and practice of reducing disaster risks through systematic efforts to analyze and reduce the causal factors of disasters. Reducing exposure to hazards, lessening vulnerability of people and property, wise management of land and the environment, and improving preparedness and early warning for adverse events are all examples of disaster risk reduction."[12] The disciplines like disaster management, disaster mitigation and disaster preparedness are included in DRR. In the 1970's the prevalent global paradigm was ad hoc disaster response. A permanent disaster management approach came into existence in 1980's which continuously deals with all aspects of disaster management cycle: preparedness, response, recovery and reconstruction. Continuous efforts are being done for the last twenty five years to address the challenges of repeating disasters and to come out of the cycle of disaster through preventive risk management (HDR, 2016).

The importance of DRR to achieve sustainable development was identified in Millennium Development Summit in September 2010. The High-level Plenary Meeting on accelerating progress towards the achievement of all Millennium Development Goals (MDGs) by 2015 identified that disaster risks are increasing globally due to increasing vulnerability to natural hazards. A risk-sensitive approach would be required for sustainability and accelerated achievement of MDGs.

The year 2015 was a landmark year for the United Nations and Global Development Agenda. The convergence of interests and global concerns for sustainable development, disaster risk reduction and climate change led to the formation of a new roadmap for a sustainable and safe world together: The Sendai framework for disaster risk reduction, The 2030 agenda for sustainable development and The Paris Climate Agreement. These agreements of global significance provide opportunities to build coherence across different but overlapping policy areas.

The Hyogo Framework for Action (HFA), 2005-2015 recognized that DRR is a cross-cutting issue in relation to sustainable development and it is an important element to achieve internationally agreed development goals, including those contained in the Millennium Declaration. Rio+20 or The United Nations Conference

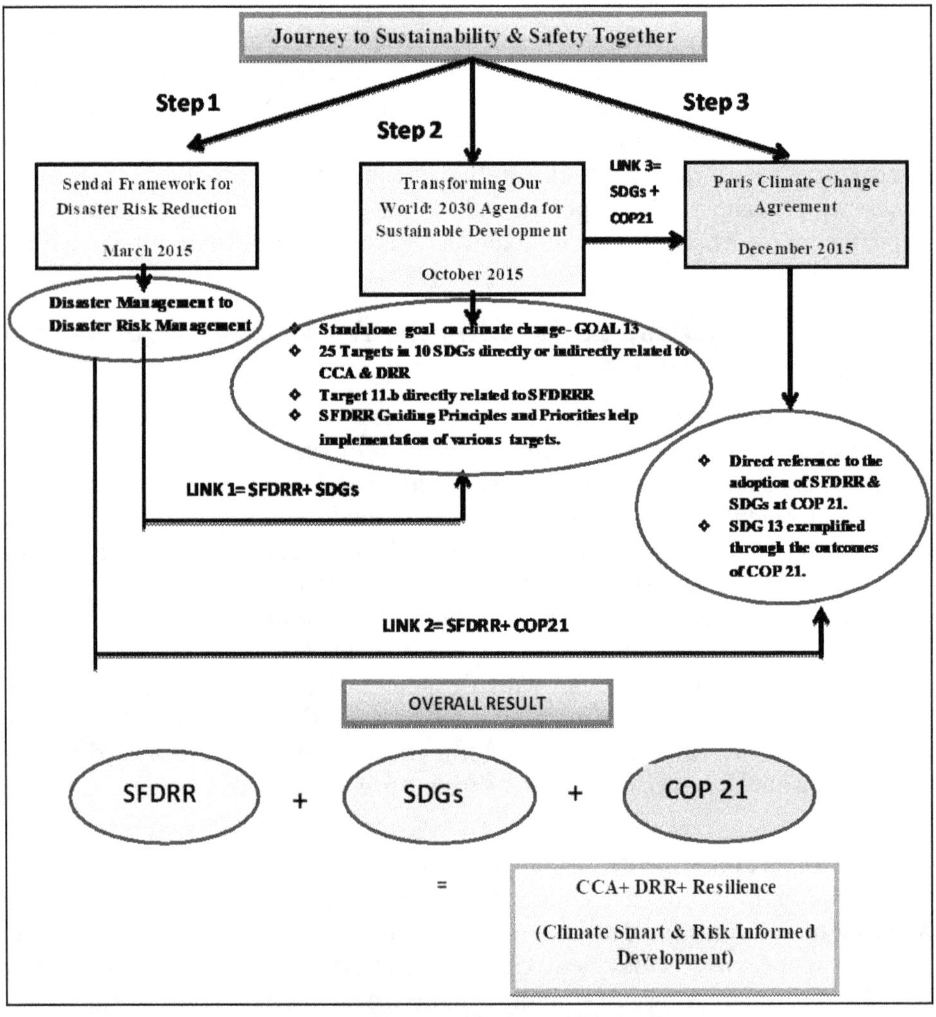

Figure 2.2: Paradigm Shift in Disaster Management through Synergies between SFDRR, Paris Agreement and SDGs.

on Sustainable Development took place in Rio de Janeiro, Brazil in June 2012. Member states of the conference launched a process to develop a set of Sustainable Development Goals (SDGs) that built upon the MDGs and converged with the post 2015 development agenda. The lessons learnt from 1.3 million deaths, 4.4 billion affected people and economic losses of worth $2 trillion due to disasters since the 1992 Rio Earth Summit called for an accelerated implementation on HFA 2005-2015.[13] Rio+20 Outcome Document- The Future We Want, contains a section on DRR which sets out a substantial foundation for consideration on post 2015 framework to guide nations after the expiry of HFA 2005-2015.

There is a growing global consensus that disaster risk reduction, climate change adaptation and sustainable development are linked to each other. Many evidences of linkages between the three agendas are observed while studying the Sendai Framework for Disaster Risk Reduction 2015-2030, Sustainable Development Goals 2030 and the Paris Climate Agreement 2015. All of them share a common aim of making the development sustainable. Commitment to the goals and their implementation must become a global priority. To ensure the achievement of SDGs, it is very important to consider current and future challenges caused by disasters and climate change.

Pre-IDNDR Disaster Management in India

Post independence (1947) and until the United National International Decade for Natural Disaster Reduction (IDNDR 1991-99) disaster management in India has been more of a contingency management oriented. A Contingency Action Plan for dealing with contingencies arising in the wake of natural calamities was developed at national level, with provision of periodic updating. The major emphasis in the plan was on providing relief in response to the occurrence of a major calamity due to any of the natural hazards beyond the coping capacity of the States or Union Territories through their own resources. The preparedness for providing relief was ensured through Contingency Plans formulated for operations at the State levels as well as at the District levels. The primary relief functions of the Central Government included the following aspects:

1. Forecasting and operation of warning systems,
2. Maintenance of uninterrupted communication,
3. Wide publicity to warnings of impending calamity, disaster preparedness and relief measures through television, radio and newspapers,
4. Transport with particular reference to evacuation and movement of essential commodities and petroleum products,
5. Ensuring availability of essential commodities at reasonable prices particularly the commodities through the Public Distribution System,
6. Ensuring availability of medicines, vaccine and drugs,
7. Preservation and restoration of physical communication links,

8. Investments in infrastructure, and
9. Mobilization of financial resources.

The High Powered Committee on Disaster Management (HPC) in its report submitted in 2001 gave clear recommendations on the preparation of Disaster Management Plan(s) and suggestions for effective mitigation mechanisms. Its recommendation included the draft of the Disaster Management Act, a National Response Plan, strategy to move from disaster response to disaster preparedness, and establishment of National Disaster Management Authority. The ten years period of post HPC paradigm shift in India was parallel and also in compliance with the Hyogo Framework of Action (2005-15).

The Yakohama Strategy for disaster management (1994) was renewed at the world conference on Disaster Reduction held at Hyogo (Japan) in January 2005. The conference laid emphasis on some crucial but neglected aspects of disaster management such as governance and policy framework, risk identification and early warning, knowledge management, reducing risk factors and preparedness for effective response and recovery. The Hyogo conference adopted the framework of Action, 2005-2015 called "Building the Resilience of Nations and Communities to Disasters." Disaster management frameworks such as the Hyogo Framework for Action (HFA) 2005 - 2015 have played an important role in advancing the agenda for DRR (Djalante *et al.,* 2012). The HFA was adopted by 168 countries (including India) and supports the efforts of nations and communities to become more resilient to, and cope better with, the hazards that threaten their development gains.

The HFA recognized *that sustainable development, poverty reduction, good governance and DRR are mutually supportive objectives* (*UNISDR, 2007*). Taking the objectives into account, and drawing on the conclusions of the review of the *Yokohama Strategy, the expected outcome for the next 10 years was resolved as "The substantial reduction of disaster losses, in lives and in the social, economic and environmental assets of communities and countries",* and adopted following strategic goals to attain the expected outcome:

a) The more effective integration of disaster risk considerations into sustainable development policies, planning and programming at all levels, with a special emphasis on disaster prevention, mitigation, preparedness and vulnerability reduction;
b) The development and strengthening of institutions, mechanisms and capacities at all levels, in particular at the community level, that can systematically contribute to building resilience to hazards;
c) The systematic incorporation of risk reduction approaches into the design and implementation of emergency preparedness, response and recovery programmes in the reconstruction of affected communities.

India has been following Five Year National Plans (under Planning Commission). The earlier five year plans did not mention on disaster management.

The Tenth Five-Year Plan 2002-2007 for the first time had a detailed chapter entitled Disaster Management: The Development Perspective. The plan emphasized the fact that development cannot be sustainable without mitigation being built into the development process. Disaster mitigation and prevention were adopted as essential component of the development strategy. This period witnessed the passage of Disaster Management Act 2005 by the parliament which installed an era of paradigm shift from 'response and relief centric' to 'prevention-mitigation, preparedness centric' holistic disaster management approach.

Disaster Management Governance and Law (India)

India is a large country and has had more than its share of major natural hazards like drought, floods, earthquakes and cyclones throughout its history of civilization. With the industrial growth and technical revolution, the risk of industrial-chemical and accidents also grew manifolds. Besides natural, industrial/ chemical, technological, fire and transport disasters, there is a list of other mishaps, resulting into disasters, totally of civil origin, those related with law and order, or homeland security, for example, stampede, sabotage, hijacking, riots, terrorist attack, serial bomb blasts, *etc.* Naturally, the country developed its own practices and strategies for dealing with various natural calamities and human-made disasters. However, the mainframe of disaster management in India has been concerned mostly with the natural, environmental, industrial and technological disasters. Management of socio-civil management is adhered primarily to the law and order enforcement agencies.

Disasters are events of environmental extremes as inevitable entities of this living world. Disaster management highlights the interdependence of the economy, environment and inclusive development. The United Nation Report titled "Living with risk" claims that though there has been decline in the loss of human lives from disaster the occurrence of disaster is on rise (UNISDR, 2004). World's biggest industrial disaster – Gas tragedy 1984, that took place in Bhopal a central city of India, shaken the world and triggered global awakening towards human environment and safety, and led to a journey of environmental jurisprudence. However, this was followed by several environmental, technical and natural disasters in India and this part of the world. Disasters, especially the ones originating from natural environmental process – geological, hydrological, meteorological or biophysical, were dealt mostly in a conventional ad hoc mode of response and relief centric approach. Most international environmental strategies starting Stockholm Conference of 1972 referred to the natural hazards as key impediment to the process of sustainable development. Agenda-21 adapted at the Rio Summit 1992 is still believed to be relevant as a milestone strategy towards sustainable development by adopting integrated approach of environment, disaster reduction, safety, and sustainable economic growth. India's commitment and activities following the Rio Summit, which coincided with the UN-IDNDR (International Decade for Natural

Disaster Reduction 1991-99), helped naturally recognize the need of integrated and holistic approach to disaster management.

Major disasters, for example, earthquakes of Uttarkashi (1991), Latur (1993), Chamoli (1999), Assam floods (1998), Orissa super cyclone (1999), created an environment of serious brainstorm on the state of disaster management in India and on the steps needed to improve the situation. India was a party to Yokohoma declaration on disaster reduction for a safer world. The constitution of High Power Committee (HPC), chaired by Mr. J. C. Pant, former secretary of Indian Government, was a key step in this direction. HPC resulted in a detailed study report with set of fundamental and practical recommendations. The Gujarat Bhuj earthquake 2001 experience triggered the proposing of Disaster Management Bill, which, however, was in dormancy until another major disaster – Indian Ocean tsunami 2004 severely struck coastal states of India, along many other nations in the region. The year 2005 yielded with passage of Disaster Management Act which legally established a paradigm shift to 'prevention-mitigation based holistic disaster management'. Interestingly, year 2005 also witnessed India's participation in Kobe world conference on disaster reduction that resulted into Hyogo Framework for Action (2005-15).

Though India's pioneering law which enabled for systemic planning and preparedness for disaster emergencies and tiered approach of authorities, was the 'Emergency Planning, Preparedness and Response Rules 1996' under the Environmental Protection Act 1996 of India, the mechanism of holistic planning for disaster management and authorities at national, state, district and local level has been brought by the 2005 Disaster Management Act. In the backdrop of current awakening on climate change, environmental extremes and ecological degradation aggravating disasters and vulnerabilities, a systemic thought process of integrating environment management, disaster risk reduction and human development as imperative facets of sustainable development has emerged following the Intergovernmental Panel on Climate Change (IPCC) Fourth Assessment Report in 2007. This is suggested as 2nd paradigm shift in disaster management (Gupta and Nair, 2010), which now coincides with the new epoch of Sendai Framework for DRR, along Paris Agreement on Climate Change contributing towards the Sustainable Development Goals (2015). Social dimensions, thus, particularly those related with livelihood and food security, shelter, water and sanitation, environmental-health, gender sensitivity, empowerment, growth opportunities, ecological security and sustainable habitat have adequate and multiple references in the current strategies and plans of disaster risk reduction at almost all levels.

The implications of climate change and ecological degradation on the occurrence and intensity of natural disasters has been a matter of research and policy studies by international communities and also in India, in this regards the 4th Assessment report of Inter-government Panel on Climate Change (IPCC) in 2007 marked a global awakening. This triggered the evolution of a new paradigm of disaster management governance that recognizes the need of integrating with framework of climate change adaptation at various levels of planning and developmental activities. The period also witnessed and complimented by India's participation in the Millennium Ecosystem Assessment (2005), Millennium Development Goals for 2015, Kyoto Protocol on Green House Gas Emission (that ended in 2012). This new policy environment given rise to a new thinking on disaster risk reduction, termed as the 2nd paradigm shift in disaster management (Figure 2.3).

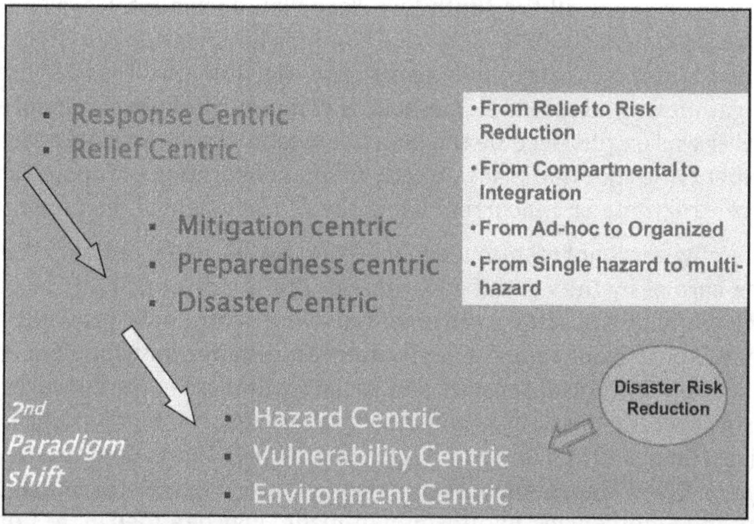

Figure 2.3: Paradigm Shifts in Disaster Management (Gupta and Nair, 2009).

Paradigm Shift in Disaster Management in India

Despite being a disaster prone country, India, traditionally had not been very proactive in managing disasters. After independence, in initial years some developments which took place with regard to disaster management practices were primarily as a result of supplementing the efforts of British Government. The Geological Survey of India (GSI), which was established during the colonial era, was modified and restructured, new specialized wings and regional offices were set up after independence. Specialization was made in seismotectonic studies and seismology. While there were only four seismological centers in 1950s, it sprang up to forty in 1989. Special cyclone warning radars were set up along the east and west coasts of India. India inaugurated its first flood forecasting station in 1958 on the river Yamuna for the national capital of Delhi. On the recommendations of various committees, a flood forecasting and warning organization was set up under the Central Water Commission (CWC) in 1969. The Famine Commission that was appointed during the British rule recommended the establishment of a separate Department of Agriculture in 1881, to deal with floods and droughts. Till 1970, the Department of Agriculture existed and it was responsible for the matters on floods and droughts. The task of disaster management, at an all India level, rested only with the Relief Commissioners in each state, who functioned under the Chief Relief Commissioner. Their role was limited to disbursement of relief material and money in the affected areas (Kanal, 2013).

The post-independence Indian Government initially did not realize the importance of disaster management, despite being aware of India's disaster history. This is evident through the fact that there had been no mention about disaster management till the Ninth Five Year Plan document. It was only in the Tenth Plan Document (2002-07) that the Planning Commission (2002) carried out a separate chapter on disaster management for the first time. Disaster prevention and mitigation was adopted as an essential component of development strategy, only after it was emphasized by the plan document. In the same manner, disaster management does not find place directly in the Constitution of India in any of the list, the central, state or concurrent list in the 7th Schedule of the Constitution.

Generally, items not mentioned in the 7th schedule of the Indian Constitutions are taken care of by the central government. But the responsibility of managing disasters lies with the state government and the centre only provides financial assistance. Apart from this, the legislation on disaster management has been related to entry 23 (Social Security and Social Insurance) in the concurrent list of the Constitution and the states would also be able to enact their own legislations on the subject (Goel, 2007). The Second Administrative Reforms Commission 2006 in its report on Crisis Management recommended that "a new entry, Management of Disasters and Emergencies, natural or man-made" may be added in the Concurrent List of the Seventh Schedule of the Constitution. However, the recommendation was not accepted and acted upon by the Government of India.

India is regularly facing severe hazards, some recent major disasters, which India has seen are: i) Orissa Super cyclone (29th October, 1999); ii) Bhuj Earthquake (26th January, 2001); iii) Indian Ocean Tsunami (26th December, 2004); and iv) J&K Earthquake (8th October, 2005) Orissa Super Cyclone of 1999, Bhuj earthquake of 2001, Tsunami on 2004, *etc.*, to name a few. Even latest incidence of Leh Cloudburst (August 6: 2010), Sikkim Earthquake (September 18:2011), AMRI hospital fire incident (December 9: 2011), Uttarakhand cloud burst (June 16-17:2013), Jammu and Kashmir flood (September:2014) are some examples of big disasters in India. These disasters have forced Indian Government to switch over from erstwhile reactive and response centric approach of Disaster Management to a proactive and holistic management of disasters. The Disaster Management Act (DM Act) was passed by the Parliament in the year 2005. The DM Act 2005 clearly spells out the institutional structures with corresponding functional responsibilities, to bring about paradigm shift and create a culture of Prevention, Preparedness and Mitigation to ensure resilience of the communities against disasters. It also ensures that in a post-disaster situation, the affected communities may be assured of sustainable livelihoods and reduced vulnerability to future disasters. Indian

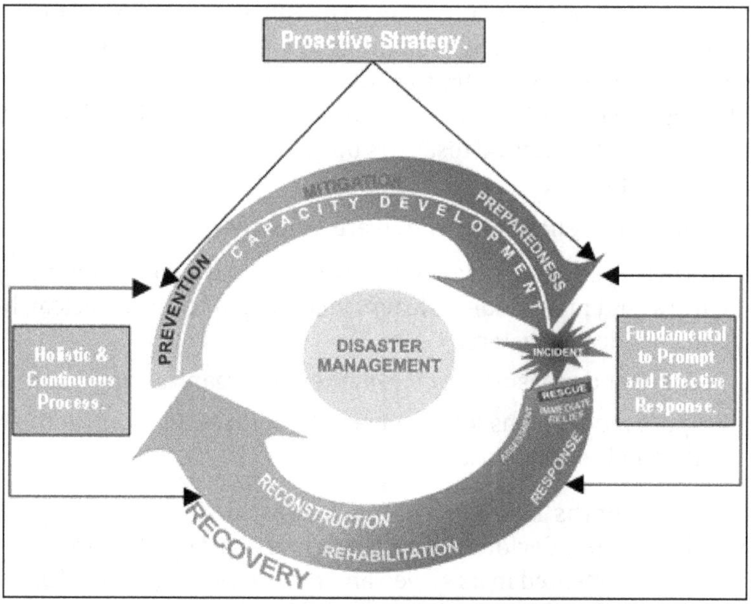

Figure 2.4: Disaster Management Continuum.

institutions, which are created to mange disasters, are expected to be accountable, participative, and transparent.

After DM ACT 2005 many national level institutes have been created. State Governments are sensitized and also being facilitated by central government for creation of similar hierarchy of institutions at state levels. A paradigm shift is now taking place at the national level from the relief centric syndrome to holistic and integrated approach with emphasis on prevention, mitigation and preparedness. These efforts are aimed to conserve developmental gains as well as to minimize losses to lives, livelihood during disasters. A typical Disaster Management Continuum as shown below, comprising of six elements that is, Prevention, Mitigation and Preparedness in pre-disaster phase, and Response, Rehabilitation and Reconstruction in post-disaster phase defines the complete approach to Disaster Management.[14] In the context of natural disasters, the holistic DM cycle has been centered on four overlapping phases: mitigation, preparedness, response, and recovery. The succeeding text attempts to analyse the measures taken by Government of India for enacting DM ACT 2005 and providing DM its due focus in policy making and governance:

High Powered Committee (HPC)

Having in pace with the international initiatives and multi-hazard vulnerability profile of the Country, a High Powered Committee (HPC) was constituted on 20 August 1999 under the Chairmanship of Mr. J. C. Pant, Former Secretary to Government of India in the Ministry of Agriculture, with members drawn from ministries, states and experts from the concerned organizations. The HPC was constituted to bring about institutional reforms in the areas of planning and management of disasters. It was also the first attempt in the country towards evolving a systematic, comprehensive and holistic approach for disaster management. While the original mandate of the HPC was confined to the natural disasters only, it was later on expanded to include man-made disasters also.

The terms of reference of the HPC were as follows:

(i) To review existing arrangements for preparedness and mitigation of natural and man-made disasters including industrial, nuclear, biological and chemical disasters;

(ii) Recommend measures for strengthening organizational structures; and

(iii) Prepare model plans for management of these disasters at the National, State and District levels.[15]

Barely two months had elapsed since constitution of the High Powered Committee, that a Super Cyclone struck the Orissa Coast on 29 October 1999. This cyclone was unprecedented in its sweep and ferocity, killing nearly 10,000 people and affecting over 15 million people across 12 districts of Orissa. Keeping in view the disastrous effects of the cyclone, the Government of India considered it as the calamity of rare severity to be handled at the national level. A High Powered Task Force was constituted after the disaster under the Chair of the Defence Minister with a view to:

(i) Prepare a comprehensive rehabilitation and reconstruction plan;

(ii) Recommend the mode of implementation including funding; and

(iii) Recommend mitigation measures that need to be taken in other areas of the country which were prone to such calamities.[16]

Empowered Group of Ministers: One year and four months later, on 26 January 2001, the country experienced an earthquake with magnitude of 6.9 on the Richter Scale in Bhuj area of Gujarat State. In this disaster, nearly 20,000 people died, over 1,55,000 were injured, and six lakh people were rendered homeless. In the aftermath of the earthquake, an Empowered Group of Ministers headed by the Home Minister was set up to monitor the situation and issue appropriate directions to the various ministries of the Central Government to coordinate relief activities. [17]

In addition to the Empowered Group of Ministers, an all-party National Committee on Disaster Management was constituted in February 2001 under the Chairmanship of the Prime Minister with the following terms of reference:

(i) To suggest short term, medium and long term steps for relief, rehabilitation and reconstruction of earthquake affected areas of Gujarat;

(ii) To deliberate on the necessary institutional and legislative measures needed for an effective and long term strategy to deal with major natural calamities in future; and

(iii) To look into the parameters that should define a national calamity.

The significant point to note in both disasters namely Orissa Super Cyclone and Bhuj Earthquake is that the reaction of the Government was response and relief centric. The Committee followed a highly process-oriented and participatory approach at the national, state and district levels involving all concerned governments, ministries, departments, scientific, technical, research and development organizations, social science institutions and covering more than a hundred non-governmental organizations. Care was also taken to consult a representative cross-section of Urban Local Bodies as well as Panchayati Raj Institutions. The HPC, after carrying out detailed deliberations over two years, submitted its report in October 2001 (HPC 2001).[18] It made wide range of recommendations that dealt with the constitutional and legal framework, organizational structures and institutional mechanism and cultures of strategic thinking, preparedness, prevention and quick response. Some of the important recommendations of the HPC relate to:

(i) Appropriate mention of Disaster Management in any one list of the Schedule 7 of the Constitution of India;

(ii) Suitable legislation at the National and State levels;

(iii) A Nodal Ministry of Disaster Management to be set up at the National level;

(iv) Integration of development plans with disaster mitigation strategies; and

At least 10 per cent of the planned funds at the National, State and District levels to be earmarked for schemes that specifically address disaster prevention and preparedness.

Working Group on Disaster Management

On the recommendations of the Vice Chairman, National Committee on Disaster Management and approval of the Prime Minister, the HPC was converted into the Working Group on Disaster Management on 27 July 2001 with the aim to provide inputs for a suitable analysis of the HPC Report. National Committee on Disaster Management in discussion with the Working Group carried out a detailed review of the recommendations of the HPC and chalked out the action points for its implementation. It was noted that there was need for building up holistic capabilities for disaster management to handle both natural and man-made disasters. It was accordingly decided that the subject of Disaster Management be transferred from the Ministry of Agriculture to the Ministry of Home Affairs (excluding drought and epidemics and those emergencies/disasters which were specifically allotted to other Ministries). The Government (Allocation of Business) Rules 1961 were accordingly amended in February, 2002 and the transfer of work took place in June, 2002. Subsequently, the Ministry of Home Affairs (MHA) proposed the Constitution of a National Disaster Management Authority (NDMA) as an overarching organization to manage and coordinate all activities relating to preparedness, mitigation, response, rehabilitation including human resource development, capacity building and long-term prevention. It was also recommended that this Authority should be a part of the DM division within MHA.

The Working Group on Disaster Management presented its recommendations to the Prime Minister in June 2003, wherein it endorsed the recommendations of the HPC regarding the legislation to be enacted on Disaster Management at the National as well as the State levels. Subsequently, an Inter-Ministerial Group (IMG) was constituted to examine the recommendations of the Working Group. IMG endorsed that each state must enact its own legislation on disaster management but there was no requirement of Progression towards DM Legislation

After Indian Ocean Tsunami of 26 December 2004, many short comings, in dealing with multi state disasters, were highlighted. In an All Party Meeting held on 09 January 2005, a need for national level legislation for management of natural and man-made disaster was accepted. Consequent to this meeting, Government of India decided to enact law on Disaster Management to provide for requisite institutional mechanism for drawing up and monitoring the implementation of the DM Plans, ensuring measures by various wings of the government for prevention and mitigation of disasters and for undertaking a holistic, coordinated and prompt response to any disaster situation. Accordingly a Bill relating to Entry 23 (Social Security and Social Insurance) in the Concurrent List of the Constitution was introduced in the Rajya Sabha on 11 May 2005.

Disaster Management Act 2005

Disaster Management Bill 2005 was passed by the Parliament on 23 December 2005. The DM Act lays down institutional, legal, financial and coordination mechanisms at the National, State, District and Local levels and integrates it with the activities of various government departments and organizations. In order to facilitate effective coordination of holistic approach to disaster management – Disaster Management Authority – at National, State, District and Local level(s) have been set up as per the provisions of the Act. Law envisaged mandates of the National Institute on Disaster Management, setting up of National Disaster Response Force, Disaster Response Funds and Disaster Mitigation Funds at the National, State and District levels. National Disaster Response Fund (NDRF) and State Disaster Response Fund (SDRF) have already been constituted by converting National Calamity Contingency Fund (NCCF) and Calamity Relief Fund (CRF). Ministry of Home affairs under guidance of Finance Ministry, issued Guidelines for constitution and administration of aforementioned funds in September 2010. However Disaster Mitigation Fund, as mentioned in DM Act is so far not created. The institutional framework as created under the provision of the ACT is expected to usher in a paradigm shift in DM from relief-centric approach to a proactive regime that lays greater emphasis on adoption of a culture of preparedness, prevention and mitigation.[19]

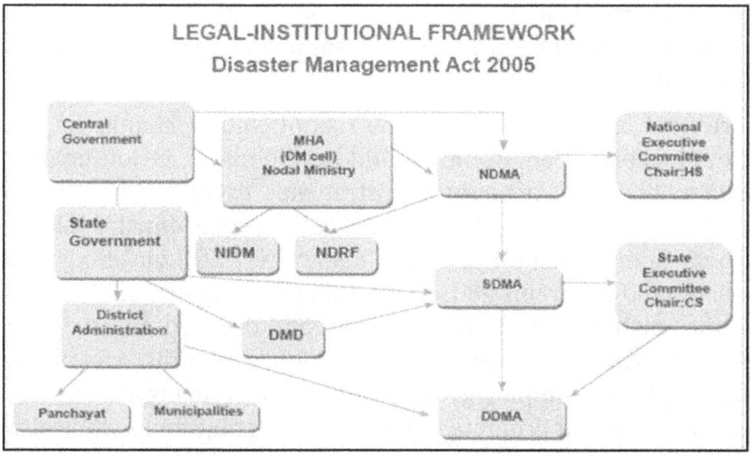

Figure 2.5: Institutional Framework for Disaster Management.

Even before enactment of National DM Act 2005, the state of Uttarakhand had its own State legislation. It was passed by the state legislature (Vidhan Sabha) and was notified as Uttaranchal Act 29 (2005) (published on 11 November 2005). After the National DM Act 2005 came into existence in December 2005, the old state DM Act was replaced.

Synergies with Sustainable Development Goals

Following the Stockholm Conference 1972, constitutional sanction was given in 1976 to environmental concerns through the 42nd Amendment, which were incorporated into the Directive Principles of State Policy (Article 48 A) and Fundamental Rights and Duties (Article 51A(g)). Indian constitution is unique in having specific provisions related to environment and human rights. Article 21 (Right to Life) of the Indian Constitution states "No person shall be deprived of his life or personal liberty except according to procedure established by law". The right to life has been employed in a diversified manner in India. Besides the mere right to survive as a species, quality of life, the right to live with dignity and the right to livelihood *etc.* also comes within the purview of Article 21. The Constitution of India provides that all are equal before the law and shall be accorded equal protection of the law. Article 14 (Right to Equality) states that "The State shall not deny to any person equality before the law or the equal protection of the laws within the territory of India". Article 14 can be used to challenge government sanctions for mining and other activities with high stakes on human rights and environmental impact, where the permissions are arbitrarily granted without adequate consideration of environmental impacts. The Constitution Act of 1976 (Forty Second Amendment) explicitly incorporated environmental protection and improvement as a part of state policy. Article 48 A, which is Directive Principles of State Policy, provides that the state shall endeavor to protect and improve the environment and safeguard the forests and wildlife of the country. Article 51A (g) imposes a similar responsibility on every citizen to protect and improve the natural environment including forests, lakes, rivers and wildlife, and to have compassion for living creatures.

Government of India, accepting the risk of climate change implications of different dimensions of environment and development, including on disasters, and foreseeing the needs of adaptation strategies, established a Prime Ministers Council on Climate Change. On June 30, 2008, the Prime Minister released India's first National Action Plan on Climate Change (NAPCC), outlining existing and future policies and programs addressing climate mitigation and adaptation. The plan identifies eight core 'national missions' and directed ministries to prepare and implement action plans accordingly. Emphasizing the overriding priority of maintaining high economic growth rates to raise living standards, the plan 'identifies measures that promote the development objectives while also yielding co-benefits for addressing climate change effectively.' Ministry of Environment, Forests and Climate Change is the Nodal Ministry on climate change related issues. National Action Plan on Climate Change identified eight missions:

- ✰ National Solar Mission
- ✰ National Mission on Sustainable Habitat
- ✰ National Mission for Enhanced Energy Efficiency
- ✰ National Mission for Sustaining the Himalayan Ecosystem
- ✰ National Water Mission
- ✰ National Mission for Green India
- ✰ National Mission for Sustainable Agriculture
- ✰ National Mission for Strategic Knowledge on Climate Change

In March 2015, the 10-year-old Hyogo Framework came to an end and was replaced by the Sendai Framework. The Sendai Framework for DRR sets out four priorities:

1. Understanding disaster risk,
2. Strengthening disaster risk governance to manage disaster risk,
3. Investing in disaster risk reduction for resilience;
4. Enhancing disaster preparedness for effective response and to "Build Back Better" in recovery, rehabilitation and reconstruction.

India was among 187 UN member countries that on signed a new protocol, replacing the Hyogo Framework for Action 2005-15, committing to a new development model, a far reaching new framework for disaster risk reduction (DRR) with seven targets and four priorities for action.

Implementing Sendai Framework in India

India is a signatory to SFDRR, and attempting to comply with it on voluntary basis. The country is making efforts to achieve the global targets by making advancement in the entire disaster management cycle by following the recommendations in the Sendai Framework and by adopting internationally accepted best practices. Recently released National Disaster Management Plan of India (NDMP), 2016, incorporates the approach articulated in Sendai Framework to achieve substantial reduction in disaster risk and losses in lives, livelihoods, and health and in the economic, physical, social, cultural, and environmental assets of persons, businesses, communities, and countries. The NDMP is aligned with the goal and priorities of SFDRR. The integration of four priorities for action under the Sendai Framework into the NDMP, 2016 is given in the Table 2.2:

Table 2.2: Incorporation of Four Priorities for Action under the Sendai Framework into the National Disaster Management of India

Sendai Framework for DRR (2015-2030) Priority	*Chapters of NDMP with the Concerned SFDRR Priority as its Dominant Theme*
1. Understanding disaster risk	Chapters 2 and 3
2. Strengthening disaster risk governance to manage disaster risk	Chapters 3, 4, 5, 6, 8, and 9
3. Investing in disaster risk reduction for resilience	Chapters 3, 4, 5, 6, 7, and 8
4. Enhancing disaster preparedness for effective response and to "Build Back Better" in recovery, rehabilitation and reconstruction.	Chapters 4, 6, 7, 8, 9, and 10

Source: NDMP, 2016, NDMA, GOI.

Conclusion

Disaster management entails a wide range of functions and skillsets, which interalia include planning and organizing day-to-day management activities, counter-disaster operations, crises management activities, recovery functions and specific tasks related to informed action communication. It calls for systematic efforts on the part of all concerned to contribute towards appropriate disaster prevention, planning, preparedness, response, relief, recovery and restoration. Of course leveraging Science and Technology for prevention, mitigation, preparedness and capacity building is the bottom line for paradigm shift to take place.It is evident from the literature and data available that substantive efforts are being made world over to cope up with disasters in an effective manner. By and large, depending upon various available resources, the initiatives and measures have been taken for more focus on preparation and planning. In some of the developed countries, it is more visible in comparison to the developing or least developed countries. Nevertheless, the paradigm shift is, by all means witnessed across the globe.

Notes

1. www.unep.org/Documents.multilingual/Default.asp
2. http://www.un.org/esa/sustdev/csd/csd15/media/backgrounder_brundtland.pdf
3. www.unisdr.org
4. www.unisdr.org
5. www.unisdr.org/we/inform/publications/
6. www.unisdr.org
7. www.unfccc.int/kyoto_protocol
8. www.un.org/millenniumgoals
9. Millennium Development Goals, India Country Report 2015 at http://mospi.nic.in/Mospi New/upload/mdg_26feb15.pdf
10. http://www.unisdr.org/2005/wcdr/intergover/official-doc/L-docs/Hyogo-framework-for-action-english.pdf
11. http://www.usicomos.org/wp-content/uploads/2016/03/Sendai-Framework-and-Cultural-Heritage-UNISDR.pdf. Accessed 17 June 2016
12. http://www.unisdr.org/who-we-are/what-is-drr [Accessed 17 June 2016]
13. http://www.unisdr.org/archive/27335
14. Major disasters, occurred before, and played crucial role in enactment of DM ACT 2005 are mentioned here.
15. www.ndmindia.nic.in/committee/hpcomm.html
16. https://en.wikipedia.org/wiki/1999_Odisha_cyclone
17. saarc-sdmc.nic.in/pdf/Earthquake3.pdf
18. http://saarc-sdmc.nic.in/pdf/india
19. http://ndma.gov.in/

Chapter 3

ICT: AN ENABLER FOR EFFECTIVE DISASTER MANAGEMENT

Introduction

Information and Communication Technologies (ICT) applications, has power to change life of individuals, communities and societies at large. Various Broadband access technologies, broadband-enabled services and ICT applications offer new opportunities for interaction among people, for transparent governance, for sharing the world's knowledge resources and expertise and thereby empowering people and transforming people's lives as a result this process contributes to inclusive and sustainable development across the world. In this chapter we can see how ICT can be of immense help in different phases of disaster management continuum.

Disasters in the contemporary society, whether of natural or technological origin, are complex phenomena that demand multi-disciplinary assessments and cross-cutting actions. Managing disasters as considered today are not simply technical issues of environment science but also as an interrelated area of study that involves and require contributions from a wide spectrum of expertise. Earth sciences, natural sciences, as well as anthropological and historical studies, economics, sociology, psychology, *etc.* have direct contributions towards the understanding of such systems. On the other hand, politics, social and physical planning, engineering, systems management, law, medicine, security sciences, and others contribute to the prescriptive capacity in building up the practice of preventive decisions. Understanding of risk and vulnerabilities from the perspectives of natural and social systems and processes, as well as their mutual interactions, with cause and effect analysis is an essential step in the study of disaster management.[1]

A complete definition of disaster by UN-ISDR[2]: "*A social crisis situation occurring when a physical phenomenon of natural, socio-natural or anthropogenic origin negatively impacts vulnerable populations and their livelihoods, production systems infrastructure and historical heritage, causing intense, serious and widespread disruption of the normal functioning of the affected social unit. The impacts and effects cannot be overcome with the resources autonomously available to the affected society. Impacts are expressed in different forms such as the loss of life, health problems, the destruction, loss or rendering useless of the totality or part of private or collective goods and severe impacts on the environment. These negative impacts require an immediate response from the authorities and from the population in order to attend the affected and to re-establish acceptable thresholds of wellbeing and life opportunities. Natural disasters not only disrupt the normal life of the affected community and the State but also put the developmental efforts in a standstill posture as the funds earmarked for new initiatives are transferred to relief, response, and rehabilitation work. History stands testimony to the fact that disasters have resulted in substantive loss of human lives, livestock, agriculture, industry, and the public utilities*".

With the pace of fast growth and conlfiting environmental challenges, development and disaster management are obviously integrated as a single issue. It is an essential component of any development framework. Effective disaster management has been recognized as a key requirement towards achieving the Millennium Development Goals (MDGs) by the specified target of 2015[3]. Total economic loss from natural calamities and man-made disasters across the world, including Nepal earthquake and heat wave in India, was at $ 37 billion in the first half of 2015, Swiss Re[4] said in a report. The Report further said that 18000 people lost their lives in first half of 2015 up from more than 4800 in the first half of 2014. Earthquake in Nepal (there were more than 9000 fatalities) and heat waves in India and Pakistan claimed these lives.[5]

The link between proactive disaster management and development began to be seriously examined from various perspectives in the 1990s. 1990s was designated as International Decade on Disasters. In this decade, international community, under the auspices of the United Nations, paid special attention to fostering international co-operation in the field of natural disaster reduction. Different perspectives bring to form a shared understanding that disaster impacts depend to a large extent on historical and current development activities[6],[7] (Nonita, 2011).

Global climatic changes enhance vulnerabilities of communities dependent on natural resources for their survival such as agriculture and fisheries. Issues related to global climate change are expected to amplify and further complicate the development challenges that perplex policy makers, practitioners, researchers. As per estimations of environmental scientist's the incidences and severity of extreme events are increasing and sea level will continue to rise. In regions where water plays a significant role in socio-economic welfare, water scarcity or excess are known to cause socio-economic and further ecological distress as those impacted by a succession of disasters resort to environmentally compromising survival tactics.[8] The shift in rainfall patterns and climate zones will result in increased pressure on freshwater sources and on hydro sensitive ecosystems. The annual economic losses for developing countries are estimated to be in excess of 10 per cent of GDP.[9]

Climate change presents two types of disaster threats, particularly in developing countries. One is the potentially devastating impacts, on vulnerable communities, of more frequent and more intense extreme weather events. This contributes to the second threat, the compounding of what are already complex development problems, potentially leading to a downward development spiral for the world's poor. The Internatinal Federation of Red Cross (IFRC) reminds us, "*Whether the changing climate with all its uncertainties, contributes to more disasters that affect more people, does not depend merely on what happens to the weather. Climate change and the accompanying risks will be superimposed on an unequal world, where vulnerability to disasters is directly linked to poverty.*"[10] Other trends further complicate the picture including population growth and weakening of governance structures. Addressing any one trend is not going to help in achieving desirable outcomes. Therefore, managing disasters need to be integrated with broader development strategies and policy frameworks that identify and address the root causes of vulnerability, and improve community resilience to disasters.

Recent incidences have shown that there is no country that does not stand the threat of a disaster, though the magnitude and level may vary. Therefore, disaster preparedness is no longer a choice; but it is mandatory irrespective of where one lives. The Second World Conference on Disaster Risk Reduction (Hyogo, Kobe, Japan 18-22 January 2005) saw participation of 4000 delegates from 168 countries who agreed and adopted the Hyogo Framework for Action (HFA) 2005- 2015 for Building the Resilience of Nations and Communities to Disasters. HFA aimed at reduction of disaster losses by 2015 by building the resilience of nations and communities to disasters. This means reducing loss of lives and social, economic, and environmental assets when hazards strike.

HFA had identified 5 important Priority Actions as following:

- ☆ **Priority Action 1**: Ensure that disaster risk reduction is a national and a local priority with a strong institutional basis for implementation.
- ☆ **Priority Action 2**: Identify, assess and monitor disaster risks and enhance early warning.
- ☆ **Priority Action 3:** Use knowledge, innovation and education to build a culture of safety and resilience at all levels.
- ☆ **Priority Action 4:** Reduce the underlying risk factors.
- ☆ **Priority Action 5:** Strengthen disaster preparedness for effective response at all levels.

It is evident from HFA, the priority actions 2, 3 and 4 demanded extensive use of science and technology for attaining desired level of preparedness, prevention and mitigation of effects of hazards from turning into disasters. During disasters, information is as much a necessity as water, food and medicine.[11] The right kind of information communicated and used at the right time can save lives, livelihoods and resources.[12] Responding effectively to disasters demands rapid access to reliable and accurate data, mechanism to convert data in to useful ground information and the capacity to assimilate, analyse, and taking timely decisions. It requires speedy communication with administration, all stakeholders,key actors, to assemble resources to mobilize, deploy and coordinate emergency personnel; and to coordinate response activities among agencies involved. Communication and information sharing are thus very critical for effective disaster management.

Similarly Sendai Framework for Disaster Risk Reduction 2015-2030 has identified following 4 priority areas:

- **Priority 1:** Understanding disaster risk;
- **Priority 2:** Strengthening disaster risk governance to manage disaster risk;
- **Priority 3:** Investing in disaster risk reduction for resilience; and
- **Priority 4:** Enhancing disaster preparedness for effective response and to "Build Back Better" in recovery, rehabilitation and reconstruction.

These priority action areas require extensive utilization of collection of scientific information, data analysis, simulations and modeling. All these requisites may only be fulfilled by use of ICT. The term Information and communication technologies (ICTs), consist of range of tools, applications, systems used, to input, store, edit, retrieve, analyse, synthesize, process and share information and data in all its forms. Thus broadly ICT encompasses: radio, television, broadband, satellite and cellular mobile phones, computer and network hardware and software, websites, portals, remote sensing systems (GPS, GIS *etc.*) satellite systems and so on, as well as the various services and applications associated with them, these are data storage, big data analysis system integration, Saas, videoconferencing and distance learning[13]. ICTs are, thus, among the lifelines during any emergency situation. The role of ICT as an enabler for all policies, strategies has already been accepted.

It has been reported that, in cloud burst (June-2013) of Uttarakhand, the major bottleneck during response was non-availability of suitable communication mechanism amongst responders, community leaders, government functionaries and victims. This Disaster has extracted a very high price in social, psychological and economic terms. Its impact was felt in trails of destruction, families torn apart, children orphaned, livelihoods destroyed, and communities traumatized. The Comptroller and Auditor General (CAG) of India in its performance audit report on the disaster management mechanism in the country, submitted to Parliament CAG had highlighted that the Uttarakhand Disaster Management Authority (UDMA), constituted in 2007, had not formulated any rules, regulations, policies or guidelines for disaster management in the state.[14] Effect of this disaster could have been substantially reduced had people been well informed timely; hazard and vulnerability analysis been properly carried out and risk communicated to key stakeholders; DM plans been in place; SOPs and well structured Incident Response System been in place. It is pertinent that proper technological interventions are much needed for addressing disaster risks issues related to changing social, economic, environmental conditions and land use, and the impact of hazards associated with geological events, weather, water, climate variability and climate change.

In the context of natural disasters, the holistic DM cycle has been centered on four overlapping phases: mitigation, preparedness, response, and recovery. Recent Trends in natural disasters reveal that there is continuous increase in the number of disasters in most parts of the world. These are considered to be major impediments to achieving human security and sustainable socio-economic development. Though natural disasters cannot be prevented, yet with the appropriate use of Science and Technology the effect of such disasters could be mitigated.

More detailed DM cycle diagram, as shown below in figure 3.1, indicates various overlapping stages of DM cycle.

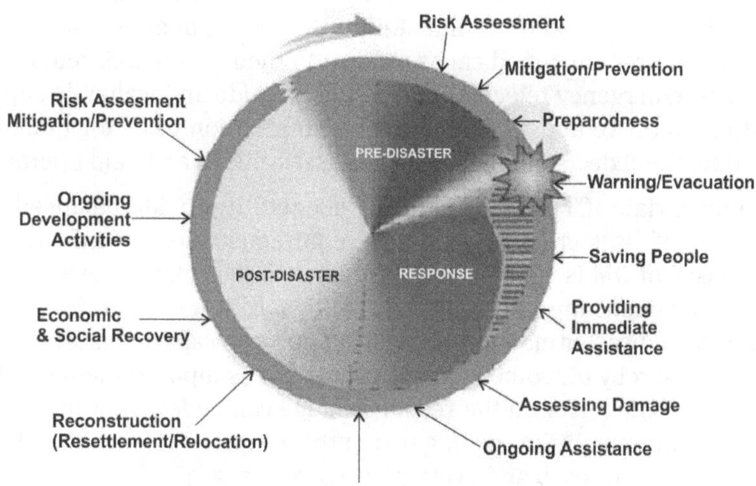

Figure 3.1: Disaster Management Cycle.

As shown in the Figure 3.1 above[15], actions taken in pre-disaster, during disaster and post-disaster phases need not be decided in isolated silos but should flow seamlessly across various stages of DM cycle. Disaster information management has traditionally been adapted in an insular environment because of which interactions between the Stakeholders involved in the four stages of DM cited above have not been sufficient enough to attain synergy and therefore DM techniques to support the various disaster phases have been improper and non-uniform in nature. The most important steps towards reducing disaster impact are to correctly analyse the potential risk and identify measures that can prevent, mitigate or prepare for emergencies. ICT can play a significant role in highlighting risk areas, vulnerabilities and potentially affected populations by producing geographically referenced analysis through, for example, a geographic information system (GIS).[16] A well developed early warning system can substantially mitigate severe effects of a disaster. Timely information in the location, progression and regression of disasters at various phases of the disaster depicted in a dynamic map is essential towards effective management of the disasters and this can be generated in Geographic Information System (GIS) linked with Global Positioning System (GPS), derived by integrating the real-time space-based imageries of the hazards with the corresponding ground information. Remote sensing and GIS technology, satellite imagery products are therefore an integral component of emergency telecommunications in all phases of assistance from preparedness to emergency response (NDMICS, 2011).

The ITU has been working in several countries to help governments and the humanitarian actors to have access to relevant Telecommunication resources. Emergency Telecommunications play a critical role in the immediate aftermath of disasters by ensuring timely flow of vital information which is much needed by government agencies, and other humanitarian actors that are involved in rescue operations and providing medical assistance to the injured. GIS, remote sensing products and emergency telecommunications provide an invaluable support for constituting objective and rapidly available information basis for humanitarian coordination, rapid decision making, needs assessments and field operations[17].

An appropriate ICT infrastructure is needed to provide balanced support to each phase of disaster continuum. The requirements for ICT infrastructure in various phases of DM is specific and particular to that phase ,nevertheless new disaster management approach calls for a faultless continuous flow of information to each phase of disaster management. Such approach calls for a seamless flow of information whereby outcome learning of one face is input to the next phase for example the product used in the response stage can be fed back into the future mitigation programme[18.] Succeeding text of this chapter explains how ICT can be utilised effectively for each and every phase of a disaster:

ICT for Disaster Risk Mitigation

Mitigation measures are pre-disaster activities necessary to be implemented in advance for reducing the impacts of disasters. Development and promulgation of disaster mapping and micro/macrozonation of disaster prone areas, ordinances building bylaws, codes ,creation of the critical baseline data for disaster events, Plans, policies, SOP preparations, web based tools *etc.* are some of important mitigation activities. Mitigation may also include implementing legislation that prevents building structures in areas prone to landslides, flood, earthquake and tsunami. Utilising existing databases linked to geographic features in GIS makes the task of monitoring easier.[19]

For Mitigation phase ICT plays essential role as a tool for planning mitigation activities, based on hazard profile and vulnerability assessment of the area. A well built communication network with proper interconnectivity with all existing networks, substantiated with large bandwidth for supporting data flow and web based data management systems. Geographic information systems (GIS)[20] are the most significant of ICT tools and useful in Disaster Mitigation. The power and strength of GIS lie in their ability to integrate spatial with non-geographic data into one encompassing system, and graphically display spatial patterns, creating a common operational picture. GIS allow real time monitoring for emergency early warning as well as modeling of possibilities. The combination of remote sensing[21] with GIS introduces more planning information and enables accurate predictions.

It is observed that satellite-based advanced technologies and GPS enabled Geographic Information System (GIS) play a very crucial role to facilitate in implementing knowledge-based efficient disaster management. The combination is frequently used for assessing and mapping of hazard and risk areas, vulnerable groups, planning of evacuation routes, location of emergency centres, as well as assessing post-disaster damage and recovery assessments. Timely information on the occurrence, progression and regression of disasters during the various phases are essential for effective management of disasters and this can be derived by integrating the real-time,aero-space imageries with the corresponding ground information (usually available on Geographic Information System (GIS) platform linked with Global Positioning System (GPS)). When all these data are integrated and mapped into a common operational picture, responses can be better targeted, and priorities established. For example, for management of floods, Digital Elevation Model (DEM)[22] is an important parameter for improving the flood forecast models and hazard zoning operation, flood damage assessment, determination of the friction coefficients on flood plains[23], coastal erosion monitoring, and topographic data input to GIS based relief, rescue and flood simulation operations *etc.* DEM can be derived from various channels such as ground survey methods, Aerial Photogrammetry[24] and Interferometric Synthetic Aperture Radar (InSAR)[25,26]. Airborne Scanning Laser Altimetry is the state-of- art technology for providing very accurate DEM up to 15cm with very less ground information support. LiDAR data can also be acquired in inaccessible moderate vegetated terrains like forests both day and nights even in semi- bad weather conditions. LiDAR[27] system broadly consists of Laser Scanning Assembly, Laser Range Finder, Global Positioning System (GPS) and Inertial Measurement Unit (IMU).

The International Charter on Space and Major Disasters (ICSMD), established in 1999, provides a unified system of space data acquisition and delivery to disaster-affected communities when so requested by member agencies. It delivers high quality satellite imagery to front-line responders generally within 24 hours[28.] In early 2009, the Charter was activated by UNDP in response to floods in the north-central and northeastern regions of Namibia which affected 17 per cent of the country's population. Satellite imagery showed the extent of the flooding along the Chobe River in Capribi as well as the flood changes over time. This dictated a phased return of the evacuees; the common operational picture facilitating consensus among the different agencies[29].

With the increasing prevalence of Earth Observation (EO) technologies being employed in development work, the World Bank and the European Space Agency (ESA) have partnered for the purpose of mainstreaming the use of EO into the World Bank's lending operations, across all of the Bank's Sustainable Development Network's sectors[30]. The first small-scale demonstrations were provided to the World Bank by ESA in 2008 contributing to coastal zone management in Bangladesh, study of marine ecosystems in Belise, urban risk management in Alexandria and Tunis, and water resources management in Peru/Bolivia. In July 2010, the collaboration was scaled up. Twelve World Bank Teams from across a variety of sectors and representing all regions of Sustainable Development Network were selected and received the pro-bono assistance of ESA to apply Earth Observation services in selected operations. Sectors benefitting include:

- Disaster Risk Management;
- Urban Development;
- Agriculture and Forest Management;
- Water Resources Management;
- Coastal Zones Management;
- Marine Environment; and
- Climate Change Adaptation.

Networking requirements to support the mitigation phase should meet key attributes, like the need to move large volume of data/information, broad connectivity among a diverse group of organisations *etc.* For mitigation phase real time data may not be that critical as it is for response and relief phases. The Mitigation data are largely archive-based, and hence several cataloged and linked repositories are to be accessed through good search engines or the directory systems should permit access to the distributed users. Data providers are responsible for data quality (timeliness and accuracy), limiting redundancy, and updating catalog/directory information. Use of modeling/prediction tools for trend and risk analysis is important. The scientific community is becoming reliant on the Internet for access to data and scientific collaboration in supporting mitigation efforts. NDMICS(2011)

ICT for Disaster Preparedness

Preparedness activities range from early warning, community development, training of the first responders and the community, logistical support, supply, and resource systems needed for disaster response to early warning and monitoring activities preceding the disasters. Dissemination of warning data through a reliable communication medium during the preparedness phase is essential but not time critical; however dissemination of actual warning data for an impending disaster is intense and time critical. A warning can be defined as the communication of information about a hazard or threat to a population at risk, in order for them to take appropriate actions to mitigate any potentially negative impacts on themselves, those in their care and their property.[31] The occurrence of a hazard does not necessarily result in a disaster. While hazards cannot be avoided, their negative impacts can be mitigated. The goal of early public warning is to ensure to the greatest extent possible that the hazard does not become a disaster. Such warnings must be unambiguous, communicate the risks succinctly and provide necessary guidance. The success of a warning can be measured by the actions that it causes people to take, such as evacuation or avoiding at-risk areas.

Early warning can help in many types of hazard warning systems are in place and have proved beneficial for a variety of hazards. In the case of tsunamis, the benefit of an internationally coordinated system was shown in the 2011 earthquake and tsunami in Tohoku, Japan, which threatened many Pacific islands, providing time for many people to evacuate to high ground: warnings were more coordinated than in the devastating Indian Ocean Tsunami in 2004. The primary information should be on the likely impact of the event in terms of the extent of the area affected, location specific details, population affected, availability of resources for evacuation of the people and relief and quick assessment of damages.

Advancements in Information and Communication Technology (ICT) have made it possible not only to forecast some of the disasters but also have made available means for quick and effective rescue and relief operation, thereby minimizing the deadly impacts of some of the worst disasters. For instance, compared to more than 10,000 killed during the cyclone which hit Andhra Pradesh in 1979, improved communication techniques limited the loss to less than 1000 during the May 1990 cyclone in the same state. Similarly when cyclone Phailin crossed the Bay of Bengal on the eastern coast provinces (states) of Andhra Pradesh and Odisha on Saturday the October 12, 2013; Red Message, the highest alert from the Indian Meteorology Department in Delhi, was used timely by the local district officials and could drastically reduced loss of human lives. The early warning messages for this cyclone prompted India's biggest evacuation in 23 years with more than 550,000 people moved up from the coastline in Odisha and Andhra Pradesh to safer places.[32,33]

It is important to note that disaster warning is indeed a system, not a singular technology, constituting the identification, detection and risk assessment of the hazard, the accurate identification of the vulnerability of population at risk, and finally, the communication of information about the threat to the vulnerable population in sufficient time and clarity so that they can take action to avert negative consequences. This last mile component of taking action on warnings, emphasize the importance of preparedness of local administrations and capacity building of community by extending education and creating awareness so that they may respond with the appropriate actions.[34]

Radio and television are two traditional media most widely used in communication of disaster warning, because they are relatively cheap, provide a reliable one-to-many communication (broadcast mode) medium and most importantly, do not require literacy. These mediums have much wider reach in comparison to other communication medium, besides warnings can be issued in vernacular language. Radio in particular is the most accessible medium to the poor, especially women in their homes, or fishermen at sea, workers out in the fields. Television, while more powerful, is comparatively more expensive and is not movable.A study on media, perception and disaster-related behaviour in Bangladesh revealed that early, easily understandable and language-appropriate warning dissemination through radio can reduce the potential death toll of catastrophic cyclone and tidal bore.[35]

Radio broadcasting is an efficient method for disseminating information at the disaster site and across communities in the affected area and the key players of response or relief operation. Thanks to rapid advancement in technology, completely portable radio station in a box is now commercially available, which can be set-up at the disaster site within few hours. Tailor-made programs as per the requirement of disaster site can be transmitted to the community over a large area. The packaged portable radio station typically contains three portable packs namely broadcast studio, transmitter rack and antenna system. The broadcast studio provides features such as audio mixer, CD players, audio recorder, cassette player, monitoring headsets, which are essential for creating the information content. In order to receive the broadcast messages, AM/FM receiver/tuner sets have also to be supplied to various officials, NGOs, Civil Defense and Home Guard personnel. AM/FM receivers are available with the provision to work on battery, AC adapter or solar power and cell phones charging facility is also available. In the event of mobiles phones functioning in the area, FM transmission can also be received by the communities, who have mobile phones with FM capability. After the Indian Ocean tsunami of 2004, many radio manufacturers considered introducing new digital radio alert systems that react even if the set is switched off. In order to trigger this alarm, a special flag integrated into the received signal from a terrestrial transmitter or a satellite would be used and the set would automatically tune to the emergency broadcast channel.[36]

The Mozambique Red Cross integrated the Freeplay Lifeline[37] radio into its cyclone and flood early warning activities. A volunteer is charged with listening to the radio. When a warning is heard, the volunteer alerts village leaders and a pre-planned response is launched. This has been credited with the dramatic reduction in devastation from cyclones and floods. There were 700 casualties in the year 2000 flood, 30 in 2007 and none in 2008. This type of utilization of radio broadcasting of alarms has also been used in hurricane preparedness in Haiti, and in RANET[38] projects in Niger, Kenya, and Uganda.[39]

Amateur radio[40] (also known as 'ham radio') operators have also assisted their communities and countries during disasters by providing reliable communications to disaster relief organizations, especially when traditional communications infrastructure breaks down. In such a situation, amateur radio operators transmit emergency messages on voice mode about the well-being of survivors and information on casualties to friends and relatives. Indian Ocean tsunami 2004 destroyed electricity and communications infrastructure in the Andaman and Nicobar Islands, in such situation, amateur radio operators were the critical link between the islands and the Indian mainland and helped in the coordination of rescue and relief operations. Amateur radio broadcasters are authorized to communicate designated predefined radio spectrum band. They require a license from the licensing authority to ensure that only competent operators use their skills. In India, Department of Telecommunication (DOT) under the Ministry of Communication and IT, in Government of India, is empowered to authorize such frequency use. Wireless Planning and Coordination Cell along with other services licensing unit of DOT decide such matter.

A satellite radio is a digital radio that receives signals broadcast by communications satellite, which covers a much wider geographical range than terrestrial radio signals. Satellite radio receives its signal from a communication satellite,[41,42] and therefore, has a wider geographical range than terrestrial radio. It is very useful when terrestrial transmission towers are damaged due to a disaster. A satellite radio, combined with fixed or mobile phones was shown to be the most effective and reliable of five ICT tools in eight combinations tested for transmitting early warning information from government agencies to 32 at risk villages in Sri Lanka. This Addressable satellite Radios for Emergency Alerting (AREA) system is being combined with GPS to issue hazard alerts in text and audio formats directly to specific at-risk communities in Bangladesh, India, Indonesia and Thailand. But satellite usage is very costly for both service set up and the purchase of a satellite enabled radio. Satellite radio functions on line of sight between the antenna and the satellite when line, thus service can be interrupted by trees, buildings and some weather conditions. Radio and television are however limited by the fact that they provide only one-way communication[43] and are generally turned off in the night.

The humanitarian activities of amateur radio and amateur satellite services are highlighted in Recommendation ITU–R M.1042 (International Telecommunication Union - Radio division is ITU-R) which encourages the development of networks within these services that are capable of providing communications during disasters and relief operations.[44]

Telephones, both fixed (Wireline) and mobile (wireless) can play a very vital role in dissemination of early warning messages. Telephones have an inherent advantage of broadcast, as well as two way communication[45]. Now a days when tele-density, particularly mobile tele-density[46,47] is rapidly increasing; this makes mobile telephone a very important medium for early warning. In some of the rare remote areas of India, which presently are not connected through any mode of communication, government, under National telecom Policy (NTP) 99, and now under NTP 2012 has made provisions for connectivity to all the villages of India. In remote rural areas of India, most of the villages were connected through Village Panchayat Telephones (VPT) using different kind of technologies. Thereafter department of telecommunication (DOT), under Ministry of Communication and Information Technology, developed various schemes, projects under Universal Obligation Fund (USO) for mobile and broadband connectivity of remote areas. In the new scenario the overall teledensity has reached **93.61 as on 31 May, 2017**. The major contribution has been from wireless segment as total subscriber base of **1,204.98 million** consists of wireless segment of **1,180.82 million** and wireline subscriber base of **24.16 million as on 31 May 20017.** Subscription in Urban Areas is **697.1 million** with a teledensity of **172.3**, whereas Rural Area subscription is **508 million** with a teledensity of ***57.6*** at the end of May 2017. Total number of broadband subscribers is **291.61million.**[48]

In such prolific growth environment for Telephones, they offer maximum opportunity to be a very useful tool for early warning and information sharing medium in all phases of disasters. Effectiveness of telephone call can be enhanced by a mechanism called telephone trees' to warn communities of impending dangers. When an individual receives a warning message (either through phone or by other means), he/she is supposed to make a predetermined number of phone calls (usually four or five) to others in a pre-prepared list. This arrangement not only ensures the timely delivery of the warning message, but also ensures the minimum duplication of efforts (NDMICS, 2011).

A very famous incident that has been quoted by various Disaster Management practitioners and by MSSRF[49] *occurred in one small coastal village of Nallavadu in Pondicherry, India. A timely telephone call, warning about the impending tsunami saved the village's entire population of 3,600 inhabitants, as well as those of three neighboring villages. Villagers of Nallavadu were involved in the M.S. Swaminathan Research Foundation's Information Village Research Project. Vijayakumar, a former project volunteer, was working in Singapore and heard a tsunami alert issued there. He immediately telephoned the research centre in Nallavadu, which issued an alert. His timely action, led to the evacuation of the four villages before the tsunami hit the coast.*[50]

Mobile phones were the main tool for implementing a community-based flood monitoring and early warning system in An Giang and Dong Thap provinces of Vietnam, established after the devastating August 2008 Mekong floods. Thirty-eight at-risk villages were provided with mobile phones, and villagers in seven villages trained to take wet season water level measurements twice a day. They report their water level readings, via text messages, to the Southern Region Hydro-Meteorological Centre in Ho Chi Minh City, the local agency responsible for flood forecasting. The information is entered into the Centre's computer and the calculated flood forecast fed back to the villagers who publicise the information on information boards at central locations. Any imminent flood threat is publicized via loudspeakers[51.] Similarly in coastal areas of Andhra Pradesh in India loudspeakers are being considered as a cost effective mean of last mile connectivity for alerts.

Short message service (SMS) is a service available on most digital mobile phones that permits the sending of short messages between mobile phones, other handheld devices and even landline telephones. There are instances when many residents of affected coastal areas are unable to make contact with relatives and friends using traditional landline phones, but could easily communicate with each other through SMS as long as the mobile network is functional (NDMICS 2011, P34). During the time of Disasters, due to anxiety of people voice traffic increases exponentially in landline and mobile telephones. Due to very high traffic, as everybody in the area tries to use telephones, telephone, exchanges have to operate on a very high traffic capacity which is beyond handling capacity of system, and thus lead to congestions/collapse of voice communication. Mobile communication have a facility,that even when it's voice channels are congested, Short Message service (SMS), may be use d through separate data channel. An example of effective use of SMS service is Wenling China where sensor detection of building integrity (in relation to cyclones) is part of the EWS52. Evacuation alerts via SMS are sent to residents before a cyclone hits. The use of SMS is limited by the fact that it can reach to only mobile users.

Most of today's wireless systems (based on CDMA and GSM technologies) support a feature called cell broadcasting, through which a public warning message in text can be sent to the screens of all mobile devices operating in any group of cells of any Size, ranging from one single cell (about 8 kilometers across) to the whole country (NDMA, 2012, p. 34). The only possible disadvantage to SMS/cell broadcasting is that not every user may be able to read a text message when they receive it, unless the warning messages are sent in local languages. Further, these messages would still be inaccessible to those who cannot read, even in their own language. Nevertheless text messages sent in vernacular language simultaneously reaching many people is very effective way of disseminating important disaster related messages.

Telefax service which carries data/information over fixed line networks may also be widely used for written communication for example for sending text messages related with warning, co-ordination plans queries *etc.*, may be sent through group fax messages to multiple places at a time. Indian National Centre for Ocean Information Services (INCOIS) and Indian Meteorological Department (IMD), widely use Telefax and e-mail service for mass dissemination of warnings.

GIS plays important role for preparedness phase as well by providing real-time monitoring of the progressive development of the emergency. Remote weather stations can provide current weather indexes based on location and surrounding areas. Wind direction, temperature and relative humidity can be displayed by the reporting weather station. Wind information is vital in predicting the movement of a chemical cloud release or anticipating the direction of radio activity released *etc.* Earth movements (earthquake), reservoir level at dam sites, radiation monitors, *etc.* can all be monitored and displayed by location in GIS. If necessary, this type of information and geographic display can be delivered over the Internet to the public.[53]

In remotest areas, where at present full national coverage is still a concern to be addressed to, through adoption of various technologies and combination of various method for creation of Public awareness, such as broadcast announcements, use of print media, nukkar nataks, capacity building through training activities making use of interactive video preparedness phase may be geared up. Besides, community participation may be enhanced through use of various ICT tools, for example a web based posters may be useful for volunteer registrations and trainings.

ICT Applications for Emergency Response

Disaster-affected communities have always been the first responders in emergencies. New technologies are greatly enhancing capacity of people for self-help. The increased access to ICT tools, such as the prevalence of mobile phones, laptops, broadband internet access, smart phones, video cameras enabled with Bluetooth and Wi-Fi connectivity are creating new ways for affected communities to organize, coordinate and respond to their own problems. Community may now connect to outside agencies, coordinate with all stakeholders, self-organize community-based actions and monitor on ground actions and making everyone more accountable. Technology can *"empower individuals and communities threatened by hazards to act in sufficient time and in an appropriate manner so as to reduce the possibility of personal injury, loss of life, damage to property and the environment, and loss of livelihoods"* (*UNISDR, 2006*), one of the fundamental principles of community-based early warning. More broadly, technology enables people-centered humanitarian action[54].

Disaster management activities, in the immediate aftermath of a disaster, can be made more effective by the use of appropriate ICT tools.[55] This time period demands very fast decision making, prompt actions and immediate deployment of search and rescue teams. Communications among response teams, search and rescue teams, coordination in charge, control rooms, medical teams for trauma care, Non Government organizations, and aid agencies and to the general public, become most vital in the immediate aftermath of a disaster. Rapid, reliable, scalable, controlled access communication is vital to efficient disaster response operations. Major challenges are presented by extreme conditions of infrastructure destruction, communication traffic peaks, mobile users, and sensitive data.

Information on GIS-platform, which helps in design of Mitigation measures, provides critical information and helps in preparation of appropriate action plans for response as well. GIS information can be transmitted to various stakeholders (For example incident commanders at Emergency Operation Centers as well as the disaster response personnel) for quick and effective implementation with excellent coordination. Disaster relief workers and the private sector (for example insurance agencies) could use these GIS information products in the post-disaster recovery phase, replacing the current paper plan approach. During response phases, GIS can accurately support better response planning in areas such as determining evacuation routes or locating vulnerable infrastructure and vital lifelines, *etc.* It also supports logistical planning to be able to provide relief supplies by displaying previously available information on roads, bridges, airports, railway and port conditions and limitations. Based on the information provided by GIS, it is also possible to estimate the quantity of food supplies, bed space, clothes, medicine, fire tenders, medical and paramedics that will be required at each shelter, based on the number of expected evacuees.

Electromagnetic waves in the lower Ghz range can penetrate most building materials, which allows for detection of hidden targets. The principle of operation is based on the discrimination between an inanimate object and a living human being can be made by the detection and decoding of reflection of electromagnetic waves (in terms of periodicity and frequency spectrum) transmitted from the trapped human being. By tracking the signal processing to the periodicity of the breathing, the trapped human being can be localised. There are many upcoming technologies working on such principle of propagation and reflection of high wavelength electromagnetic waves, and are very useful for rescue of people who are trapped and still alive.

The importance of telecommunication during emergencies resurfaced when bomb-blasts incident occurred in Mumbai on 13th July 2011, where it was reported that the response was affected due to congestion in the cellular mobile network of Mumbai. In order to facilitate a mechanism wherein important functionaries engaged in 'response and recovery' work during emergencies get the calls on priority, TRAI[56] had issued a pre-consultation paper on 8th November, 2011 on 'Priority call routing' in Telecom networks for persons engaged in 'response and recovery' work during emergencies', in which stakeholder's inputs/comments/ views/papers were solicited on:

a. *Any aspect, in general, of making Indian telecom networks resilient so as to be effective during disasters/emergency situations.*

b. In particular, the aspect of 'Priority call routing' in telecom networks for persons engaged in 'response and recovery' work during emergencies.

Subsequently, TRAI released its final recommendation on priority routing of calls of those who are involved in response and relief on 26th November, 2013. These recommendations have been accepted by the Ministry of communication and IT GOI, in April 2015. Services are yet to roll out.

During response phase of DM continuum, Radio, to some extant may be used for dissemination of official communication to affected community. The Internet is acknowledged to be one of the most reliable information infrastructures even under adverse physical conditions, and electronic mail, its most widely used application. This was a critical tool during the 1997 Cambodia floods[57]. The floodwaters had washed up venomous snakes and people were being bitten. The local WHO field offices did not have the antiserum in stock nor the taxonomic information on Cambodian snakes. The field officers sent e-mails to members of the Global Health Disaster Network (GHDNet) who forwarded it to several mailing lists. Specialists and institutions in the region were identified resulting in a speedy sourcing of the antiserum.

Normally radio and television receivers are available in most of the critical locations such as police and fire stations, hospitals, government buildings, auditoriums, stadiums and public shelters. These stations/repeaters/receivers are often installed with backup generator power. Thus, both citizens and emergency responders will often benefit from the distribution of key information by these local broadcasters. Television broadcasting is particularly powerful when it

During response most critical part is information collection about victims and their location updates. Many times it happens that victims are dislocated and are not able to contact their family and friends; an ICT open source platform may be very useful in such situations. During Uttarakhand Disaster (June 2013), many pilgrims, who were from different parts of India got trapped in distant locations. Search and Rescue (SAR) teams of Indian Army, Paramilitary forces, NDRF and NDMA officers were making arrangements to rescue trapped people from various means which inter-alia included Helicopters for very ill persons/ old people/women/small children. kIt was becoming very difficult to manage database of trapped people in different locations. Excerpts from post response report of Mr Vinod kajla officer in charge from NDMA, during Uttarakhand Disaster (June 2013), is given below:

"On 24th June there were heavy rains in Uttrakhand and very little flight movement was possible except on the Harshil – Chinliyasaur Sector. I went to the Police Control Room and enquired in detail about how the details of rescued, missing, dead being put on the website. They had put up a large number of pdf files for various dates at various places. These were all in different formats as some of them mentioned name or name and age. Some were in more details with name age and place. Some mentioned name and state. This should have been standardised at the start."

displays maps of emergency areas annotated with critical information prepared by the broadcasters' news and meteorological personnel, and first responders and emergency agencies. Hundreds of thousands (or millions) of television broadcast viewers, whether fixed or mobile, can each examine the same maps and determine their own location relative to the areas in crisis. In this way they are able to help each other by drawing evacuation routes, whether to shelter in their present location, or measures for property protection and the safety of family members and others.[58]

As mentioned above it is evident that it could be much easier to manage rescue operation with the help of a web based missing person registry system; a web based tools for data management, to avoid ambiguity and bring more clarity and consistency in deployment of search and rescue teams and Helicopters, allocation of medical teams and relief materials *etc.*

In the similar chaotic situation of Indian Ocean Tsunami (2004), "*Sahana*", free and open source software (FOSS)-based system was developed by Lanka Software Foundation. In Sri Lanka, which was severely hit by the tsunami, ICT volunteers put together the *Sahana* Disaster Management System to help track families and coordinate work among relief organizations during and after the tsunami disaster. *Sahana* consists of a series of integrated web-based disaster management applications aimed at facilitating the management of missing people and disaster victims; managing and administering various organisations; managing camps; and managing requests and assistance in the proper distribution of resources. Subsequently, *Sahana* was also deployed to manage the earthquake disaster in Northern Pakistan (2005), the Guinsaugon landslide in the Philippines (2006) and the earthquake in Yogjakarta, Indonesia (2006).

Sahana was aimed to handle post disaster trauma of victims:

- ☆ *By effectively finding missing persons;*
- ☆ *Coordinating and balancing the distribution of relief organisations by connecting relief groups;*
- ☆ *Registering and tracking all incoming requests for support and relief;*
- ☆ *Maintaining link between helping donors and relief distributers; and*
- ☆ *Tracking the location and numbers of victims in the various camps and temporary shelters set up all around the affected area.*

The following examples show how *Sahana* assisted disaster victims during the Indian Ocean tsunami in 2004 and the Pakistani earthquake the following year (2005).[59],[60],[61]

Example 1: Tracing Missing Persons

After a devastating Tsunami many people were misplaced or were missing. Many children, women and old age persons got separated from their families. People were desperately waiting for their relatives and friends to be found or for latest information, about the condition of their loved ones. *Sahana* was designed with Missing Person Registry, an electronic version of a bulletin board of missing and found people. It can capture information not only on the people missing, but also about those who sought details about the missing, thus increasing their chance of reuniting. Even if the victims or families did not have access to this information themselves, it was quite easy for any authorized NGO or civil society group to connect to the central portal and provide that service in the areas they were working.

Example 2: Coordinating Donor Groups

In the immediate aftermath of the 2004 tsunami in Sri Lanka, many NGOs, International organizations and community groups came forward for all kinds of support. Though they all had intentions to support but coordinated support system was required. In an environment where resources are in short supply, it is essential that response efforts should not be duplicated. Otherwise, such duplication could result in congested supply routes, competition between helping organisations, saturation of support provided to some areas while other affected areas are neglected. Consequently, goodwill can be lost. *Sahana* has developed such an organization registry. It keeps track of all the relief organisations and civil society groups working in the disaster region. It captures information on both the places where they are active and the range of services they are providing in each area to ensure that, there is no overlap.

Example 3: Recording the Locations of Temporary Camps and Shelters In a disaster situation, there are usually no pre-planned locations for camps and shelters.

In the aftermath of a disaster it is important to have a track record of available temporary shelter or camps in the area with locations, capacity and extant populations staying therein. This is paramount to distributing aid effectively and ensuring that no affected areas are inadvertently ignored. A sub-application of the *Sahana* system keeps track of the location of all the camps in the region. If necessary it can provide a GIS view to plot the location of the camps in the disaster affected area. The *Sahana* Disaster Management System from LSF has shown how ICT can play an important role in times of disasters to manage and coordinate the relief and humanitarian efforts. The use of FOSS for developing *Sahana* and its availability as FOSS is significant in that it is easily and freely available for any country or community to use both in times of disaster and also in preparing for disasters.[62]

There are many more such examples where people are collaborating and developing ICT tools by leveraging technologies to mitigate effects of disasters. A very pragmatic example of community based early warning system in India is as given below:

Example 4: Community-Based Flood Early-Warning System[63,64]

One of the effects of climate change is the formation of melt water lakes on the lower sections of glaciers in the Himalaya region. Such lakes are inherently unstable as a result they are prone to catastrophic flood surges. To enhance the resilience

There are similar examples available where ICT has been very effectively used for the benefit of community in distress. Besides, it is not essential to reinvent the wheel every time and everywhere. It is therefore very important that knowledge management tools be available and accessible to disaster prone developing countries so as to keep track of such FOSS available globally and countries affected by disasters may openly and effectively use them for the benefit of all. Example of Uttrakhand flash flood previously given in this chapter calls for such mechanism which provides information on globally available ICT tools open for use during disasters. UN organizations may help to facilitate development of such programmes on capacity building through knowledge and technology transfers to least developed and developing nations.

of 45 vulnerable communities in the Indian Himalayan region to flood hazards, a team of experts from International Centre for Integrated Mountain Development (ICIMOD) jointly with an NGO called ***Aaranyak***[65] have promoted the Community-Based Flood Early-Warning System. The Community-based Flood Early-warning System is an ICT-enabled system to detect and respond to flood emergencies that are prepared and managed by the communities. The wireless system manages flood or flash flood risk by providing early warnings to downstream communities and enhances cooperation between upstream and downstream communities in the sharing of flood information. This ICT solution consists of two units a transmitter and a receiver. The transmitter is installed along the riverbank, and the receiver is installed at a house near the river. A flood sensor attached to the transmitter detects rising water levels. When the water reaches a critical level, a signal is wirelessly transmitted to the receiver. The flood warning is then disseminated via mobile phone to concerned agencies and vulnerable communities downstream. Critical flood levels are set with the help of the local community.

Nowadays social media, with ever growing internet technologies, are coming up in a big way. People are using social networking for coming together for socio-political causes. This is being considered as most powerful tool for connecting people for a cause.

Example 5: Use of Skype and YouTube in Syria

"Mamdouh[66] (name changed to preserve real identity), who works with an international non-governmental organization (NGO) that preferred not to be named to preserve their identity. Mamdouh, a Syrian who was forced to flee the country, has helped to establish encryption tools and virtual private network (VPN) accounts, to create secure ways for Syrians inside the country to communicate via the internet (Mamdouh, personal communication, 2013)." Mamdouh says "Syrians who manage to get online head to YouTube to share footage of the humanitarian disaster they are witnessing. They also flock to Skype to communicate with family members and to get information and requests to the humanitarian community".

Role of Crowd Sourcing During Emergency Response and Relief

ICT is widely used for crowd sourcing.Use of 'crowd sourced' data is getting momentum with better broadband Internet connectivity and use of smart mobile phones, smaller and handy ICT devices such as Ipads, notepads smaller palmtops *etc.* Crowd sourcing was used extensively in the response to the 2010 Haiti earthquake, allowing local people, mapping experts and other stakeholders to communicate what they saw and heard on the ground, and to produce information that could be used by humanitarian workers. This was particularly useful in locating survivors who needed assistance, but it is increasingly recognized that crowd sourcing could also help with pre-disaster activities, specifically risk identification and early warning.

Later in succeeding text use of crowdsourcing for recovery with the help of web based platforms shall be described.

Public-Private Partnerships

Public Private Partnerships (PPP) model is very good method of pooling in resources and expertise for larger good of society.

Example 6: Emergency.lu [67] [68]

In the event of a natural disaster or an emergency situation where human lives are involved, there is an immediate challenge to coordinate the relief effort, at a time when connectivity is often interrupted or damaged or where responses are needed across wide spread physical areas and often challenging operating environments. Until there is communication through network connectivity, effective aid cannot take place. Local infrastructure is often unable to provide any of these coordination and response prerequisites in the short time frame needed. Emergency.lu was formed as a public-private partnership to fill the need for a rapid response communication and coordination system with global capacity. The public-private partnership provides multi-layer support for the first hours and days following a large-scale disaster, and integrates into existing communications infrastructures used in humanitarian operations. A Luxembourg-initiated solution, emergency.lu leverages technology and expertise of a consortium of companies and organizations to support the efforts of existing humanitarian groups: SES TechCom, Hitec, and Luxembourg Air Rescue have joined in a public-private partnership with the Ministry of Foreign Affairs of Luxembourg.

"emergency.lu" works with the support of its operational partners "Administration des services de secours" (ASS), World Food Programme (WFP) and the Emergency Telecommunication Cluster (ETC). The "Administration des services de secours" (ASS) is the governmental agency for civil protection and rescue services in Luxembourg. ASS uses its Humanitarian Intervention Team (HIT) which is based on volunteers support for civil protection and communal fire brigades. ASS sends different types of emergency aid units to countries that are struck by natural or man-made disasters

which includes Urban Search and Rescue Team (USAR), a Water Purification unit, a Deep Water Rescue unit and a Technical Assistance and Support Team (TAST). World Food Programme is humanitarian agency fighting hunger worldwide. WFP is part of the United Nations system and is voluntarily funded for, saving the lives of victims of war, civil conflict and natural disasters.

emergency.lu offers a suite of local and remote services to facilitate communication and coordination services on site and with remote headquarters or teams.[69]

- ☆ *Voice communication via Voice over IP technology*
- ☆ *Instant Messaging*
- ☆ *Tracking and Tracing of people and equipment*
- ☆ *Map assessment*
- ☆ *Food distribution and asset management*
- ☆ *Situational reports*
- ☆ *Integration of extensive local and remote data sources*

This first humanitarian mission took place in January 2012 with the support of technical staff from the Luxembourg Civil Protection.

Example 7: The International Telecommunication Union (ITU) with support from FedEx, IGO global communication, Inmarsat, Iridium, TerraStar Global, Thuraya and Namelyada, helps in deploying mobile satellite terminals and other equipments to establish communication links for coordinating relief operations.

"Certainly ICT plays prominent role in various aspects of disaster response such as :resource information management and tracking; communication under emergency situations; collecting essential items for the victims; national and international fundraising; management of property and casualty status; and prioritization of response requirements".

ICT in Post-Disaster Recovery

Immediately after a disaster, it is essential to restore vital services and systems. This may include providing temporary food, water and shelter to those who have lost homes in the disaster. Medical services are needed for those who are injured. GIS can play several roles in this process. It can identify the damage and begin to establish priorities for action (triage). GIS can also ensure uniformity in the distribution of supplies (medicine, food, water, clothing, *etc.*) to emergency distribution centers. They can be assigned in proper amounts based on the extent and type of damage in each area. GIS facilitates online monitoring of the status of ongoing work in the recovery phase. As work is progressing and/or completed, GIS can visually display current project status by quickly updating the data. Current status can be easily viewed and accessed at the centralized GIS interface in the Emergency Operation Centre (EOC).

Post disaster recovery requires huge amount of on ground data collection related to Post Disaster Need Assessment (PDNA) for recovery projects planning and implementation, rebuilding, claims processing, and documentation of lessons learnt. ICT offers logistical support option for providing details on damage surveys, visuals and maps, equipment, human resources, and funds availability. ICT plays very important role in Post Disaster Need assessment (PDNA). PDNA is a credible, rigorous and globally recognized standard methodology for assessment of losses during course of a disaster. PDNA enables to take stock of physical damage to property, infrastructure, and productive assets as well as to account for losses resulting from the disruption of economic flows brought on by the disaster. This significant loss dimension often did not feature in previous assessments. Having a better understanding of both of these measures and how they affect each sector provides the Government with a stronger knowledge base to plan and budget. Further, the PDNA looks to do justice to the human recovery aspect of the disaster and build a solid disaster risk management strategy into the plan to support recovery. Priority areas are set out in a comprehensive framework to inform the Government's budgeting priorities and to engage with donors and partners where necessary.

Based on detailed damage assessment, the government finalizes estimation of funds requirements and fund generation mechanisms. After normalcy returns, the damage assessment team carries out a detailed study of the damaged houses, livelihood assets and crops *etc.* In India, usually the Panchayat level Officer is made responsible for carrying out such assessment assisted by the damage assessment teams. Assessments must be carefully planned and managed. A sequence of activities is involved and each of them must be planned in detail.

The following activities inter-alia constitutes the assessment process:

- ✰ Identify information needs and sources of reliable data;
- ✰ Collect data;
- ✰ Analyze and interpret data; and
- ✰ Report conclusive forecast and provide solution based alternatives to planners and decision-makers

Therefore, for appropriate damage assessment, making plans and strategies for disaster recovery; it requires collection, interpretation and analysis of a big volume of data. Without support of ICT tools, it is not possible to effectively do this exercise. Manually collected data tend to be erroneous and subject to various kinds of sampling and judgmental errors. Internet along with database management tools are ideal for such information creation processing and sharing. Different types of tools are now available for systematic data collection, storage, and analysis. These tools are not only important for disaster response, relief and recovery phases but are also useful for future planning, Disaster risk reduction, capacity building and improvement of community resilience.

From 1994, the creation of a common conceptual and methodological framework was begun *in Latin America*[70] *by groups of researchers, academicians, and institutional actors linked to the Network of Social Studies in the Prevention of Disasters in Latin America (Red de Estudios Sociales en Prevención de Desastres en América Latina - LA RED). These groups conceptualized a system of acquisition, collection, retrieval, query and analysis of information about disasters of small, medium and greater impact, based on pre-existing official data, academic records, newspaper sources and institutional reports in nine countries in Latin America.* This effort was then picked up by UNDP and UNISDR who sponsored the implementation of similar systems in the Caribbean, Asia and Africa. "*The developed conceptualization, methodology and software tool is called Disaster Inventory System - DesInventar (Sistema de Inventario de Desastres)*[71]. *The development of DesInventar, with its conception that makes visible disasters from a local scale (town or equivalent), facilitates dialogue for risk management between actors, institutions, sectors, provincial and national governments. DesInventar is a conceptual and methodological tool for the generation of National Disaster Inventories and the construction of databases of damage, losses and in general the effects of disasters".*

A major disaster generally triggers an outpouring of technical and financial assistance from Government, international donors, non-profit non-government organisations, and individuals. The potential for waste, misappropriation and misuse of these resources is high. Lack of transparency and accountability can lead to irreversible loss of goodwill and generosity. Keeping transparency during recovery phase is very important for sustaining faith of various stakeholders. "*Self-organising, self-managed social networking tools with free and open source platforms such as Ushahidi are good in meeting this requirement. Ushahidi combines SMS, Twitter and Google Maps to crowd source crisis information. It is a free, open source; decentralised platform developed by Kenyan bloggers in the aftermath of the 2008 Kenyan elections but is now used in other disasters. Location-specific information is communicated directly to subscribers. Ushahidi has developed Swift River as a rapid verification system for crowd sourced information by crosschecking tagged information from different sources*"[72].

Crowdsourcing information from mobile phone, email, RSS feeds, the web, and feeding Crisis mapping. Through initiatives such as Ushahidi and Google Crisis Response, crisis mapping utilizes Crowd sourcing as well as satellite imagery, participatory maps and statistical models to power more informed and effective real-time information on an upcoming crisis in times of uncertainty and confusion. The vast amount of data that can be produced from such systems can be analyzed through networks of stakeholders (such as Crisis Mappers).

Long-term recovery, is a long drawn process and it inter alia includes restoration of all services, it involves replacement of existing damaged/temporary infrastructure and building new ones with better facilities. GIS tools can be used to track the progress of these activities. It is also possible to priorities restoration investments with the help of a GIS and decision support system. A GIS can also be very helpful for account keeping for recovery activities. Long term recovery and rehabilitation[73] often requires going back to planning tools adjudge the impact of disaster in order to understand about sectors of the economy were hit the hardest; strata of society which are worst affected with demographic statistics; what could be assessed accurately about cost of lost production in the manufacturing or agricultural sector, disruption to localized small and medium business, or loss of environmental assets; infrastructure losses *etc.*On the basis of such analysis post disaster recovery plans of action and their successive calibrations may be implemented. ICT plays major role for such planning by providing data collection, processing, simulation, resource mapping and optimization techniques for designing and implementation of such plans.

Internet of Things (IOT) - M2M Communication

With rapid technological growth in the field of Machine to Machine popularly known as M2M communication ICT is envisioned to play major role in all facets of disaster management continuum. Broadly, Machine to machine (M2M),[74] refers to technologies that allow both wireless and wired systems to communicate with other devices (machines) of similar abilities. M2M uses a device (such as a sensor/meter/ actuator[75]) to capture an event, which is relayed through a network (wireless, wired or hybrid) to an application, that translates the captured event into meaningful information. There are various definitions given by Global Standardization organizations for M2M. European Telecommunications Standards Institute (ETSI)[76] has defined M2M2 as 'Physical telecommunication based interconnection for data exchange between two ETSI M2M compliant entities, like: device, gateways and network infrastructure.' Accordingly M2M may be described as machines are connected to the Internet, using a variety of fixed and wireless networks and communicate with each other and with the wider world.

M2M is also being inter-changeably used with various other terms - Internet of Things (IoT), Internet of Everything (IoE), Smart systems (Homes, Cities, Meters, Grids *etc.*). IoT refers to addition of communications and sensing capabilities to a wide range of physical objects. In the coming decade it is expected that millions of IoT devices will be deployed in electric meters, parking meters, thermostats, car components, roads *etc.* IoT devices will send data directly using protocols such as Wi-Fi and Bluetooth and via mobile networks, specialized networks and over the global network.[77]

Advancement M2M technology will add entirely a new dimension to disaster early warnings. Technologically empowered sensors, gauges, meters will take measurements, *viz.* hydrological measurements will talk to their respective devices at back end which in turn will inputs to weather prediction models and more accurate forecast/nowcast/early warnings will be generated. This machine trails would be extended to provide necessary triggers to mitigation and risk aversion mechanical models as well.

International Telecommunication Union (ITU) had established various Groups (Foucus Groups, study Group) with the objective of developing recommendations relevant to M2M from telecom/ICT perspectives. Study Group 20 (SG 20) is currently working to address the standardization requirements of Internet of Things (IoT) technologies, with an initial focus on IoT applications in smart cities and communities (SC and C). Mandate of SG20 is to develop international standards to enable the coordinated development of IoT technologies, including machine-to-machine communications and ubiquitous sensor networks. SG20 core objectives inter alia include standardization of end-to-end architectures for IoT, and mechanisms for the interoperability of IoT applications and datasets employed by various vertically oriented industry sectors.[78]

Climate Change and Adaptation: Role of ICT

Land and people, their resources and ecosystems, whether in developed or in a developing country, are prone to environmental challenges. The prevalent environmental challenges include: climate change, land-use changes and natural resource degradation. These changes are particularly important from human security and sustainability point of view as these often determine or contribute to the intensity of disaster related natural or anthropogenic events, besides aggravating the underlying factors of vulnerability. This, issues pertaining with the energy efficiency, waste management, air pollution, water quality and scarcity, natural habitats and biodiversity, attain central position in the science of sustainability along the social, economic and cultural aspects of vulnerability to the impact of climatic changes – including shocks and stresses.

Communities in developing countries particularly in small island states, arid and high mountain zones, and in densely populated coastal areas are considered to be particularly vulnerable. Vulnerability of socio-economic systems also depends on type and costs of assets and infrastructure located in such hazard-prone areas. India is a large country with transition of developmental stages, with the Great Himalayas, the world's third largest ice mass in the north, 7500 km long and densely populated coastline in the south and east. Nearly 70 percent of its population still living in rural or semi-urban areas with livelihood directly dependent on climate-sensitive occupations (agriculture, forests, and fisheries, ecosystems resources such as water, biodiversity, mangroves, coastal zones, grasslands and related businesses). Heat wave, floods (land and coastal), tropical cyclones and droughts are known to hit commonly.

The terms 'mitigation' and 'adaptation' refer to two alternatives, but at the same time complimentary, approaches of dealing with climate change. **'Mitigation'** deals with the *causes* of climate change and works to reduce man-made effects on the climate system. In contrast, **adaptation** makes changes to prepare for and negate the *effects* of climate change, thereby reducing the vulnerability of communities and ecosystems. By adapting to cope with the effects of climate change, communities, enterprises and institutions can build up their **'resilience' to climate change**. There has been a dichotomy for long time and the 'mitigation' and 'adaptation' strategies travelled keeping abeyance from each other, but however, there is a paradigm shift and the two are finding synergies and co-benefits. This is particularly significant when these are being explored with broader aim of sustainable development and incorporating disaster risk reduction objectives.

Mitigation Vs Adaptation and Role of ICTs

Mitigation – slowing down (and reversing) the climate change process – involves reducing the flow of heat-trapping greenhouse gases (GHG) into the atmosphere, either by reducing sources of these gases (for example, the burning of fossil fuels for power, heat or mobility) or enhancing the "sinks" that accumulate and store these gases (such as the oceans, forests and soil). The goal of GHG mitigation is to avoid significant human interference with the climate system, and "stabilize greenhouse gas levels in a timeframe sufficient to allow ecosystems to adapt naturally to climate change, ensure that food production is not threatened and to enable sustainable economic development process" (from the 2014 report on Mitigation of Climate Change from the United Nations Intergovernmental Panel on Climate Change, page 4).

Adaptation – adapting to life in a changing climate – involves adjusting to actual or expected future climate scenario of events. The goal is to reduce our vulnerability to the harmful effects of climate change (like sea-level encroachment, more intense extreme weather events or food insecurity). It also encompasses making the most of any potential beneficial opportunities associated with climate change (for example, longer growing seasons or increased yields in some regions). Throughout history, people and societies have adjusted to and coped with changes in climate and extremes with varying degrees of success. Climate change (drought in particular) has been at least partly responsible for the rise and fall of civilizations. Modern life is tailored to the stable climate we have become accustomed to. As our climate changes, we will have to learn to adapt. The faster the climate changes, the harder it could be.

Climate Change Adaptation (CCA) is a response to global warming and climate change, that seeks to reduce the vulnerability of social and biological systems to relatively sudden change and thus offset the effects of global warming.[79] *Thus "CCA is adjustment in natural or human systems in response to actual or expected climatic induced outcomes, which moderates harm or exploits beneficial opportunities".*[80] A typical definition of climate change adaptation is[81]: "Adjustments in ecological, social, or economic systems in response to actual or expected climatic stimuli and their effects or impacts. It refers to changes in processes, practices, and structures to moderate potential damages or to benefit from opportunities associated with climate change."

As the climate change process seems to be irreversible, it became urgent to develop sound adaptation mechanisms to the current and future scenarios of the climate. It is likely that the biggest impacts of change will be on livelihood and health resources through effects on agricultural, food systems, water and energy (Brown and Funk, 2008). In general the more mitigation (GHG emission reduction) there is, the less will be the impacts to which the communities will have to adjust, and the less the risks for which risk management or emergency prepared would be required. Conversely, the greater the degree of preparatory adaptation, the less may be the devastating impacts resulting from any given degree of climate change.

The predicted manifestations of global climate change are diverse and extensive. Information and Communications Technologies (ICTs) offer great potential to enable and enhance climate change adaptation projects, programmes and activities. As yet these roles have received relatively little systematic consideration (Sabajee *et al.*, 2014)[82]. Impacts of climate change are like a set of interlocking causal chains. Starting with a primary set of effect(s), *e.g.*, changes to weather patterns, leading to physical consequences, generally on a fairly large scale, which in turn lead to human and ecosystem level impacts and on to consequences for individuals and communities. Figure 3.2 reveals an example for understanding the scope of ICT application in climate change adaptation context. Current and potential roles for ICTs in the contexts of adaptation are diverse. Some of these are directly related to meeting challenges associated with climate change, such as:

- Provision of early warning systems for extreme weather
- Sharing knowledge of adapted farming practices
- Awareness raising of climate-related risks
- Co-ordinating disaster recovery information
- Supporting consultation and participation in developing adaptation policies
- Provide training in flood management
- Sensor networks providing data to aid adaptation decision making
- Gathering and analysing information for vulnerability assessments

Adapting to climate change involves cascading decisions across a landscape made up of agents from individuals, firms and civil society, to public bodies and governments at local, regional and national scales, and international agencies. A broad distinction can be drawn between action that often involves creating policies or regulations to build adaptive capacity and action that implements operational adaptation decisions. Within a strategically defined framework, it is possible to categorize interventions more precisely in a number of ways[83]:

- By sector: A number of development sectors are particularly susceptible to the impacts of climate change, notably agriculture and water resources, forestry and fisheries, health and livelihoods.
- By geographic scale: Climate change is a large-scale phenomenon but will also have impacts that vary greatly between individual locations.

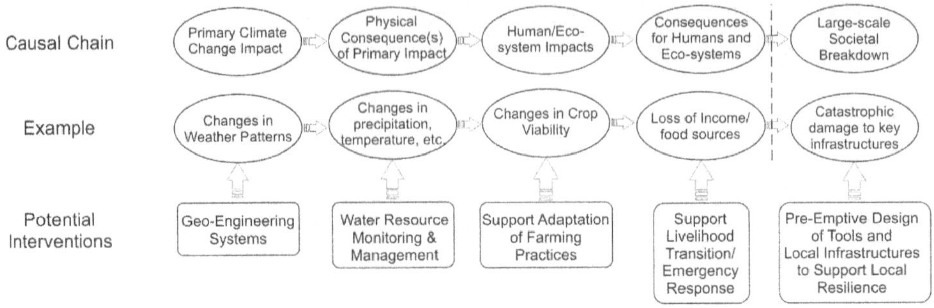

Figure 3.2: Model of Interplay of Factors in Identifying Adaptation Opportunities for ICTS (*Source*: Sabajee *et al.*, 2014).

Interventions therefore range from those at regional or national level down to specific actions to meet the needs of individual communities.

- By technology: ICTs are highly diverse and can be implemented on different scales. At one end of the scale they include expensive one-off applications such as sensor networks, satellite earth stations and meteorological systems. At the other, they include the mobile phones that individuals can use to access information, report problems or share experiences.

The multiple infrastructure systems in cities are in fact a "system of systems," or a network of systems that support interlocking operations or functions. They have become more integrated using ICT, leading to the "*internet of things*" (IoT) and enabling integrated management of operations. Harnessing the potential of these networks for sustainable urbanization is a crucial feature of smart sustainable cities. ICT based operational and governance can help reduce energy footprint, as witnessed in OECD countries (Urban *et al.*, 2014)[84], which ultimately contributed to reduction in the carbon footprint.

Role of ICTs in Adaptation through Key Environmental Strategies

Adaptation to climate change has twin coverage to integrate objectively, *i.e.* the climatic understanding, projections and impact assessment, with the sound observations and management of environmental settings for effective planning and governance. The relationship between ICTs and the environment is complex, as ICTs can play both positive and negative roles. Positive impacts can come from dematerialization and online delivery, transport and travel substitution, a host of monitoring and management applications, greater energy efficiency in production and use, and product stewardship and recycling. Negative impacts can come from energy consumption and the materials used in the production and distribution of ICT equipment, energy consumption in use directly and for cooling, short product life cycles and e-waste, and exploitative applications, for *e.g.*, remote sensing for unsustainable fishing (Berkhout *et al.*, 2009).

Looking at ICTs to provide tools for dealing with environmental issues from a developing and emerging country perspective, following application categories are identified (ITU, 2008):

(i) Environmental Observation

Terrestrial (geophysiography, land, soil, water), ocean, climate and atmospheric monitoring and data recording technologies and systems (remote sensing, data collection and storage tools, telemetric systems, meteorological and climate related recording and monitoring system), as well as geographic information systems (GIS).

(ii) Environmental Analysis

Once data have been collected and stored, various computational and processing tools are required to perform the analysis. This may include land, soil, water and atmospheric quality assessment tools, including technologies for analysis of atmospheric conditions including GHG emissions and pollutants, and the tracking of both water quality and availability. The analysis of data may also include correlating raw observational data with second order environmental measures, such as biodiversity, energy, *etc.*

(iii) Environmental Planning

Planning makes use of the information from environmental analysis as part of the decision making process for the purpose of policy formulation and planning at the international, regional and national levels. Planning may include classification of various environmental conditions for use in agriculture and forestry and other applied environmental sectors, and is often focused on specific issues such as protected areas, biodiversity, ecotourism, housing, industrial or GHG emissions. Planning may also include the anticipation of environmental conditions and emergency scenarios, such as climate related natural disasters and their socio-economic or health consequences.

(iv) Environmental Management and Protection

This involves activities of managing and mitigating impacts on the environment as well as on the communities and their resources, by helping adaptation to given environmental conditions or events. This includes resource and energy management systems, GHG emission management and reduction systems and controls, pollution control and management systems and related methodologies, including mitigating the ill effects of pollutants, disaster emergencies and other human-induced hazards.

(v) Impact and Mitigating Effects of ICT Utilization

Sustainable ICT resource management includes environmental aspects of materials, energy and waste, including toxic waste in the form of heavy metals. ICT use can mitigate the environmental impacts directly by increasing process efficiency and as a result of dematerialization, and indirectly by virtue of the secondary and tertiary effects resulting from ICT use on human activities, which in turn reduce the human induced risks on the environment and humans themselves.

(vi) Environmental Capacity Building

Improvement of environmental conditions would depend on the actions of individuals and organizations. Capacity building includes efforts to increase knowledge and skills, and mass awareness of environmental issues and priorities, the development of professionals, and integrating environmental and climate change content into formal education.

ICT Enabled Adaptation Framework

Climate adaptation refers to the ability of a system to adjust to climate change (including climate variability and extremes) to moderate potential damage, to take advantage of opportunities, or to cope with the consequences. Mapping of the relationships between ICT characteristics/affordances and adaptation context is critical in understanding the many roles that ICT may play in adaptation in any practical situation. The attributes of paradigm shift from traditional emergency response based approach to adaptation based risk reduction approach are shown in the Figure 3.3. There are evolving perspectives on ways and means for ICTs intersection with adaptation to climate change. Based on a number of analytical approaches, a new framework to guide future policy and programming interventions for ICT enabled climate change adaptation approach has been recommended.

This framework categorizes a range of interventions at four points along a continuum of adaptation, as following:

I. Addressing underlying drivers of vulnerability – interventions which are concerned with the precursors or contributing factors making land, people and their resources vulnerable to the impacts of climate change, rather than being concerned with those impacts themselves.

II. Building the response capacity of local and regional systems and communities – interventions which help communities and local governments to develop resources needed in responding to the impacts of climate change.

III. Reducing and managing risks related to climate variability and climate change – interventions which provide information and facilitate communities in changing lifestyle, occupational practices, and economic behaviours in ways that make them more sustainable in new climate conditions.

IV. Confronting climate change – interventions which directly enable protection from the physical impacts of climate change such as rising sea

General Differentiating Concept or Attribute	Examples of Terms Used
Purposefulness	Autonomous ↔ Planned Spontaneous ↔ Purposeful Automatic ↔ Intentional Natural ↔ Policy Passive ↔ Active Strategic
Timing	Anticipatory ↔ Responsive Proactive ↔ Reactive Ex ante ↔ Ex post
Temporal Scope	Short term ↔ Long term Tactical ↔ Strategic Instantaneous ↔ Cumulative Contingency Routine
Spatial Scope	Localized ↔ Widespread
Function/Effects	Retreat - Accommodate - Protect Prevent - Tolerate - Spread - Change - Restore
Form	Structural - Legal - Institutional - Regulatory - Financial - Technological
Performance	Cost - Effectiveness - Efficiency - Implementability - Equity

Figure 3.3: Objective Differentiation of Conventional 'Disaster Response' to 'Adaptation Based Risk Reduction' (Shabajee *et al.*, 2014).

levels, landfall of tropical cyclones and the spread of malarial mosquitoes into newly favourable regions.

ICTs also have a complex but interesting relationship with sustainability and with the underlying cause of climate changes. This relationship can be understood through the effects of ICTs (Creech *et al.*, 2012):

I. First order (direct) effects related to impacts ICTs have on climate change, in particular the CO_2 emissions from the production, use and disposal of communications equipment and services, accounting for between 2 and 2.5 percent of global emissions.

II. Second order (indirect) effects relate to the role ICTs can play in reducing emissions resulting from other industrial sectors, by having them adopt ICTs to improve efficiency and production. A complication arises from rebound effects which may eliminate the gains resulting from apparent reductions, such as increased power consumption resulting from lower energy prices achieved through greater energy efficiency.

III. Third order (societal) effects result from large-scale changes in social, occupational, cultural and economic behaviour resulting from widespread use of ICTs, including changing patterns of trade, production and consumption, and global to local engagement of citizens in decision-making.

Community Based Adaption Approach with ICT Applications

Communication for development can effectively support an ICT-enabled adaptation strategy, from data collection to implementation, monitoring and evaluation phases. Both Communication and Information are essential when promoting community-based adaptation, which mainly targets small communities in the developing world that are vulnerable to climate change due to their dependence on natural resources and/or their location in areas prone to climate-related shocks (Huq and Reid, 2007). Adaptation initiatives at the community-level can build on existing natural resources management or disaster risk management practices, but there is often a need to broaden the set of stakeholders involved in the development and implementation of an appropriate strategy in order to deal with climate change issues. This is due to the complexity of the phenomenon, which requires different forms of expertise and capacities to be involved for developing a sound adaptation process (Sala, 2010).

Promoting community resilience through sound natural resources management is definitely a win-win strategy, but trying to directly target climate change implies the participation of various professionals from different sectors and institutions. Information technologies and 'communications for development' can make a difference by linking different capacities and ensuring that communities are at the centre of adaptation planning and implementation. Supporting multi-stakeholder action through the application of ICTs in several ways, utilized fundamental principles of communication for development, such as:

- Facilitate data collection at local level with tracking tools, handheld devices, *etc.*
- Integrate local, traditional and modern scientific knowledge through knowledge bases, databases, Geographic Information Systems, *etc.* for developing user friendly improvised data products (*e.g.* weather data products, like in case of agro-advisories).
- Facilitate participatory approach to problem analysis and adaptation planning utilizing downscaling tools, modelling software and decision support systems, among others.
- Facilitate the translation of uncertainty into practical policies through support systems, collaborative project management tools, monitoring systems, *etc.*
- Ensure information flows among concerned groups through information systems and knowledge networking tools.
- Channel information at the right people at the right time through appropriate media through community-based early warning systems, mobile phones, radios, *etc.*
- Empower local people to manage their adaptation efforts, providing them with guidelines, media, e-learning modules, monitoring system, *etc.*
- Promote local information sharing through web portals, virtual

consultations and other traditional tools that are useful to discuss and define protocols and alternatives for adaptation, good practices, programmes and benchmarks.

☆ Link rural knowledge base and information systems through intermediate nodes and tools, to promote data integration across different domains (*i.e.* GIS for spatial integration, community based mapping, *etc.*).

Role of ICT in Urban Climate Adaptation

Urban growth and associated industrial commercial conglomeration, in addition to the housing intensity sprawl, is often seen as contributing to high GHG emissions and environmental degradation, but the fact remains that cities form epicentre of economic activities and infrastructure concentration often highly vulnerable to the effects of climate stresses and shock events. Many cities started addressing climate change by efforts to reduce emissions, putting in place several mitigation actions such as sustainable transport strategies, waste management systems, establishing building codes or by promoting enhanced use of renewable energy. However, GHG emissions continue to rise and cities are starting to feel the effects of climate change impacts, evidencing the need for them to adapt to both current and future manifestations (UN-HABITAT, 2012).

Sectors in cities are interconnected, and therefore, a failure in one sector (*e.g.*, in the case of extreme weather events) could have a 'domino effect' on other sectors in the city and lead to an overall economic loss for a country or region

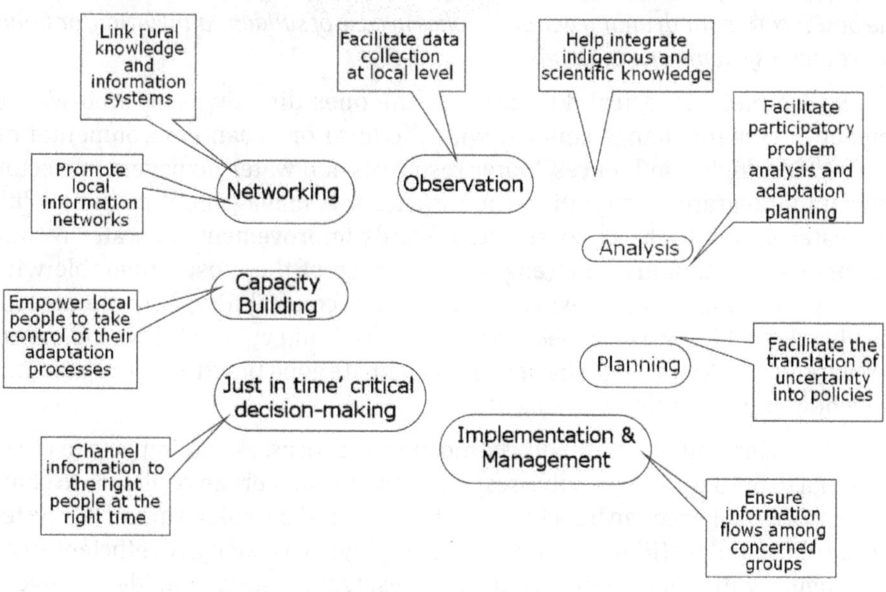

Figure 3.4: A 'Communication for Development' Approach for ICT Applications in CBA (Sala, 2010).

(GTZ, 2009). As reported, windstorms and floods in Asia between 1996 and 2005 caused over 70,000 deaths, with an estimated economic loss of around US$ 190 billion. Such large scale loss is due to the fact that infrastructure, including ICT infrastructure were neither fully resilient nor adequately redundant or flexible to meet the demands in stressed situation.

Several cities around the world and in India as well, are in the process of becoming smart and sustainable, and have an enormous opportunity to include ICT infrastructure and ICT solutions as part of their climate change adaptation strategies to respond more effectively to both current and future climatic risks. Most of the urban infrastructure that will exist in 40 to 50 years has not been built yet (ITU, 2012). The ICT sector can be an enabler of climate change adaptation in cities, but the sector itself has to adapt to climate change in order to ensure the continuity of essential ICT-related services provision. ICTs' use for adaptation can be grouped in three strategic categories: (1) ICTs for enhanced disaster risk management (Table 3.1); (2) ICTs to improve city resilience and adaptive capacity; and (3) ICT for informed adaptation decision making.

Nexus on ICTs, climate change and development initiative (NICCD, 2011), suggested that -

"Social media tools can also support public awareness and education campaigns, as well as foster participative processes. ICT applications such as participatory videos, photo-diaries or the use of mobile phones for collective mapping/monitoring exercises, could be used to foster greater involvement of low-income urban dwellers in climate change and risk-reduction initiatives, involving them in decisions such as the best location for drinking water supplies in case of sudden salinization, or failures in drainage systems due to floods."

Hydro-meteorological disasters are the ones directly associated with the impact of climate change coupled with the local or urban environmental heat and meteorological influences. Water resources and water management sector is crucial in integrating adaptation and disaster risk management in cities. Within the water sector, ICTs can contribute towards improvements in water resource management techniques[85]; strengthen the voice of the most vulnerable within water governance processes; create greater accountability; provide access to locally relevant information needed to reduce risk and vulnerability; and improve networking and knowledge sharing to disseminate good practices and foster multi-stakeholder partnerships, among others.

Observing and understanding atmospheric processes and impacts in cities is essential to prepare a sustainable response, and to build urban resilience to climate change. City resilience can be achieved in two ways: (i) by making urban ICT systems more robust and by (ii) increasing the city's adaptive capacity. An efficient system is "designed with safety margins that allow residents to survive sudden change and to quickly bounce back". It is important to mention that the social, economic and

Table 3.1: Examples of ICT Use for DRM Based Adaptation in Cities

Disaster phases	*ICT services involved*	*Major tasks*
DISASTER PREVENTION: Prediction and detection	☆ Meteorological services (meteorological aids and meteorological-satellite service) ☆ Earth exploration-satellite service ☆ Geographic Information Systems (GIS) ☆ Blogging, web 2.0 and social networking	☆ Weather and climate prediction ☆ Detection and tracking of hurricanes, typhoons, forest fires, among others ☆ Providing warning information
EMERGENCY PREPARDNESS and RESPONSE: Alerting	☆ Amateur services ☆ Broadcasting services terrestrial and satellite (radio, television, *etc.*) ☆ Fixed services terrestrial and satellite ☆ Mobile services (land, satellite, maritime services, *etc.*) ☆ Blogging, web 2.0 and social networking	☆ Receiving and distributing alert messages ☆ Disseminating alert messages and advice to large sections of the public ☆ Delivering alert messages and instructions to telecommunication centers for further dissemination to public ☆ Distributing alert messages and advice to individuals
RECOVERY: Relief	☆ Amateur services ☆ Broadcasting services terrestrial and satellite (radio, television, *etc.*) ☆ Earth exploration-satellite services ☆ Fixed services terrestrial and satellite ☆ Mobile services (land, satellite, maritime services, *etc.*) ☆ Blogging, web 2.0 and social networking	☆ Assisting in organizing relief operations in areas (especially when other services are still not operational) ☆ Coordination of relief activities by disseminating information from relief planning teams to population ☆ Assessment of damage and providing information for planning relief activities ☆ Exchange of information between different teams/groups for planning and coordination relief activities ☆ Exchange of information between individuals and/or groups of people involved in relief activities

Source: ITU-R (2013) and LIRNEasia (2008).

physical elements of a city; including ICTs, can become more robust. It is critical that any form and means of communication (*e.g.*, broadcasting radio receivers, Internet, mobile phones) and network infrastructure are not compromised by a disaster. Remote monitoring and data collection using ICT-equipped sensors (telemetry) is essential for climate change-related research.

Conclusion

ICT is of paramount importance in various phases of Disaster Management continuum. Utilization of ICT is no more a choice but a mandatory requirement for attaining paradigm shift in Disaster management. Task force report on Millennium Development Goals (MDG), 2013 has given specific mention of access to ICT technologies for all countries and Disaster Risk Management. There is broader concern all over the world, about sustainable development through mainstreaming of Disaster Risk Management and Climate Change Adaptation in developmental plans and strategies of developing and developed world. ICT is an imperative for overall sustainable development. Hence, the international strategies, namely, Sendai Framework, Paris Climate Agreement and Sustainable Development Goals (2015-30) and their call for a greater synergy in implementation, needs a systemic ICT base for their planning, implementation and monitoring frameworks.

It is also evident that, the requirement for communication/broadband access, safety, security, privacy, and bandwidth varies among disaster phases. The very reason for the existence of so many technological solutions is that none of them is perfectly fit for every situation. One medium that might fit best under a certain set of circumstances might be of little use under another. Thus, use of the best technological combination depending upon situational requirement is suitable. Use of ICT tools require proper planning at all levels by all stakeholders, which in turn require huge amount of planned interventions undertaken by Government. Besides, having pertinent policy framework in place at every level of Governance is the basic yardstick for efficacy of use of ICT as critical tool in disaster management.

Notes

1. http://eqs.metu.edu.tr/node/119.
2. Living with Risk: A global review of disaster reduction initiatives, UN/ISDR.
3. The eight Millennium Development Goals (MDGs) – which range from halving extreme poverty to halting the spread of HIV/AIDS and providing universal primary education, all by the target date of 2015 – form a blueprint agreed to by all the world's countries and all the world's leading development institutions. They have galvanised unprecedented efforts to meet the needs of the world's poorest
4. Swiss Re is a leading wholesale provider of reinsurance, insurance and other insurance-based forms of risk transfer.
5. http://economictimes.indiatimes.com/news/international/world-news/
6. Allen 2006; Berke *et al.,* 1993; Drabek and Key 1976; Lewis; Paton and Johnson 2001; Tobin 1999; Wisner *et al.,* 2003.
7. NONITA T. Y. 2011
8. Polsky and Cash, 2005
9. Stern 2007, NONITA T. Y. 2011 op.cit.
10. IFRC 2009

11. Disaster management developing country communities and climate change The Role of ICT:NONOTA T. YAP 2011.
12. Coyle and Meier 2009; Denning 2006; Gunawardene and Noronha 2007.
13. TechTarget 2004; UNESCAP 2008; IBRD 2010; NONITA T. Y. 2011 op.cit.
14. http://www.oneindia.com/2013/06/24/u-khand-tragedy-cag-submits-report-1244367.html
15. Picture Source: healthsysteminfo.blogspot.com
16. Chanuka Wattegama op.cit
17. http://www.itu.int/ITU-D/emergencytelecoms/tampere.html
18. National Disaster management Guideline on Information and communication system By NDMA, February 2012.
19. NDMICS (2011)
20. GIS refers to an organised collection of hardware and software for storing, indexing, retrieving, processing, spatially analysing and displaying data.
21. Remote sensing is an information gathering technique using sensors and other instruments aboard vehicles (for example satellites, aircraft, spacecraft) not in physical contact with and at distance from the object being observed. NONITA T. Yap op.cit
22. Paper on LIDAR DEM Application in India, by K. Kalyanaraman, V R. Venkataraman, ASRKV Murali Mohan *et al.*
23. Used for creating Inundation models required for flood early warnings.
24. Aerial Photogrammetry is the science of making measurements from aerial photographs, especially for recovering the exact positions of surface points. Moreover, it may be used to recover the motion pathways of designated reference points located on any moving object, on its components and in the immediately adjacent environment. Photogrammetry may employ high-speed imaging and remote sensing in order to detect, measure and record complex 2-D and 3-D motion fields.
25. Interferometric Synthetic Aperture Radar, abbreviated InSAR, is a radar technique used in geodesy and remote sensing. This geodetic method uses two or more synthetic aperture radar (SAR) images to generate maps of surface deformation or digital elevation, using differences in the phase of the waves returning to the satellite or aircraft. InSAR can be used to measure tectonic deformation, for example ground movements due to earthquakes. It was first used for the 1992 Landers earthquake, but has since been utilised extensively for a wide variety of earthquakes all over the world. InSAR can be used in a variety of volcanic settings, including deformation associated with eruptions, inter-eruption strain caused by changes in magma distribution at depth, gravitational spreading of volcanic edifices, and volcano-tectonic deformation

signals.

26. https://en.wikipedia.org/wiki/Interferometric_synthetic_aperture_radar
27. LiDAR(Laser Imaging, Detection and Ranging) originated in the early 1960s, shortly after the invention of the laser, and combined laser-focused imaging with radar's ability to calculate distances by measuring the time for a signal to return. LiDAR is popularly used as a technology to make high-resolution maps, with applications in geomatics, archaeology, geography, geology, geomorphology, seismology, forestry, remote sensing, atmospheric physics, laser altimetry, and contour mapping.
28. Non-members may make the request through members (Laituri 2010).
29. Charter was also successfully activated during the Senegal and Burkina Faso floods of September 2009. NONITA T. Y. op.cit.
30. Web.worldbank.org/Winning Applications for Earth Observation Services
31. Wattegama Chanukya (2007) op.cit.
32. http://www.ndtv.com/article/india/cyclone-phailin-triggers-india-s-biggest-evacuation-operation-in-23-years-431466
33. https://en.wikipedia.org/wiki/Cyclone_Phailin
34. Wattegama Chanukya (2007) op.cit
35. Ahmad 2007, Wattegama (2007) Op.cit
36. Dunnette, 2006, Wattegama (2007) Op.cit
37. WDR 2009
38. RANET is an initiative using RAdio and InterNET for the communication of hydro-meteorological information for rural development. NONITA T. Y. op.cit
39. IFRC op.cit
40. Wattegama (2007) op.cit
41. The Role of ICT:NONOTA T. Y op.cit
42. A satellite is essentially a radio-frequency repeater that is launched into orbit (circular or elliptical) around the earth. It transmits data (from the simplest digital data to the most complex television programmes) to and from the earth (with a latency period depending on how high their orbit is) via ground stations and hubs connected to the backbone core network.
43. One way: as transmission of information takes place only from central radio/Television transmission station, (downward) People/masses receive information through suitable receiving instruments tuned to that frequency. Whereas public can not transmit message to central station (upward).
44. https://www.itu.int/itunews/manager/display.asp

45. Message broadcast from a central office/warning centre/emergency operation centre, may be done through Short Message Service (SMS) on wireless phones, both GSM and CDMA. Community may also send relevant and important information through voice/SMS back to central place.
46. Number of telephone connections per 100 people in a specified geographic area. Teledensity is often used to compare the level of access to voice and data communications services between metropolitan and rural areas, or between one country and another. The governments of many emerging economies are focused on increasing teledensity as an economic enabler.
47. The telecommunications services market which includes fixed and mobile, data and voice services will continue to be the largest IT segment in India with IT spending forecast to reach $30 billion in 2014, to according to Gartner, Inc. (October 2013)
48. Source: TRAI report on The Indian Telecom subscription data (13 July 2017) op.cit
49. M S Swaminathan Research Foundation MSSRF, is a non-profit research organization and was established in 1988.It inter alia intensively work on coastal system research and use of information education and communication technology for sustainable development in rural India.
50. Subramanian, 2005
51. IFRC 2009, NONITA T. Y. 2011.op.cit
52. UN ISDR 2008.
53. NDMA guideline on Disaster Communication op.cit
54. World Disaster Report 2013 by International Federation of Red Cross and Red Crescent Societies.
55. NDMICS 2011
56. TRAI consultation paper on Telecom Network Failure during Emergencies/ Disasters Priority Routing of calls of Persons engaged in Response and Recovery May 2012
57. NONITA T. Y.op.cit
58. Draft Final Report for Question 5/2 https://www.itu.int/md/D14-SG02.RGQ-C-0247/en
59. http://wiki.sahanafoundation.org/
60. https://en.wikibooks.org/wiki/ICT_for_Disaster_Management/ICT_for_Disaster_Response
61. Wattegama Chanuka (2007)
62. http://wiki.sahanafoundation.org/, http://www.sahana.lk/
63. http://unfccc.int/secretariat/momentum_for_change/items/8688.php

64. http://www.icimod.org/?q=10925
65. http://www.aaranyak.org/Aranyak is working for conservation of biodiversity in Northeast India through research, environmental education, capacity building and advocacy.
66. Chapter 2 : Technology and community-centered humanitarian action, World Disasters Report 2013
67. Document RGQ22-1/2/28 Emergency.lu Rapid Response Communications Solution (SES World Skies).
68. http://emergency.lu/index.php/about, www.wfp.org, http://ictemergency.wfp.org
69. http://emergency.lu/index.php/system-elements/services
70. http://www.desinventar.net/what_is.html
71. http://www.desinventar.org
72. www.ushahidi.com
73. PDNA report of Fiji cyclone hit in December 2012.
74. http://www.dot.gov.in/sites/default/files/Nationalper cent20Telecomper cent20M2Mper cent20Roadmap.pdf
75. An actuator requires a control signal and a source of energy and is responsible for moving or controlling a mechanism or system. The control signal may be low energy electric voltage or current, pneumatic or hydraulic pressure, or even human power. (https://en.wikipedia.org/wiki/Actuator)
76. TR 102 725 V1.1.1 (2013-06)
77. http://trai.gov.in/sites/default/files/Consultation_Paper_M2Mper cent20_18_October_2016.pdf
78. Taken from http://www.itu.int/en/ITU-T/about/groups/Pages/sg20.aspx
79. https://en.wikipedia.org/wiki/Climate_change_adaptation
80. http://unfccc.int/essential_background/glossary/items/
81. B. Smit and O. Pilifosova, "Chapter 18: Adaptation to Climate Change in the Context of Sustainable Development and Equity," in Working Group II: Impacts, Adaptation and Vulnerability, IPCC, Ed. 2001.
82. Sabajee *et al.* (2014). ICT 4 Climate Change Adaptation: Systemic and Generative Perspectives and Tools. 2nd International Conference on ICT for Sustainability (ICT4S 2014), Pp 248-257.
83. Creech *et al.* (2012). Use of ICTs for Climate Change Adaptation in Africa, International Institute for Development. www.eTransformAfrica.org. In: The Transformational Use of ICTs in Africa.
84. Urban, B. *et al.* (2014). Energy consumption of consumer electronics in U.S.

homes in 2013. Final report to the consumer electronics association (CEA). Fraunhofer USA center for sustainable energy systems. June 2014.

85. Finlay and Adera (2012). Application of ICTs For Climate Change Adaptation in the Water Sector: Developing country experiences and emerging research priorities. Association for Progressive Communications (APC) and the International Development Research Centre (IDRC).

Chapter 4

EARLY WARNING SYSTEMS

Introduction

India's Disaster Management Act, 2005, defines 'disaster' as "catastrophe, mishap, calamity or grave occurrence in any area, arising from natural and human-induced causes, or by accident or negligence which results in substantial loss of life or human suffering or damage to, and destruction of property or damage to, or degradation of, environment and in of such a nature or magnitude as to be beyond coping capacity of the community of the affected areas". Disasters induced as natural calamities such as floods, cyclones, earthquake, Tsunami, landslides, tidal waves *etc.* are considered as natural disasters and those caused by chemical, industrial, transport related accidents, mass congregation and security related threats including terrorism, biological and nuclear attacks are classified as human-induced disasters. Nationally and internationally, the frequency and intensity of natural as well as human-induced disasters is increasing, and these disasters certainly cause significant personal, material and economic strain on individuals, communities and to every level of governments.[1]

As per trend analysts ninety per cent of the global population growth is taking place in least developed countries (LDCs).[2] In these countries, vulnerability to hazards is already very high through dense concentrations of population in largely unsafe human settlements and poor socio-economic and environmental conditions. All disasters, due to having its effects on services, natural resources and production systems,agriculture, water bodies, Soil, horticulture, forest, animals, increase risks and decrease a country's ability to reduce its vulnerability and strengthen risk governance capacities. However with the use of science and technologies in recent decades, countries in all regions have strengthened their capacities to combat risks associated with major hazards. Technology is changing preparedness, response and recovery mechanisms. More importantly, it is changing how local communities, who inevitably are the first responders, react and improve their actions. Both the directly affected populations and the institutions working to support them are creating new mechanisms to connect people at risk; facilitating to prevent catastrophes, save lives and rebuild communities Paradigm change is now a buzz word for Disaster Management tactics. *Enhancing disaster risk governance, managing disaster risk for disaster risk reduction, reducing vulnerability and enhancing resilience are key priorities of Sendai Framework for DRR.*[3] Use of science and technology inter alia for early warning systems for preparedness prevention mitigation and capacity building of community is increasingly gaining momentum.

Early Warning Systems for Disaster Management

Following[4] the ten-year review of the progress made in the area of disaster reduction, the World Conference on Disaster Reduction (WCDR), held in Kobe, Japan in January 2005, adopted an important policy document, the Hyogo Framework for Action (HFA) 2005-2015: Building the Resilience of Nations and Communities to Disasters. It was resolved that *"disasters have a tremendous detrimental impact on efforts at all levels to eradicate global poverty; the impact of disasters remains a signicant challenge to sustainable development"*. The Hyogo Declaration added that *"it is critically important that the Hyogo Framework for Action 2005–2015: Building the Resilience of Nations and Communities to Disasters be translated into concrete actions at all levels and that achievements are followed up through the International Strategy for Disaster Reduction, in order to reduce disaster risks and vulnerabilities."* The international community accordingly recognised the need to develop indicators to track progress on Disaster Risk Reduction (DRR) activities as appropriate to particular circumstances and capacities as part of the effort to realise the expected outcome and strategic goals set in the HFA.

HFA identified 5 important Priority Actions as given under:

- ☆ *Priority Action 1*: Ensure that disaster risk reduction is a national and a local priority with a strong institutional basis for implementation.
- ☆ *Priority Action 2*: Identify, assess and monitor disaster risks and enhance early warning.
- ☆ *Priority Action 3*: Use knowledge, innovation and education to build a culture of safety and resilience at all levels.
- ☆ *Priority Action 4*: Reduce the underlying risk factors.
- ☆ *Priority Action 5*: Strengthen disaster preparedness for effective response at all levels.

Priority Action -2 of HFA, as mentioned above, identified early warning as one of the major elements required for disaster-risk reduction which could save lives and help protect livelihoods and national development gains. Early warning systems have been accepted as an effective tool to reduce vulnerabilities and improve preparedness and response to natural hazards.

The importance of early warning has been underlined in various UN General Assembly resolutions as a critical element of disaster reduction. When the International Strategy for Disaster Reduction (ISDR)[5] was established in 2000 as the successor to the International Decade for Natural Disaster Reduction (1990-1999), the promotion of people-centered early warning systems was clearly underlined and included in its mandate. The significance of early warning for disaster reduction has been repeatedly emphasized in major international agendas, including the Yokohama Strategy, Agenda 21[6], the Barbados Plan of Action[7] for Small Island Developing States, the Johannesburg Plan[8] of Implementation, the Mauritius Strategy and the G8 Summit in Gleneagles,[9] as well as major multilateral environmental agreements, including the UN Framework Convention on Climate Change and the Convention to Combat Desertification.

The UN Secretary-General in a report " *In Larger Freedom: Towards development, security and human rights for all",* proposed that *"the United Nations system should take a leadership role in developing comprehensive global capacities for systematic people-centered early warning systems, which would cover all hazards for all countries and communities"*. Subsequently, he requested that a global survey be undertaken, with a view to advance the development of a global early warning system (GEWS) for all natural hazards. The survey report, coordinated by the ISDR secretariat, concluded that *"while some warning systems are well advanced, there are numerous gaps and shortcomings, especially at the local community level in developing countries, for effectively reaching and serving the needs of those at risk."* The report also recommended *"the establishment of a globally comprehensive early warning system, rooted in existing early warning systems and capacities, as well as a set of specific actions toward building people-centered systems at the national level. It also proposed filling in the main gaps in global capacities, strengthening the scientific and data foundations for early warning, and developing the institutional foundations for a global early warning system."*

In the report on the *International Strategy for Disaster Reduction ISDR* (*2006*), the UN Secretary-General encouraged Member States and organizations to develop a GEWS for all hazards and communities, based on existing systems. It was mentioned in the report that *"there was immediate need to address the associated technical and organizational gaps and needs, as recommended in the Global Survey of Early Warning Systems. For GEWS to work, coordinated planning and recommendations are needed to define priorities and practical objectives to be achieved, and engage the attention and participation of all relevant stakeholders. These actions should take place through existing relevant mechanisms, particularly the International Early Warning Programme (IEWP)"*[10].

Early warning received due attention in India and Asia-Pacific after the 26 December 2004 tsunami, when it became clear that a tsunami warning system and associated public education could have saved thousands of lives. Need of an effective early warning system in India was emphatically established during 2004 Indian Ocean tsunami. Early warning can help in many types of hazards. In the case of tsunamis, the benefit of an internationally coordinated warning system could be seen in the 2011 earthquake and tsunami in Tohoku, Japan, which threatened many Pacific islands: warnings were more coordinated than in the devastating Indian Ocean Tsunami in 2004, providing time for many people to evacuate to high ground. Similarly, effective early warning systems significantly helped during Cyclone Phailin in Andhra Pradesh and Odisha[11]. The Red Message, the highest alert issued by the Indian Meteorology Department in Delhi, was usable to a large extent by the local district officials. The alert also enlisted where and what type of damage was expected to shelter and infrastructure. Early warning messages sent out through cell phones and TV channels helped saving lives even in the most remote rural areas. People knew about the coming storm and were able to take some measures to reduce their exposure. Due to timely early warnings and having disaster preparedness and mitigation tools in place, local government ministry for disaster Management and Orissa State Disaster Management Authority managed the largest-ever evacuation exercise in the state and planned large-scale relief preparations. As a result nearly 500,000 people were evacuated in time and moved to higher grounds and safer cyclone shelters.

Another example is Bangladesh, which now has a 48-hour early warning system in place that allows people to evacuate to safe shelters hours before cyclones make landfall, resulting in reduction in deaths in cyclone aftermath. In 1970, about 3 million people died due to Cyclone Bhola, compared to a death toll of 3,000 in 2007 during Cyclone Sidr, which authorities were able to track as it grew in strength. A lead time of a few seconds can make a difference, and some countries are working with the limited information available.[12]

Early Warning Systems: Essential Component of Disaster Preparedness

Early warning systems are increasingly considered to be an integral component of disaster preparedness and involve a broad spectrum of stake holders. A timely warning can save lives. Several countries have significantly reduced deaths by developing effective early warning systems. In Mexico City, for example, technical systems can identify the first seismic wave following the start of an earthquake that may have happened more than 100 kilometers away. Such system allows authorities to use this information to shut down critical systems such as gas supply lines.[13] Similarly, the Ferny Creek Bushfire Alert System (FCBAS) in Australia is an emergency communication system, meant to issue alerts for fire dangers, broadcast via three independent strategically-located sirens. Operational during declared fire danger periods, its purpose is to sound the alarm only to alert residents when an emergency call fulfils predetermined criteria pointing to a potential threat.

The Third International Conference on Early Warning (EWC III) held in Bonn, Germany from 27-29 March 2006 provided the opportunity to present new and innovative early warning projects and to discuss natural hazards and risks around the world and how their impacts can be minimized through the implementation of people-centered early warning. The document "*Developing Early Warning Systems: A Checklist*" was developed as a conference outcome, to both inform and draw upon the discussions and practical examples raised during the conference, and to support the implementation of the early warning components of the Hyogo Framework for Action. Early warning systems do not exist in every part of the world. A quarter of the countries assessed in the 2011 Global Assessment Report for Disaster Risk Reduction reported that communities did not receive any timely warnings for impending hazards.[14]

Early Warning Systems: Procedural Characteristics

Early warning (EW)[15] is "*the provision of timely and effective information, through identified institutions, that allows individuals exposed to hazard, take necessary actions to avoid or reduce their risks and prepare for effective response, and is the integration of four main elements*"

As per International Strategy for Disaster Reduction (ISDR), United Nations (UN), 2006[16] Early Warnings as an outcome oriented process can be divided in following important aspects:

1. Risk Knowledge: Risk knowledge and its fair assessment provide critical information to build mitigation and prevention strategies and designing early warning systems.
2. Monitoring and Predicting: Early Warning Systems (set of equipments, sensors, simulators, communication links, ICT tools, technical human resources) with monitoring and predicting capabilities provide timely estimates of the potential risk faced by communities, economies and the environment.
3. Disseminating Information: Communication systems are needed for delivering warning messages to the potentially affected locations to alert local people and local and regional governmental agencies. The messages need to be reliable, systematic, consistent and simple to be understood by authorities and public. It requires a lot of social and cultural awareness to spread appropriate information to specific recipients.
4. Response: Coordination, good governance, Standard Operating Procedures, and appropriate action plans are key points in effective early warning. Likewise, public awareness and education are critical aspects of disaster mitigation.

Failure of any part of the system will imply failure of the whole warning system, therefore an effective early warning system embrace aforesaid all four elements.

Early warnings may be disseminated to targeted users (local early warning applications) or broadly to communities, regions or to media (regional or global early warning applications). This information gives the possibility of taking action to initiate proactive measures before a disaster actually strikes. The main goal of an early warning system is to take action to protect or reduce loss of life or to mitigate damage and economic loss, before the disaster occurs[17].

Early Warning System: Important Technologies

Forecasting and Modelling Technology[18]

Several countries have built up early warning systems based on seasonal-to-interannual climate forecasts. [19]These systems are based on using monitoring data, including temperature and rainfall values, and state-of-the art climate models. Climatologists analyse the observations and model-based predictions to predict climate anomalies one or two seasons ahead.

Remote Sensing and Geographic Information Systems (GIS)

Remote sensing and GIS applications have significant roles to play in early warning systems. Timely information in the location, progression and regression of disasters at various phases of the disaster illustrated in a dynamic map is essential towards effective management of the disasters and this can be generated in Geographic Information System (GIS) by linking it with Global Positioning System (GPS). Linking of GIS and GPS is achieved by integrating the real-time space-based imageries of the hazards with the corresponding ground information.

Satellite Communication Technology

Satellite imageries are key inputs for all types of simulation mode development of early warning systems for meteorological, hydrological predictions. Indian forecasting institutions such as IMD, CWC, ISRO *etc.* are depending upon these imageries sent from a series of Indian and international satellites.Improvements in satellite communications have helped decrease the lag time between data collection and warning. For example, the Pacific Tsunami Warning System[20] works by a recorder on the seabed relaying data on anomalies to a buoy on the surface. This data is then transmitted via satellite to ground stations every 15 seconds. Similarly improved technologies and state of art simulation models have extensively enhanced credibility of warning predictions.

Mobile Phone Technology

With the global spread of mobile phones and networks, this technology is now increasingly used to communicate warnings and coordinate preparation activities particularly SMS alerts for disseminating mass messages. Short message service (SMS) is a service available on most digital mobile phones that permits the sending of short messages between mobile phones, other handheld devices and even landline telephones. There are instances when many residents of affected coastal areas are unable to make contact with relatives and friends using traditional landline phones, but could easily communicate with each other through SMS as long as the network is functional. This is because SMS works on a different channel and can be sent or received even when phone lines are congested. SMS also has another advantage over voice calls is that one message can be sent to a group simultaneously. For example, upon detection of p-waves that precede earthquake shaking, Japanese agencies send out SMS alerts to all registered mobile phones in the country. However, during disasters such as earthquake, cyclone, cloudburst telecom towers and terrestrial telecom infrastructure may get affected thus may result in non-availability of this service. Similarly, surge in traffic during emergency can overburden network and resulting in breakdown of communication services.

With advancement in wireless broadband connectivity/evolution of mobile technology to support high bandwidth data, it is possible to avail real-time information and early warnings on mobile. Besides with mobile technology evolution to 3G/4G and now advancing to commercial rollout of 5 G, utilization of mobiles as multimedia communication tools for disaster reporting is hugely enhanced.

Internet Technologies

The Internet is a global network of many large medium and small networks enabling computers of all kinds to remain connected and to directly and transparently communicate and share services with each other all across the world. The internet provides rapid and very cost-effective means of making an intra-national and international disaster warning communications. Due to internet, information is now available in a near-real-time mode from a variety of sources at global and local levels. Now a day with undersea optical fibres internet connectivity and bandwidth offerings have increased manifolds with new technological advances, internet is facilitating the global distribution of data and information at all levels.

ICTs for Crowdsourcing[21]

The use of 'crowdsourced' data is gaining momentum with increasing internet connectivity and bandwidth with the help of ICT. *Self-organising, self-managed social networking tools with free and open source platforms such as "Ushahidi" are good in meeting this requirement. Ushahidi combines SMS, Twitter and Google Maps to crowdsource crisis information.* It is a free, open source; decentralised platform developed by Kenyan bloggers in the aftermath of the 2008 Kenyan elections but is now used in disasters. Location-specific information is communicated directly to subscribers. *Ushahidi has developed Swift River as a rapid verification system for crowd sourced information by crosschecking tagged information from different sources[22]. Crowdsourcing was also used extensively in the response to the 2010 Haiti earthquake, allowing local people, mapping experts and other stakeholders to communicate what they saw and heard on the ground, and to produce information that could be used by crisis managers. This was particularly useful in locating survivors who needed assistance, but it is increasingly recognised that crowdsourcing could also help with pre-disaster activities, specifically risk identification and early warning.*

Crisis Mapping[23]

Through initiatives such as Ushahidi and Google Crisis Response, crisis mapping utilises crowdsourcing as well as satellite imagery, participatory maps and statistical models to power more informed and effective early warning. It can provide real-time information on an upcoming crisis in times of uncertainty and confusion. The vast amount of data that can be produced from such systems can be analysed through networks of stakeholders (such as Crisis Mappers).

Other Important Technologies

Communication media, which are helpful for dissemination of early warnings are Radio, Television, Telephone, Facsimile, Public Address Systems, Email, cell broadcasting, Amateur Radio *etc.*

Social media has increasingly becoming a very important tool for community connection. Wide spread of internet and mobile has made facebook, twitter, instagram, whatsapp *etc.* important tools of mass communication.

Protocol for Dissemination and Communication of Early Warning

Dissemination of warnings often follows a cascade process, which starts at international or national level and then moves outwards or downwards in the scale, reaching regional and community levels (Twigg, 2003). Early warnings may activate other early warnings at different administrative levels, flowing down in responsibility roles, as it is necessary to follow the chain for effective early warnings. Standard protocols play a fundamental role in addressing the challenge of effective coordination and data exchange among the actors in the early warning process and it aids to the process of warning communication and dissemination. The Common Alerting Protocol (CAP), Really Simple Syndication (RSS) and Extensible Markup Language (XML) are examples of standard data interchange formats for structured information that can be applied to warning messages for a broad range of information management and warning dissemination systems. [24] The advantage of standard format alerts is that they act as technology agnostic and are compatible with all information systems, warning systems, media, and most importantly, with new technologies such as web services. CAP defines a single standard message format for all hazards, which can activate multiple warning systems at the same time and with a single input. This guarantees consistency of warning messages and would easily replace specific application-oriented messages with a single multi-hazard message format. CAP is compatible with all types of information systems and public alerting systems (including broadcast radio and television), public and private data networks, multi-lingual warning systems and emerging technologies such as Internet Web services. CAP uses Extensible Mark up Language (XML) language. It contains information about the alert message, the specific hazard event, and appropriate responses, including urgency of action to be taken, severity of the event, and certainty of the information. The Common Alerting Protocol allows technology developers and sponsors to expand, upgrade or even replace existing facilities without disrupting entire systems[25].

Early Warning: Other Important Facts

Improving the effectiveness of early warning systems does not, in itself; guide to risk reduction for disaster-prone communities unless it is followed by (early) action. Warnings are still not effectively communicated, and not sufficiently acted upon, even as agencies in developed and developing countries are now more aware of the nature, frequency, locations and intensity of various hazard types, and have advanced technical capabilities for monitoring such as climate models and remote sensing.[26] '*A good example is one of the most devastating disasters in history, the 2004 Indian Ocean Tsunami. The Pacific Tsunami Warning Centre in Hawaii picked up the earthquake. But despite the phone calls made by the centre to government agencies in countries such as Indonesia and Thailand, the emergency communication and early warning infrastructure was missing and so the warning was not disseminated to local communities*'[27]

Uncertainties of Events

The uncertainty inherent in scientific information is one of the reasons for failing to act on disaster warnings. Thus the message should at the same time communicate the level of uncertainty and expected cost of taking action but also be simple so as to be understood by those who receive it. Most often, there is a communication gap between EW specialists who use technical and engineering language and the EWS users, who are not acquainted with scientific jargons. To avoid this, early warnings need to be reported concisely, in layman's terms and without scientific jargon. Information from forecasts is often in a language and format that is not easily understood by the local communities who are target recipient for such information. If information is not interpreted well by its recipients it cannot be acted upon.

Risks Prioritisation

When a warning is issued, it may not help a recipient to take decision on prioritising impending risk over his/her routine higher priorities. For example, a farmer may stay looking after their cattle rather than evacuate because they judge the risk of flood to be lower than the risk of losing their livelihood. Therefore proper awareness of local community about risks and vulnerabilities of the area along with capacity building in terms of their quick response to warnings, preparation of safety kits and family safety plans, dissemination of information on Dos and Don'ts *etc.* may help. Thus early warnings can work more effectively by taking into account how people behave in that crucial period after they receive a warning particularly how they prioritise different risks.

Reducing False Alarms

As early warning systems grow in geographical coverage and sophistication, false alarms are rising too. While some believe that they provide invaluable practice, high false alarm rates can undermine public confidence, dilute the impact of alerts and reduce the credibility of future warnings. One approach to reducing false alarms is to use reliable local hazard indicators, such as animal behaviour or vegetation changes, to verify scientific indicators of upcoming hazards. Another approach is to work with the media to avoid inaccurate, exaggerated or misleading information about potential events.

Coordinating Response

Insufficient coordination and collaboration between organisations can hold back efforts to encourage early action because the organisations that produce warnings may not necessarily be same to those that disseminate them. For example in India, CWC monitors rain gauges and issues warning through IMD networks. There should be proper coordination between such agencies, in case of India, it requires good coordination between Ministry of agriculture and IMD for handling drought or CWC and IMD in case of flood warnings. For major disasters which have effects across more than one country/region, such as Tsunami, Earthquake *etc.* sharing of information becomes more important. Better communication channels and proper pacts/memorandum of understandings/policy for regional cooperation may help to address this.

Forecasting and Warning Networks in India

Primary Nodal agencies for monitoring and early warning of disasters in India are:

- ☆ IMD (Cyclones, Floods, Drought, Earthquakes)
- ☆ Central Water Commission under Ministry of Water Resources (Floods)
- ☆ Geological Survey of India (Landslides)
- ☆ Ministry of Earth Sciences through Indian Tsunami Early Warning Centre at the Indian National Centre for Ocean Information Services (INCOIS), Hyderabad. (Tsunami)

IMD Warning and Forecasting Network

The India Meteorological Department (IMD) was established in 1875. It is the National Meteorological Service of the country and the principal government agency in all matters relating to meteorology, seismology and allied subjects. IMD is mandated:[28]

- ☆ To take meteorological observations and to provide current and forecast meteorological information for optimum operation of weather-sensitive activities like agriculture, irrigation, shipping, aviation, offshore oil explorations, *etc.*
- ☆ To warn against severe weather phenomena like tropical cyclones, norwesters, dust storms, heavy rains and snow, cold and heat waves, *etc.*, which cause destruction of life and property.

- ☆ To provide meteorological statistics required for agriculture, water resource management, industries, oil exploration and other nation-building activities.
- ☆ To conduct and promote research in meteorology and allied disciplines.
- ☆ To detect and locate earthquakes and to evaluate seismicity in different parts of the country for development projects.

According to mandate given to IMD it offers observational, data collection, monitoring and forecasting services across various sectors namely Monsoon, Hydrology, Agriculture, Health, Aviation, Transport, Shipping, Cyclone, Climatology, Mountaineering, Disaster Management *etc.* Web based forecast services for example weather forecast, meteorological information, warnings, normal *etc.* are provided through IMD Head Quarters at Delhi and various offices of IMD. Website details of IMD HQ, Delhi, Pune and other RMCs are :

- ☆ IMD HQ http://www.imd.gov.in/, http://www.mausam.gov.in/
- ☆ IMD Chennai http://www.imdchennai.gov.in/
- ☆ IMD Delhi http://www.amssdelhi.gov.in
- ☆ IMD Guwahati http://www.imdguwahati.gov.in/
- ☆ IMD Kolkata http://www.imdkolkata.gov.in/
- ☆ IMD Mumbai http://www.imdmumbai.gov.in/
- ☆ IMD Nagpur http://www.imdnagpur.gov.in/
- ☆ IMD Pune http://www.imdpune.gov.in/

National Weather Forecasting Centre (NWFC) has been set up at Mausam Bhawan, Delhi, for Weather Forecasting Services of IMD. It brings all forecasting services of IMD under one umbrella. It provides forecast guidance to six Regional Meteorological Centres (RMCs) and State Meteorological Centres (SMCs). Synoptic method is overlaid on numerical models supported by modern graphical and GIS applications to produce high quality analysis and forecasts and automated dissemination of forecast and warning. Backup support of NWFC is set up in the office of DDGM (Weather and Forecasting) at IMD Pune. Salient features of NWFC are:

- ☆ Real time reception and assimilation of multi sensor observations;
- ☆ Seamless software tools for various meteorology;
- ☆ Climatology and public weather service applications;
- ☆ An integrated and fully automated forecasting environment.

IMD is receiving and processing meteorological data from two Indian satellites. These satellites have Very High Resolution Radiometer (VHRR) for imaging the earth in visible (0.55-0.75 m), infrared (10.5-12.5m) and water vapour (5.7-7.1m) channels having resolution of 2X2 km in visible and 8X8 km in water vapour and infrared channels. In addition these satellites also have Charge Coupled Device (CCD) payloads for imaging the earth in visible (0.62-0.69m), near infrared (0.77-0.86m) and short wave infrared (1.55-1.77m) bands. Imaging from CCD is done about 5-6 times during daytime only. All the received data from the satellite is processed and archived in National Satellite Data Centre (NSDC), New Delhi.[29]

Warnings issued by IMD for different hazards are categorised according to predefined values. For example warning categories for various disasters are as shown in the Figure 4.1.

Roles and Responsibilities of different Cells of IMD

IMD Forecasting Cell

IMD[30] issues quantitative District Level Weather Forecast (DLWF) to all districts of India. The products comprise of quantitative forecasts for 7 weather parameters namely rainfall, maximum and minimum temperatures, wind speed and direction, relative humidity and cloudiness. In addition, weekly cumulative rainfall forecast is also provided. IMD, New Delhi generates these products based on a Multi Model Ensemble Technique using forecast products available from numerous models of India and other countries. These include: T-254 model of NCMRWF, T-799 model of European Centre for Medium Range Weather Forecasting (ECMWF); United Kingdom Met Office (UKMO), National Centre for Environmental Prediction (NCEP), USA and Japan Meteorological Agency (JMA). The products are disseminated to RMCs and SCs. These offices undertake value addition to these products and communicate to 130 AgroMet Field Units (AMFUs) located at State Agriculture Universities (SAUs) and Institutes of Indian Council of Agriculture Research (ICAR) *etc.*

Main functions of forecasting division are:

i. Monitoring of weather situation over the country;
ii. Issuance of short to medium range forecast;
iii. Issuance of forecast for mountain region, sports, tourism and VIP movements *etc.*;
iv. Issuance of meteorological sub-division wise weather warnings;
v. Co-ordination with RMCs and SCs for location specific forecasts; and
vi. Preparation of significant weather charts for analysis purpose.

Aviation Cell

Aviation cell does monitoring of aviation weather through about 17 Aerodrome Meteorological (AM) offices, 51 Aeronautical Meteorological Weather Offices and Meteorological Watch Offices. Meteorological information for the use of aviation activities are[31]:

i. Current weather observations (METAR/SPECI, MET REPORT/SPECIAL);

WARNING CATEGORY

Rain

Heavy: 64.4 to 124.4 mm
Very Heavy: 124.5 to 244.4 mm
Extremely Heavy: more than 244.4 mm

Snow

Heavy: 35.6 to 64.4 mm
Very Heavy: 64.5 to 124.4 mm
Extremely Heavy: more than 124.4 mm

Heat Wave

a). **When normal maximum temperature of a station is less than or equal to 40° C**

Heat Wave: Departure from normal 5° C to 6° C.
Severe Heat Wave: Departure from normal 7° C or more

b). **When normal maximum temperature of a station is more than 40° C**

Heat Wave: Departure from normal 4° C to 5° C.
Severe Heat Wave: Departure from normal 6° C or more

c). **When actual maximum temperature remains 45°C or more**

Hot Day

Plains: The maximum temperature remains 40°C or more and minimum temperature departure from normal 5° C or more.
Coastal: When maximum temperature departure is 5° C or more from normal.

Cold Wave

a) **When normal minimum temperature is equal to 10°C or more.**

Cold Wave: Departure from normal -5° C to -6° C.
Severe Cold Wave: Departure from normal -7° C or less

b) **When normal minimum temperature is less than 10°C**

Cold Wave: Departure from normal is -4° C to- 5° C.
Severe Cold Wave: Departure from normal is -6° C or more

Cold Day

Plains: The maximum temperature remains 16°C or less.
Coastal: When maximum temperature departure is -5° C or more from normal.

Fog

Phenomenon of small droplets suspended in air and the horizontal visibility is one kilometer or less

Moderate: When the visibility between 500-200 metres
Dense: when the visibility less than 200 metres
Very Dense: when the visibility less than 50 metres

Thunderstorm

Sudden electrical discharges manifested by a flash of light (Lightning) and a sharp rumbling sound (thunder)

Moderate: Wind gusts are between 31-74 kmph.
Severe: Wind gusts reach more than 75 kmph.

Dust/Sand storm

An ensemble of particles of dust or sand energetically lifted to great heights by a strong and turbulent wind.

Moderate: Wind speed 42-79 kmph and visibility 200 – 500 m
Severe: Wind speed ≥ 80 kmph and visibility < 200 meter.

Hailstorm

Solid precipitation in the form of balls or pieces of ice.

Frost

Ice deposits on ground

Air temperature ≤4°C

Squall

Sharp increase in wind speed last atleast for 1 minutes

Moderate: Wind speed 52-61 kmph
Severe: Wind speed 62-87 kmph
Very Severe: Wind speed >87 kmph

Sea State

Effect of various waves in the sea over specific area

Rough to very rough: Wave height 2.5-6 metre & Wind speed 41-62 kmph
High to very high: Wave height 6-14 metre &Wind speed 62-117 kmph
Phenomenal: Wave height >14 metre & Wind speed >117 kmph

Cyclone

Cyclonic Storm: Wind speed 32-90 kmph
Severe Cyclonic Storm: Wind speed 90-220 kmph
Very Severe Cyclonic Storm: Wind speed 90-220 kmph
Super Cyclone: Pressure Deficient >65.6 hPa & associated Wind speed >220

Figure 4.1: IMD Warning Categories.

ii. Forecasts (Terminal Aerodrome Forecast (TAF), Area/Local Forecast, Route Forecast, Take-off and Landing (TREND) Forecast);
iii. Warnings (Aerodrome warnings, Warning for Light Aircrafts, Wind shear warnings, SIGMET[32]); and
iv. Climatology (Climatology of aerodromes, Climatological summary, Climatology of upper wind and temperature)

Responsibilities of Aeronautical Meteorological Stations (AMS)[33]

Aeronautical Meteorological Stations are responsible for Supply to aeronautical users, of current weather observations of their own stations and those of other stations as required, by obtaining them from the concerned other stations. AMS are:

i. Providing documentation for flights originating from their stations after obtaining the forecasts from their associated Aerodrome Meteorological Office(s).
ii. Supply of forecasts and warnings to aeronautical users of their aerodrome after obtaining relevant information from their associated Aerodrome Meteorological Office.
iii. Supply of information received on pre-eruptive volcanic activity, volcanic eruption or volcanic ash cloud to aeronautical users.
iv. Supply of SIGMET[34] as and when received.

Responsibilities of Meteorological Watch Offices (MWO)

The major responsibilities of the MWOs are:

i. Maintain continuous watch of meteorological conditions affecting flight operations within its Flight Information Region (FIR);
ii. Prepare and disseminate SIGMET and other information relating to its FIR;
iii. Exchange SIGMET information with other MWOs in the neighbouring countries;
iv. Supply SIGMET and other meteorological information to associated air traffic services units, including important messages from other MWOs.
v. Supply information received on pre-eruptive volcanic activity, a volcanic eruption and volcanic ash cloud for which a SIGMET has not already been issued to its associated Flight Information Centres (FIC)/Area Control Centres (ACC).

World Area Forecast System

World Area Forecast System (WAFS) products are used for briefing all international flights. It is a worldwide system by which World Area Forecast Centres (WAFC) provide aeronautical meteorological en-route forecasts in uniform standard formats.

Satellite and Radar Support Cell

To facilitate the interpretation of satellite and radar products for forecasting interpretation of radar products and assistance to forecasters in NWFC in issuance of weather advisories and nowcasting[35] of severe weather. Satellite application cell monitors and interprets satellite products and functions in the same manner as the radar cell. Besides, this cell also does generation and dissemination of customised weather products from the NWFC.

Basically used for Nowcasting and very short range weather forecasting A Doppler Weather Radar (DWR) provides velocity and spectrum width data (in addition to reflectivity). This capability permits the forecaster to view weather events like thunderstorms, hailstorm, heavy rainfall *etc.* in greater detail. DWR provides various meteorological, hydrological and aviation products which are

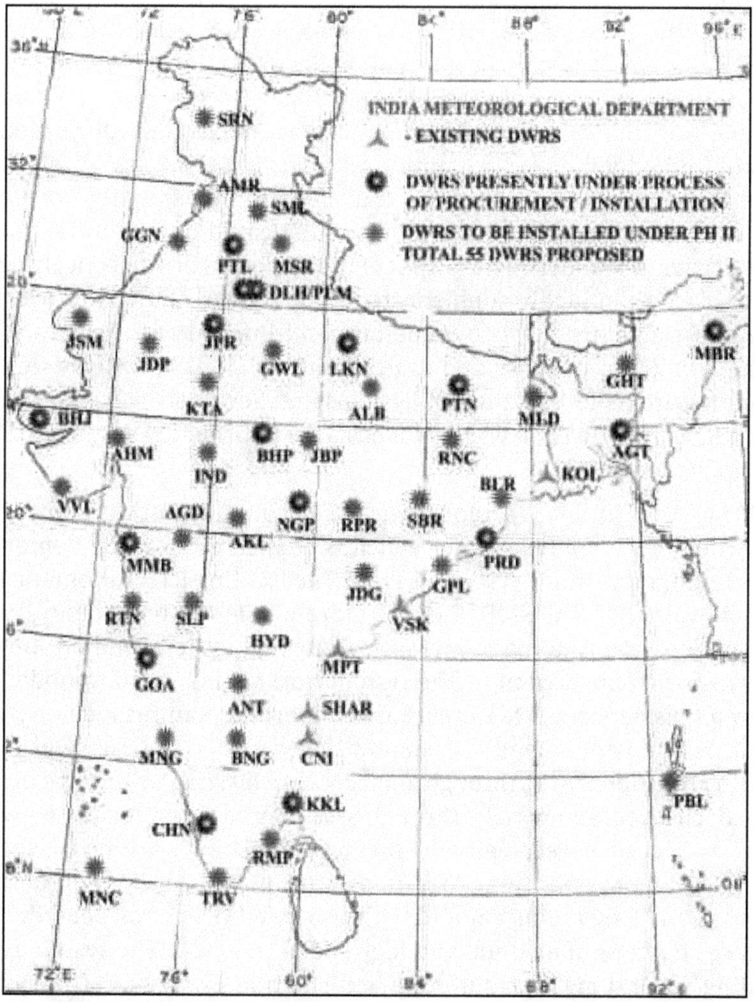

Figure 4.2: Existing and Proposed RADAR Networks of IMD.

very useful for forecasters in estimating the storm's center, fixing its position and predicting its future path. In Numerical Weather Prediction Models the DWR data, which is in digital form, can directly be assimilated, to have better rainfall estimation, better prediction of the storm track and point of landfall for tropical cyclones. Raw data from DWRs is received online at central server and is converted to the format required by modelers and shared through IMD-Telecom (TRANSMET) and also used to create a national radar mosaic.

Information System and Services Division

The Directorate of Tele-Communication was set up in IMD[36] at New Delhi in 1969 to cater the needs of National Meteorological Service (NMS) and strengthen the Meteorological Telecommunication in India. Since its inception, IMD maintains an extensive telecommunication network for speedy collection of meteorological information, both basic data and processed products, over the globe. The main telecommunication hub of the IMD's telecommunication network is located in IMD HQ, New Delhi and it is known internationally as Regional Telecommunication Hub (RTH) under the aegis of World Meteorological organization (WMO). Earlier its working was manual or semi-automatic, now manual decoding, plotting and analysis of limited volume of synoptic/upper air observations have given place to automatic decoding and analysis and forecast by using high performance computing systems running on highly sophisticated and complex Numerical Weather Prediction Models. The output of these models are objectively visualised by the forecasters in highly sophisticated graphic workstations to overlay with satellite and radar products to value-add and infer about the future state of atmosphere in different space and time resolution. Thus, focus was shifted from telecommunications to Information Systems and Services and consequently, the Telecommunication Division was renamed as the Information System and Services Division in January, 2009.

IMD has its National Meteorological Telecommunication Centre (NMTC) with a switching computer system which is connected to WMO Centers on the Global Telecommunication System (GTS)[37]. The existing RTH Automatic Message Switching System "TRANSMET" is the state-of-the-art technology system. It consists of Separate Automatic Message Switching System (AMSS) for national and international data exchange. The system works in hot standby mode. The new RTH system has enormous features. Some important features include : ability to retrieve message from e-mail and submit that message to GTS; send the whole message to predefined users through e-mail when a message with particular header is received; customised message through SMS to predefined recipients whenever a particular header is received which can be used for warning dissemination and message submission; transmission of warning messages to predefined FAX numbers; and file switching capability. The whole system has been designed to handle large number of medium and high speed channels. The system is capable of handling several gigabytes of data/information every day. The system has many advanced facilities like handling VSAT links and handling sea area bulletins under Global Maritime Distress and Safety System (GMDSS). NMTC New Delhi is connected to Super computer of NCMRWF through National Knowledge Network/ Virtual Private Network link and the HPCS (High Power Computing System) at Northern Hemispheric Analysis Centre (NHAC) New Delhi through Gigabit LAN for instantaneous transmission of global observational data and processed information received via GTS. TRANSMET is core system for data exchange with systems. Automatic Message Switching Systems (AMSS) are also operational at the major international airports namely Mumbai, Delhi, Calcutta, Chennai and Guwahati.

The Meteorological Telecommunication in IMD consists of an integrated network of point-to-point and point to multipoint (MPLS VPN) links and meteorological centers within the country and the world for receiving data and relaying it selectively. It is mainly organised on a two level basis, namely

i) The Meteorological Telecommunication Network (MTN) within the Global Telecommunication System (GTS) of World Weather Watch (WWW) program of World Meteorological Organization (WMO), and

ii) The National Meteorological Telecommunication Network (NMTN).

Nowcasting Cell

Nowcasts require high resolution of spatial and temporal meteorological data to detect and predict the occurrence of an event. Lack of data of the mesoscale imposes limit on ability to diagnose and predict an event. Nowcasting in India has benefited from major developments in observational meteorology and computer-based interactive data processing and display systems in IMD. There has been tremendous improvement in monitoring and forecasting due to introduction of (i) digital and image information at 10 minutes interval from a network of 14 Doppler Weather Radars, (ii) dense automatic weather station (AWS) network, (iii) half hourly satellite observations from Kalpana and INSAT satellites, (iv) better analysis tools in synergy system at forecaster's workstation and (v) availability of mesoscale models, (vi) computational and communication capabilities, IMD implemented nowcasting of thunderstorms, squalls and hailstorms. Considering the importance and reliability of DWR based information for nowcast of thunderstorm and associated weather, in the first phase, major stations/cities which come under the coverage of DWR were included for nowcasting of convective weather.[38]

Agro-met Advisory Service Units

To meet the farmer's need in real-time and to have a state-of-art Agro-met Advisory Service (AAS), an integrated AAS has been started from April 2007. This service involve many organisations related to the field namely Indian Council of Agricultural Research (ICAR), Ministry of Agriculture (Centre and State), State Agricultural Universities (SAUs) *etc.* National Agro-met Advisory Service Centre (NAASC) has been set up in Agro-met Division, Pune, to acts as the nodal centre and Head Quarters of state AAS units of IMD. NAASC prepares all India AAS bulletins as input to weekly meetings of Ministry of Agriculture, Ministry of Earth Sciences and Crop Weather Watch (CWWG) Group. Based inputs given by NAASC, important decisions on short time crop planning in various parts of the country are taken in this meeting. Besides, these bulletins are also uploaded on real time basis in the website (http://imdagrimet.gov.in) of the Agricultural Meteorological Division, IMD, Pune. Monthly and seasonal reports on AAS are also prepared.

State Agro-met Advisory Service Units (23) of IMD issue Agro-met Advisory Bulletins twice a week (Tuesday and Friday) on operational basis in consultation with the Director of Agriculture of the respective States. The bulletins are disseminated in English as well as local languages in through All India Radio, Doordarshan, Newspaper, Internet *etc.* The State Level bulletin are meant to identify the distressed district of the state as well as plan the supply of appropriate farm inputs such as seeds, irrigation water, fertiliser, pesticides *etc.* These bulletins are jointly prepared by State Meteorological Centers and Agro-met field units. State bulletins are useful for state government functionaries, crop weather watch group, fertiliser industry, pesticide industry, irrigation department, seed corporation, transport and other agriculture related organisations. There are about 130 observation units across India catering information to this division.

Hydrometeorology

The Hydromet Division of IMD was established in 1971 at New Delhi[39]. The division was established for providing necessary technical and operational support to various central/state government organisations and other agencies operational in the field of Hydromet Design Flood Forecasting, Hydromet, and Glaciological Studies.

The main activity in this discipline pertains to rainfall statistics and hydro meteorological studies for different river catchment. Hydromet Division provides data on storm, probable maximum precipitation, time distribution of rainfall storm, intensity-frequency analysis of rainfall which is used by design engineers for construction of dams, railways and road bridges, culverts *etc.*, for different central and state organizations, public undertakings and private agencies. This division also provides meteorological support for flood warning and flood control operation of the field units of Central Water Commission. Hydromet Division also participates regularly in glacier expedition to take meteorological observations to conduct snowmelt studies.

Seismological Observations Network of India

Bureau of Indian Standards [IS-1893 – part – 1: 2002], based on various scientific inputs from a number of agencies including earthquake data supplied by IMD, has grouped the country into four seismic zones namely., Zone-II, -III, -IV and -V. Of these, zone V is rated as the most seismically prone region, while zone II is the least. The Modified Mercalli (MM) intensity, which measures the impact of the earthquakes on the surface of the earth, broadly associated with various zones, is as follows[40 41]:

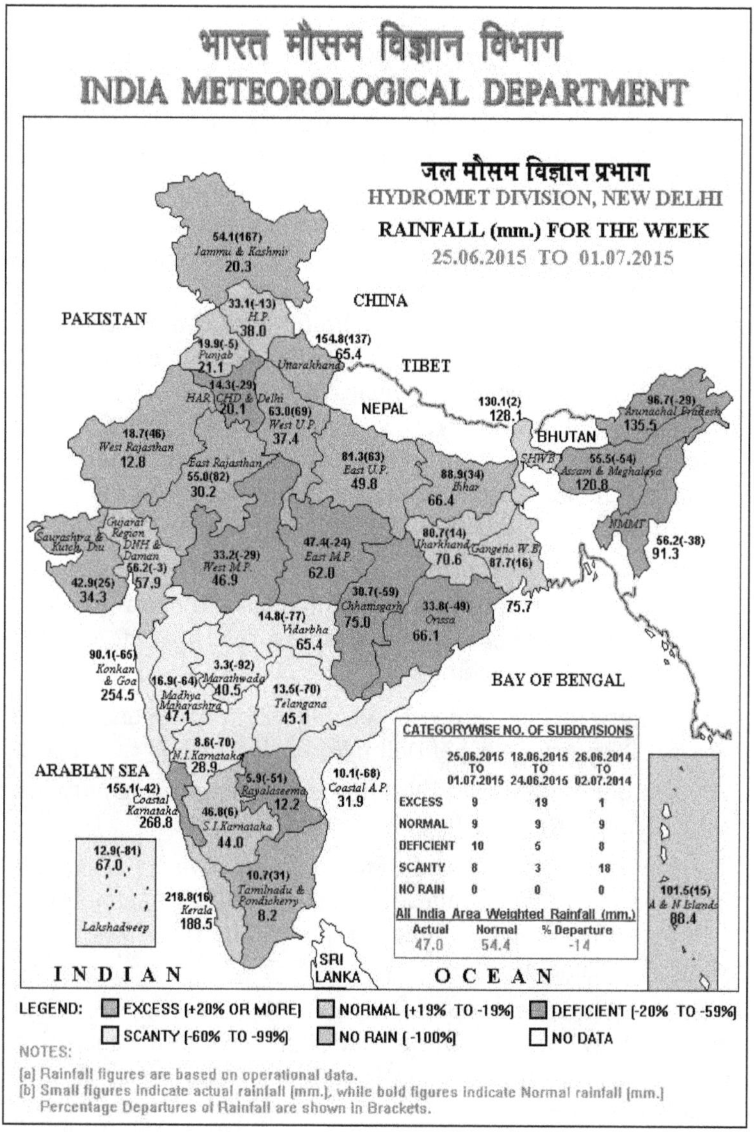

Figure 4.3: Weekly Rainfall Data.
(*Source*: http://www.imd.gov.in/section/hydro/dynamic/weeklly-rainfall.htm).

Table 4.1: Seismic Zones of India(by BIS)

Seismic Zone	*Intensity on MMI Scale*	*Per cent of Total Area*
II (Low intensity zone)	VI (or less)	43 per cent
III (Moderate intensity zone)	VII	27 per cent
IV (Severe intensity zone)	VIII	18 per cent
V (Very severe intensity zone)	IX (and above)	12 per cent

Seismological observations in the country are made through a national network of Seismic Stations operated by the IMD, being the nodal agency[42]. IMD is the nodal agency responsible for monitoring seismic activities in and around the country. IMD has rendered more than a century of seismological service to the nation with the first seismological observatory of the country having been set up by the department at Kolkata in 1898. The operational task of the department is to quickly estimate the earthquake source parameters immediately on occurrence of an earthquake and disseminate the information to all the user agencies including the concerned state and central government agencies responsible for carrying out relief and rehabilitation measures. The information relating to under-sea earthquakes capable of generating tsunamis on the Indian coastal regions is also disseminated to all concerned user agencies including the Indian National Centre for Ocean Information Services (INCOIS), Hyderabad for issue of tsunami related messages and warnings. The earthquake information is transmitted to various user agencies including public information channels, press, media *etc.* using different modes of communication, such as SMS, fax, email and also posted on IMD's Website (www.imd.gov.in).

IMD is maintaining a country wide National Seismological Network (NSN),[43] consisting of a total of 84 seismological stations, spread over the entire length and breadth of the country. This includes: [44]

i. 16-station V-SAT based digital seismic telemetry system around National Capital Territory (NCT) of Delhi;

ii. 20- station VSAT based real time seismic monitoring network in North East region of the country;

iii. 17-station Real Time Seismic Monitoring Network (RTSMN) to monitor and report large magnitude under-sea earthquakes capable of generating tsunamis on the Indian coastal regions. The remaining stations are of standalone/analog type;

iv. A Control Room is in operation, on a 24X7 basis, at IMD Headquarters (Seismology) in New Delhi, with state-of-the art facilities for data collection, processing and dissemination of information to the concerned user agencies.

The available seismological data from all the network stations including those operated by other agencies is compiled, processed, analysed and archived systematically at the National Seismological Database Centre (NSDC) on a regular basis. Monthly National Seismological Bulletins containing the phase data and the processed information on source parameters of all earthquakes located by the seismological network of IMD are prepared regularly. India, represented by IMD, is a permanent Member of the International Seismological Centre (ISC), UK. Seismological Bulletins of IMD are shared regularly with ISC, for incorporation in the ISC's Monthly Seismological Bulletins, which contain information on earthquakes occurring all across the globe. Seismological data and earthquake related information is also supplied to agencies dealing with relief and rehabilitation measures, earthquake disaster mitigation and management related matters, seismic zoning, *etc.* Earthquake data is shared with various scientific, academic and R&D institutions for research purposes. Towards early warning of tsunamis, real-time continuous seismic waveform data of three IMD stations, namely Portblair, Minicoy and Shillong, is shared with global community, through IRIS (Incorporated Research Institutions of Seismology), Washington D.C., USA.

Monitoring of Earthquakes

IMD is the nodal agency of government of India, responsible for real time monitoring of earthquakes and their aftershocks through dedicated network of seismological observatories spread across the countries, determination of the magnitude of the earthquake within 30 minutes of its occurrence.

IMD's seismological network has 84 stations. State of the art Digital Broadband Seismograph Systems. A control room at the headquarters is operational 24/7. The available seismological data from all the network stations including those operated by other agencies is compiled, processed, analysed and archived systematically at the National Seismological Database Centre (NSDC) on a regular basis. Seismology division supplies earthquake data and seismicity reports of specific regions to various user agencies. The National Center for Seismology (NCS) has been set up by bringing together all Seismology related activities of IMD (including those of EREC) under one umbrella. As part of systematic archival of historical analog charts, state-of-the-art facilities have also been established for raster scanning and vector digitization of seismic analog charts. These facilities have enabled raster scanning of old analog charts and vector digitization of significant earthquake waveforms. Seismology division supplies earthquake data and seismicity reports of specific regions to various user agencies such as, insurance companies, industrial units, power houses, river valley projects *etc.* on payment basis[45].

National Early Warning System for Tsunami and Storm Surges

In the aftermath of the great Sumatra earthquake of 26th December, 2004, Ministry of Earth Sciences[46] has set up an Indian Tsunami Early Warning System (ITEWS) at the Indian National Centre for Ocean Information Services (INCOIS), Hyderabad. The Center is mandated to provide advance warnings on Tsunamis likely to affect the coastal areas of the country. The ITEWS has the responsibility to provide tsunami advisories to Indian mainland and the island regions. Acting as one of the Regional Tsunami Advisory service Providers (RTSPs) for the Indian Ocean Region, ITEWS also provide tsunami advisories to the Indian Ocean rim countries along with Australia and Indonesia.

INCOIS is using numerical modeling to determine the potential run-ups and inundation from a local or distant Tsunami.This is very important tool to plan future disaster mitigation and management action plans and strategy to respond. These Models can also be initialized with smaller sources to understand the severity of the hazard for the less extreme but more frequent events. This information then forms the basis for creating Tsunami evacuation maps and standard operating procedures. Sufficiently accurate modeling techniques have been developed in the recent years, and these models require proper inputs on detailed bathymetry and topographic data for the area being modeled.

Tsunami Modeling can be divided into three stages[47]:

- Generation
- Propagation and
- Run-up (inundation)

The parameters and type of model employed at three stages are different and depend on the site conditions.

Generation

The generation stage of Tsunami evolution includes the formation of an initial disturbance at the ocean surface due to the earthquake triggered deformation at the sea floor. This initial water surface disturbance is transformed into a long gravity wave radiating from the earthquake source. Modeling of the initial stage of the Tsunami generation is closely linked to the studies of earthquake mechanisms. The basic parameters required for these models include the fault area (length and width), angle of strike, dip, slip, depth of fracture, dislocation and moment magnitude of the earthquake. The wave form generated varies with these source parameters and, hence, the information on above parameters is essential.

Propagation

Tsunamis travel outward in all directions from the generating area, with the direction of the main energy propagation generally being orthogonal to the direction of the earthquake fracture. Their speed depends on the depth of water, so that the waves undergo acceleration and deceleration in passing over an ocean bottom of varying depth. In the deep and open ocean, they travel at speeds of 500 to 1000 km/hr. The distance between successive crests can be as much as 500 to 650 kilometers. However, in the open ocean, the height of the waves is generally about 30 to 40 cm even for the most destructive distant Tsunamis, and so the waves pass unnoticed. Variations in Tsunami propagation result when the propagation impulse is stronger in one direction than in others because of (i) orientation or dimensions of the generating area, (ii) regional bathymetric and topographic features and (iii) rate of advance. Specifically Tsunami waves undergo a process of wave refraction and reflection throughout their travel. Tsunamis are unique in that the wave disturbance extends through the entire water column from sea surface to the ocean bottom.

Run-up (Inundation)

The propagating Tsunami wave from the deep water undergoes a change, causing increase in the wave height at the coast due to the near shore bathymetry and coastal morphology such as inlets, sand dunes, water bodies *etc.* The run up of the Tsunami on land is the most undeveloped part of the Tsunami model, primarily because of lack of two major aspects high quality field measurements for testing the models and fine resolution bathymetry/topographic data (Titov and Synolakis, 1997).

INCOIS Studies have been carried out to validate the model results with December 26, 2004 Sumatra earthquake that indicated an 80 per cent match. This model has been run for 5 historical earthquakes and the predicted inundation areas are being overlaid on cadastral level maps of 1:5000 scale. The maps for most vulnerable parts of the coastline have been provided to the central and state level departments that are involved in Disaster Management.

At the time of any earthquake only the hypo central parameters and magnitude are available in near real time. The information about fault geometry will be available very much later, which cannot be used for real time tsunami prediction for near source regions. In addition, tsunami modeling, especially the coastal inundation, cannot be run in real time for generating operational tsunami advisories as the model run takes large computing time. For operational quantitative tsunami forecast, there needs to be a method to quickly estimate the travel times and run up based on the quickly available earthquake parameters (based on magnitude and hypocenter of earthquake only). For the above reasons, the best way to generate quantitative operational tsunami advisories is to create a database of pre-run scenarios.

A "scenario" is a single tsunami model simulation that is calculated from the required initial seismic deformation condition with the pre-defined input fault geometry parameters of earthquake rupture, *i.e.*, Fault Location, Depth, Length, Width, Displacement, Strike angle, Dip angle and Slip angle. Each Scenario output contains the expected tsunami wave travel times, run-up heights and directivity maps. The fault geometry parameters are carefully selected so that they are likely to represent an actual Tsunamigenic earthquake based on knowledge of the historical earthquakes in two Tsunamigenic source regions of Indian Ocean. The Strike angle is assigned based on those of the pre-historical earthquakes which actually occurred near the simulation point and triggered Tsunami in the past. In case of the parameters of historical earthquakes are uncertain, the strike angle is assigned in such way that it will represent the worst case, *i.e.*, parallel to the coast or along nearby trench.

This scenario database of pre-computed scenarios is generated with the objective use of available real time seismic information as well as sea-level data. Each simulation covers the entire Indian Ocean domain with 15 hours simulation time and a time step of 5 seconds. The tsunami profiles of 15 hours for every 15 seconds are saved at coastal forecast points for each scenario. The coastal forecast points are selected at 30 m bathymetry assuming that till such depth, the computation is linear. Arrival times and wave heights at specific coastal locations for each scenario are stored in a database. Travel times and Surge heights on 30 m bathymetry are interpolated to get the values at Coast. Travel Times to coast are calculated by considering the speed of the wave at different depths (30, 20, 15 and 10 M). The distance to coast is divided by the average speed to get the travel times at the coast. Tsunami Wave height (H) at coast is calculated by Greens function which is fourth root of bathymetry depth (h1=30 m) multiplied by the wave height (H1) at 30 m depth.

Whenever an earthquake occurs, the closest scenario to the event is extracted from the scenario database based on magnitude and hypocenter location to identify the regions at risk. Real time observations of sea level will be used to for slip parameter. This information is used to update the forecast. Tsunamigenic zones that threaten the Indian coasts have been identified by considering the historical tsunamis, earthquakes, their magnitudes, location of the area relative to a fault, and also by tsunami modeling. The east and west coasts of India and the island regions are likely to be affected by tsunamis generated mainly by subduction zone related earthquakes from the two potential source regions, namely the Andaman-Nicobar-Sumatra island arc and the Makran subduction zone north of Arabian Sea.

The ITEWS comprises a real-time network of seismic stations, Bottom Pressure Recorders (BPR) and tide gauges to detect tsunamigenic earthquakes and to monitor tsunamis. The ITEWS comprises a real-time network of seismic stations, Bottom Pressure Recorders (BPR), tide gauges and 24 X 7 operational warning centre to detect tsunamigenic earthquakes. To monitor tsunamis and to provide timely advisories Standard Operating Procedure (SOP) are in place and are followed. Vulnerable community is kept in touch by means of latest communication methods. Tsunami advisories are issued when all relevant information is processed with back-end support of a pre-run model scenario database and Decision Support System (DSS). The warning centre is capable of issuing tsunami bulletins in less than 10 minutes after any major earthquake in the Indian Ocean thus it gives a response/ lead time of about 10 to 20 minutes for near source regions in the Andaman and Nicobar and a few hours in the case of mainland.

As part of this, a 17-station Real Time Seismic Monitoring Network (RTSMN)[48] has been set up by IMD. The network is capable of monitoring and reporting, in least possible time, the occurrence of earthquakes capable of generating tsunamis likely to affect the Indian coastal regions. The data from the 17 broadband seismic field stations is transmitted simultaneously in real time through V-SAT communication facilities to the Central Receiving Stations(CRS) located at IMD, New Delhi and INCOIS, Hyderabad for processing and interpretation. The CRSs are equipped with state-of-art computing hardware, communication, data processing, visualisation and dissemination facilities. For providing better azimuthal coverage towards detecting earthquakes of tsunamigenic potential, the RTSMN system has been configured to include about 100 global stations of IRIS (a consortium of Incorporated Research Institutions in Seismology), whose data is available freely through internet. The earthquake information is disseminated through various communication channels to all the concerned user agencies in a fully automated mode. The National Early

Warning Centre [49] generates and disseminates timely advisories to the control room of Ministry of Home Affairs (MHA) New Delhi, for further dissemination to the Public. For the dissemination of alerts to MHA a satellite-based virtual private network for disaster management support (VPN DMS) has been established. This network enables early warning centre to disseminate warnings to the MHA, as well as to the State Emergency Operations Centres. In addition, Messages are also sent by Phone, Fax, SMS and e-mails to authorised officials. In case of confirmed warnings, the National Early Warning Centre is being equipped with necessary facilities to disseminate the advisories directly to the administrators, media and public through SMS, e-mail, Fax, *etc.*

Cyclone Forecasting and Warning Network

IMD is responsible for monitoring, tracking and forecasting the movement of cyclones and depressions of North Indian Ocean. Chennai centre of IMD, namely Regional Meteorological Centre, Chennai (RMC Chennai) digitized the tracks of depressions and cyclones that occurred over the Indian seas. Software, "Cyclone eAtlas – IMD"[50], has been developed to generate tracks statistics in maps, charts and tables, on various aspects of cyclones and depressions of Indian seas.

Regional Specialized Monitoring Centre, New Delhi- monitors cyclonic disturbances over Bay of Bengal and Arabian Sea, runs numerical models for cyclone track prediction and storm surge prediction models, issue of tropical weather outlook/cyclone warnings to the panel countries, issue of tropical cyclone advisories for aviation. A Cyclone Warning Division (CWD) at HQ has been set up to coordinate and supervise the totality of cyclone warning programme in the country. *Doordarshan* (India National Television Channels) and All India Radio (AIR) stations at New Delhi are provided cyclone warning bulletins by this CWD for inclusion in the national broadcast/telecast. The CWD has to assess the totality of cyclone situations both in the Bay of Bengal and the Arabian Sea including Gulf of Oman and Gulf of Aden irrespective of the area of responsibility of individual Cyclone Warning Centres and then advice the Government of India at the Apex level. There has also been installation of high power cyclone Detection Radars (CDRs) along with coastal belt of India. Cloud pictures received from weather satellites are extensively used in locating the movement of cyclones over oceanic regions particularly when they are beyond the range of the coastal radars.

Cyclone Warning and Marine Forecasting Cell of IMD is responsible for monitoring and prediction of cyclone disturbances over the north Indian Ocean. This division is responsible for all cyclone warnings. The division issues warning and advisory to national and international disaster management agencies. This division acts as Regional Specialized Meteorological Centre (RSMC) to issue tropical cyclone advisories to World Meteorological Organization (WMO)/Economic and Social Co-operation for Asia and Pacific (ESCAP) Panel countries. The division acts as Tropical Cyclone Advisory Centre (TCAC) to provide advisories for International Civil Aviation Organization (ICAO). India Meteorology Department has established

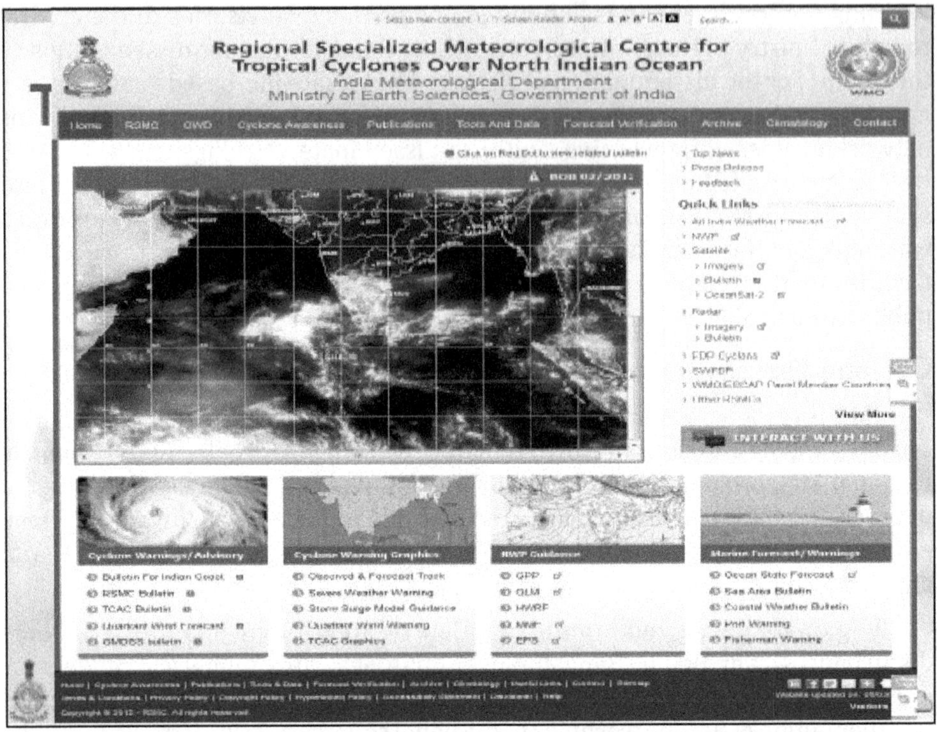

Figure 4.4: Tropical Cyclone Centre of IMD[51].

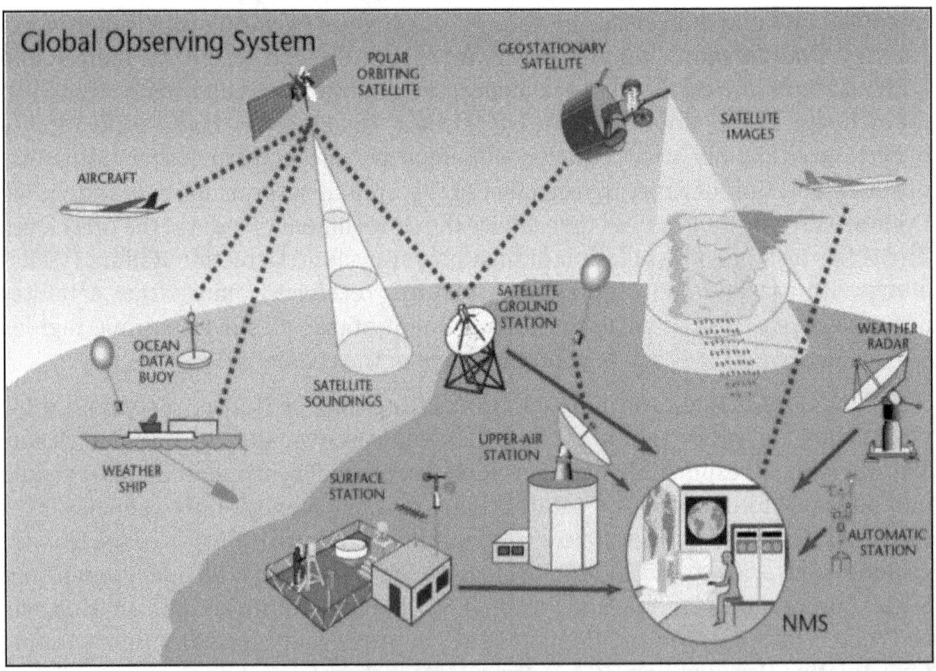

Figure 4.5: Monitoring Network for Cyclone Early Warning in India (*Source*: IMD).

a specialized centre on Tropical Cyclone Studies (Figure 4.4) which provides real time information and updates including track of tropical cyclones over north India ocean, and forewarns about the location of landfall. The approach of monitoring network for cyclone tracking and early warning is shown in Figure 4.5.

Drought Early Warning and Monitoring Network

Drought is a slowly evolving disaster connotes a situation of scarcity and distress caused by prolonged failure of rains leading to loss of agricultural production, drinking water shortages, reduction in fodder supply *etc.* It is difficult to exactly demarcate the onset and end of drought and its effects accumulate for a considerable period of time. Agricultural drought is a situation when rainfall and soil moisture are inadequate during the crop growing season to support healthy crop growth to maturity, causing crop stress and wilting. Monitoring and assessment of drought conditions at different scales and timely dissemination of information constitute the most vital part of drought management system. India Meteorology Department (IMD) is the designated agency for providing early warning and forecasting of the weather. Agricultural Meteorology Division, Pune provides timely advice on the actual and expected weather and its likely impact on the various day-to-day farming operations. Drought Research Unit, Pune provides Crop Yield Forecasts. The National Centre for Medium Range Weather Forecasting, a constituent unit of the Department of Science and Technology provides, in consultation with IMD, ICAR and State Agricultural Universities, agro-meteorological advisory service at the scale of agro-climatic zones to the farming community, based on location specific medium range weather forecasts.

In India, National Agricultural Drought Assessment and Monitoring System (NADAMS) was initiated towards the end of 1986, with the participation of National Remote Sensing Agency (Now National Remote Sensing Centre, Hyderabad) under the Government of India in Department of Space as nodal agency for execution, with the support of IMD and various state departments of agriculture. NADAMS was made operational in 1990. From the year 2012, the NADAMS project is being carried out at Mahalanobis National Crop Forecast Centre (MNCFC), Ministry of Agriculture (GOI), with the technical support of NRSC[52]. NADAMS now covers 13 agriculturally important and drought vulnerable states of the country and provides information on prevalence, spatial extent and intensity of agricultural drought at state, district and sub-district level. The project Monitoring of agricultural drought is carried out for the kharif season (June-Oct/November) as this season is agriculturally more important and rainfall dependent. Drought assessment is carried out on fortnightly basis and reporting of information to the User departments is done on monthly basis.[53]

A prototype regional South Asia Drought Monitoring System (SADMS) was developed in 2015 with support from the World Meteorological Organizations (WMO), Global Water Partnership (GWP), CGIAR's consolidated research program of CCAFS (Climate Change, Agriculture and Food Security) and Ministry of Agriculture,

Forestry and Fisheries (MAFF), Japan which is being led by International Water Management Institute. It is a weekly map of drought conditions as regional monitor for South Asia covering Bangladesh, Bhutan, India, Nepal, Pakistan and Sri Lanka. It involves the national authorities, the Country Water Partnership and other interested partners. Numerous drought indices - including the Integrated Drought Severity Index, Standardized Precipitation Index, and Soil Moisture Index - have been developed to provide advanced drought monitoring and assessment information for various purposes. In tandem, these indices not only create an accurate picture of any particular drought episode, but provide invaluable decision-making tools. Through the SADMS website (http://dms.iwmi.org/), the International Water Management Institute provides a wide array of precipitation and related information garnered primarily from the freely available satellite imagery to improve current capabilities in drought monitoring and prediction and provide regional to district scale information about drought's effect on agriculture. The drought severity maps that are produced, deliver continuous geographic coverage over large area for the first time, and have inherently finer spatial detail (500m resolution) than other commonly available global drought products such as NESDIS NOAA, MODIS Global Terrestrial Drought Severity Index using different data and approaches (Figure 4.6).

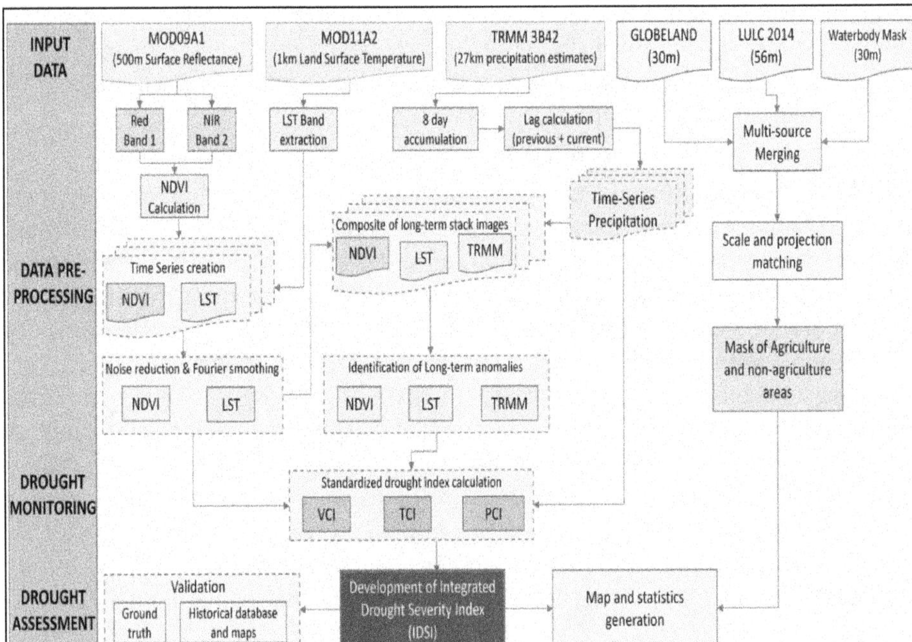

Notes: MOD09A1 – MODIS Surface Reflectance of every 8-Day product at 500m resolution; MOD11A1 – MODIS Land Surface Temperature (LST) daily product at 1,000m resolution; TRMM – Tropical Rainfall Measuring Mission; LULC NRSC – Land Use and Land Cover from National Remote Sensing Centre; Water body mask from Landsat images; NDVI – Normalized Difference Vegetation Index; VCI – Vegetation Condition Index; TCI – Temperature Condition Index (TCI); Precipitation Condition Index (PCI), IDSI – Integrated Drought Severity Index

Figure 4.6: Drought Monitoring Approach (under SADMS).

Flood Forecasting and Warning Network

Floods, heavy or catastrophic ones, are bound to occur periodically. They cannot be prevented or controlled. Increasing green cover in the catchment area, extensive water harvesting, ground water recharging, and so on, may perhaps slightly reduce the incidence of floods. However, floods continue to happen from time to time flood management systems require to minimize harm and damage and maximize benefits.

Type of floods in India:

- ☆ Snow-melt floods
- ☆ Flash floods/cloudburst floods
- ☆ Monsoon floods of single and multiple events
- ☆ Cyclone floods
- ☆ Floods due to dam bursts/failure
- ☆ Urban floods.

Flood Forecasting

Central Water Commission CWC is responsible for issuance of flood forecasting in India. As per the recommendation of Reddy Committee to manage flooding of Delhi, in the year 1958, CWC commenced the flood forecasting service in a small way by establishing flood forecasting unit for issuing water level forecasts of the Yamuna for the National Capital, Delhi. On the recommendation of various committees/ panels, a "Flood Forecast and Warning Organization" was set up in CWC in 1969 to establish forecasting sites on inter-state rivers at various flood prone places in the country. The network has been gradually expanded over the years to 175 stations including 28 inflow forecast located in 17 States/UTs covering nine major inter-state flood prone river basins comprising of 71 sub-river basins. [54] The Inflow Forecasts at 28 reservoirs are used by the dam authorities in timely operation of reservoir gates for safe flood discharges downstream as well as to ensure adequate storage in the reservoirs for meeting irrigation and hydropower generation demands during non-monsoon period. Such services are normally available during flood period from 1st May/1st June to 31st October/31st December every year. The forecast are disseminated using all types of communication means such as fax, wireless, phone, mob, SMS, email, electronic media, print media, social media, website, *etc.* Annually, over 6000 flood forecasts and advance warnings are issued by CWC regional offices across the country to the user agencies during floods. The overall accuracy of forecasts issued by CWC over the past years is around 97 per cent.[55]

CWC, through its twenty flood forecasting divisions issues forecasts through Fax, Email, SMS, Website, *etc.* to the various user agencies like civil/engineering agencies of the states/central governments such as irrigation/revenue/railways/ public undertakings and dam/barrage authorities/district magistrates/sub divisional officers besides the defense authorities involved in the flood loss

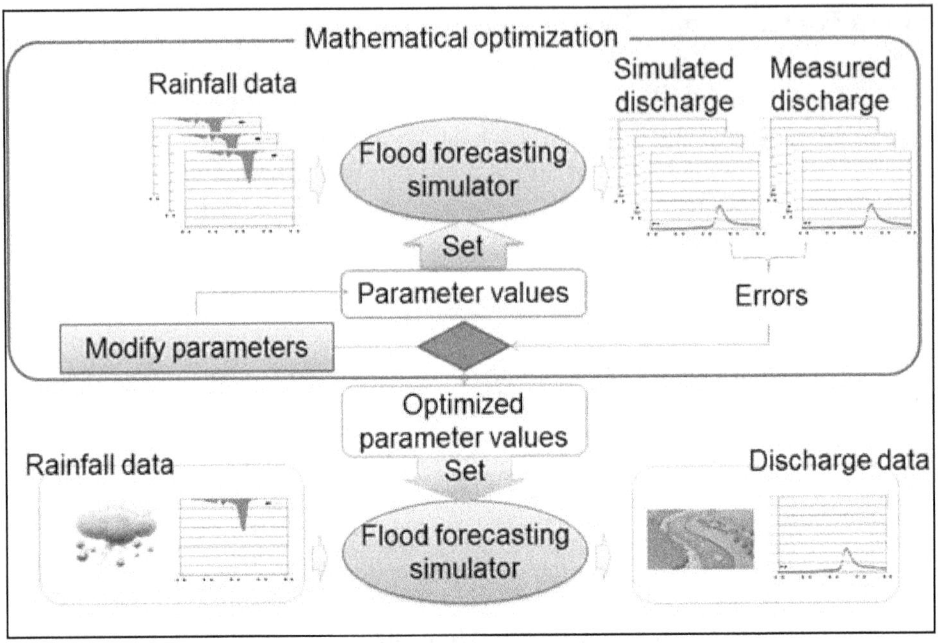

Figure 4.7: Approach of Flood Forecasting Based on Discharge Predictions.

mitigation work. CWC's flood forecasting network sites with spread across many river basins and states are tabulated in Tables 4.2. and 4.3.

Table 4.2: State-wise CWC Flood Forecasting Network

Sl.No.	*State*	*Number of Forecasting Stations*		
		Stage Forecasting	*Inflow Forecasting*	*Total*
1	Andhra Pradesh and Telangana	9	7	16
2	Assam	24	0	24
3	Bihar	32	0	32
4	Chhattisgarh	1	0	1
5	Gujarat	6	5	11
6	Haryana	0	1	1
7	Jharkhand	1	4	5
8	Karnataka	1	3	4
9	Madhya Pradesh	2	1	3
10	Maharashtra	7	2	9
11	Orissa	11	1	12
12	Tripura	2	0	2
13	Uttarakhand	3	0	3
14	Uttar Pradesh	34	1	35

Sl.No.	State	Number of Forecasting Stations		
		Stage Forecasting	Inflow Forecasting	Total
15	West Bengal	11	3	14
16	Dadra and Nagar Haveli	1	0	1
17	NCT of Delhi	2	0	2
	Total	147	28	175

Table 4.3: Major River Basin-wise CWC Flood Forecasting Network

Sl.No.	Major Interstate River Systems	Type of Forecasting Sites		Total
		Level Forecasting	Inflow Forecasting	
1	Ganga and its tributaries	77	10	87
2	Brahmaputra and its tributaries	27	00	27
3	Barak System	05	00	05
4	Eastern Rivers	08	01	09
5	Mahanadi	03	01	04
6	Godavari	14	04	18
7	Krishna	03	06	09
8	West Flowing Rivers	09	06	15
9	Southern River System (Pennar)	01	00	01
	Total	147	28	175

The flood forecasts are formulated by concerned field division after collection of data on real time basis using various communication modes. These are generally formulated as per the procedure given in the "Manual on Flood Forecasting", brought out by Central Water Commission for both stage and inflow forecasts (Figure 4.7). The flood forecasts are formulated using the conventional method as well as mathematical using MIKE-II. The conventional method uses the past data of river stages to draw a correlation diagram between the levels at the base station and the levels at the forecasting station. Correlation diagrams are of two types, namely, simple correlation and multiple correlation.

Simple Correlation

The simplest of all types of correlation is the correlation between N^{th} hour gauge of base station and $(N+T)^{th}$ hour stage of forecasting station; where T is the Travel Time between two stations. This type of correlation gives a rough estimate of the peak level likely to be attained and only linear correlation is normally adopted which is of the form:

$$Y = mX + c$$

where Y is the (N+T)th hour water level at forecasting station

X is the N th hour water level at base station

m and c are constants.

This type of graph can be developed and used for a reach of the river where no major tributary with considerable discharge joins the main river in the intervening catchment, the intervening catchment area is comparatively less, that the effect of intermediate rainfall is negligible and the travel time from the base station to the forecasting station is fairly constant for various stages. Following figure shows the typical simple correlation.

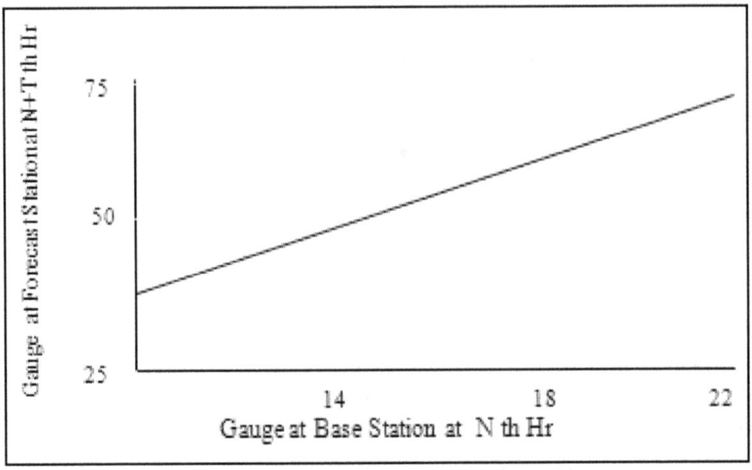

Figure 4.8: Simple Correlation for Flood Forecast.

This is the simplest of the correlation curves where one can get the stage/gauge of the forecast station simply on the basis of the stage/gauge at the base station after a particular period of time called the Travel Time. This is not true in all cases as the catchment area of the river and the flow of the river within its banks are unsteady. Hence this type of simple correlation curves has to be improved upon by considering additional parameters which controls the flow in the river.

Variation in Travel Time (Travel Time Curve)

The time taken by a flood peak to travel from the base station to the forecasting station is called the Travel Time of the flood wave. It is always not constant and depends on the river stage. During high stages travel time from base station to forecast station may reduce/increase substantially depending on the flood water of the river are within the river section or spilling into the flood plains. In order to account for these variations 'best fit' curve known as travel time curve between the peak value of water level and the time taken to reach the forecast is drawn for working out travel time for various stages in base station.

Multiple Correlations

The relationship obtained from simple correlation along with variation in travel time can be improved by considering the following additional factors and adopting multiple correlation techniques as explained below:

i. Varying conditions during rising and falling stage of the flood
ii. Rainfall in the intermediate catchment
iii. Effect of Major Tributary

Varying Condition during Rising and Falling Stages

In this diagram, the variation of the gauge at base station during the preceding time interval equal to travel time from base station to forecast station is correlated with the variation of the gauge in the forecasting station. Two different sets of lines one for the rising limb and another for the falling limb are plotted for the various gauge differences. This is normally the first quadrant correlation curve which basically considers the gauge level along with its trend at the base and forecasting stations and gives an approximately good forecast if other effects such as intermediate rainfall or contribution from major tributary are negligible.

Rainfall in the Intermediate Catchment

In order to account for rainfall in intermediate catchment the first quadrant correlation curve is updated with the likely rise due to the rainfall in the intermediate catchment. The effect of mean areal rainfall derived from the point rainfall observed in the catchment is plotted in one of the quadrant say second quadrant of the correlation curve by finding out the increase in water level at the forecasting station for various mean areal rainfall over the catchment. This gives qualitative improvement to the forecast. When there is no tributary between the base and forecasting stations, the correlation curve gives fairly accurate forecast.

Effect of Major Tributary

If a tributary joins the main river in between the base and forecasting station, the contribution from that tributary is taken into account in one of the remaining quadrant say the third quadrant of the correlation curve. If a gauge station is existing on the tributary, the travel time from the station to the forecasting station is found out and the variation in the gauge at the gauge station and forecast station during the travel time is correlated and sets of lines for both rising and falling limbs are drawn. If a reservoir is located in the tributary and if it releases water then that contribution has also to be found out and given due accordance in forecasting.

Mathematical Model

The modeling centres established under the IX, X Plans have been provided with mathematical models. The model has three modules, rainfall – runoff which converts rainfall to runoff, hydrodynamic module which routes the flow in the catchment and a flood forecasting module which formulates the flood forecast based on the Quantitative Precipitation Forecast for rainfall and boundary estimates of discharge and water level in the water level and discharge boundary stations. The model is being used in some of the modeling centres for few stations.

Telemetry for Data Collection

Under the IX, X and XI five year Plans, the use of satellite based Telemetry for Data acquisition, transmission and reception at Earth Station and Modeling Centre has been taken up in Central Water Commission. Presently about 223 Stations installed under IX and X plan schemes are under operation in Mahanadi, Chambal, Krishna, Godavari, Yamuna, Damodar, Brahmaputra river basins. These systems work on the principle of Pseudo Random Burst Mode (PRBM) in which the data collected in the Data Collection Unit (DCU) is transmitted to Satellite in three bursts in random manner during the time slot allotted for data transmission. The data thus transmitted are received in the Earth Stations or Digital Direct Readout Ground Stations (DDRGS) set up at Jaipur and Burla. The data received in DDRGS are further retransmitted to various modeling centres through VSAT. The PRBM mode of data transmission is having its inherent problems such as Data clash and window time.

Flood Forecasting Network[56]

There is a network of about 582 wireless (Flood Forcasting and Warning Network Performance Apprisal 2008, February 2010) stations for near real-time communication of hydrological and hydro-meteorological data between various base and forecasting sites and their subdivisions; and divisions.The flood forecasts are being formulated in the divisional offices (sub divisional offices in a few cases) and disseminated by these wireless stations to the users and also from the divisions to the head quarters of the CWC at New Delhi where "Flood Forecast Monitoring Directorate" is monitoring the flood situations and forecasts at the National level. In addition, there are about 220 + telemetry stations monitoring real time hourly water levels, 15 minutes rainfall as well as hourly rainfall, and other important parameters, established in Krishna, Godavari, Mahanadi, Chambal and Brahmaputra Basins. Flood forecasting has been recognized as one of the most important, reliable and cost-effective non-structural measures for flood management. Recognizing the crucial role it can play, CWC has set up a network of forecasting stations covering all important flood prone interstate rivers. The forecasts issued by these stations are used to alert the public and to enable the administrative and engineering agencies of the States/UT's to take appropriate measures.

CWC website gives all details on flood forecast, hydrograph views and site digital maps.[57,58]

Use of Additional Meteorological Inputs

The India Meteorological Department (IMD) is providing meteorological inputs such as Numerical Weather Prediction products which can provide forecasts for rainfall up to 5 days in advance. These forecasts are accurate up to 3 days in advance and Quantitative Precipitation Estimates based on Satellite Cloud top temperatures which can provide forecasts for rainfall up to 2 days in advance as well as from precipitation intensity from Doppler Weather Radar Products within a radius of

100 km. These can be utilised as boundary estimates in mathematical modeling for extending the warning time at the cost of accuracy of flood forecast. Advisory Flood Forecasts can be issued based on these types of inputs.

Role of IMD in Flood Forecasting

The formulation of flood forecast involves hydrological and meteorological data in the upstream catchment of flood forecast station in the basin for which CWC has established its own network of river gauge and rain gauge stations. Most of these stations have been established along the course of the river. In order to account for the catchment contribution meteorological data such as rainfall is being collected from IMD in catchment areas. Further, IMD provides Synoptic situation and weather forecast both qualitative and quantitative precipitation forecast (QPF) to various field divisions of CWC through their Flood Meteorological Offices (FMO). In order to provide QPF, IMD required real-time rainfall from all stations in the catchment area. Previously, the QPF were issued only when the river levels crossed warning levels after information from CWC. However from the year 2010, daily QPF is being generated by FMOs for various basins and sent to CWC field divisions. This information is also updated on the website for reference. From 2012 onwards, the telemetry data of CWC is being uploaded in the India-WARIS website[59] and user name and password were generated for IMD and supplied to them for viewing the real–time telemetry data. IMD AWS rainfall data is available on their website for downloads and further use in Flood Forecasting by CWC.

Thus process of Flood Forecasting involves following important steps:

- ✰ Coordination through 10+ flood Meteorological offices of the IMD
- ✰ 175+ Flood Forecasting Centres of Central Water Commission (CWC)
- ✰ IMD- Observes and collects Hydro-meteorological data, transmission of data to forecast centers, analysis of data and formulation of forecast (quantitative precipitation forecast) and issue of heavy rainfall warnings
- ✰ Dissemination of forecasts/warnings to Flood Forecast Centers of CWC

Frequency of Flood Forecast is broadly defined as under:

i. Major rivers (Travel time >24 hours)
   - ✰ Forecasts are being formulated based on 0800 hrs/0900 hrs
   - ✰ Water level data and forecast issued once in a day at 1000 hrs with advance warning time from 24 hrs to 36 hrs

ii. Medium rivers (Travel time 12-24 hours)
   - ✰ Forecasts are being formulated based on 0600 hrs and 1800 hrs
   - ✰ Water level data and forecast issued twice in a day at 0700 hrs and 1900 hrs with advance warning time from 12 hrs to 24 hrs

iii. Flashy rivers (Travel time < 12 hours)
   - ✰ Forecasts are being formulated based on any hr

☆ Water level data and forecast issued multiple times (more than twice) in a day with advance warning time less than 12 hrs

To overcome the limitations of existing system, the CWC has undertaken various expansion and modernization schemes to cover more areas and to make forecasting more efficient and reliable. The IMD has also taken up the expansion of its network of Automatic Rain Gauges.

Hydrological data from various river basins that are collected by IMD, CWC, and State Governments (collected by using Bureau of Indian Standards approved automatic sensors for rainfall and river flow measurements) are planned to be stored in centralized mechanism for distribution and archival. Similarly, Computer-based comprehensive catchment scale hydrological and hydrodynamic models, interfaced with flood plain inundation mapping tools, will be developed.

Landslide Risk Management in India

The devastating Malpa tragedy[60] resulting from a landslide that occurred along the Kailash- Mansarovar route in the Kumaon Himalayan region in August 1998[61] acted as an eye-opener for the government of India. Indian government decided to set up task forces for Landslide Hazard Zonation (LHZ)[62], Geotechnical Investigations and Land Use Zonation and Regulation. The Geological Survey of India (GSI) was identified as the nodal agency for LHZ while the Department of Science and Technology (DST) and the Ministry of Environment and Forests (MoEF) were identified as nodal agencies for the other two task forces, respectively. Consequently, the Department of Mines (DoM) constituted a task force to review the existing methodologies for LHZ study, to prioritize areas/belts for its study and to recommend a plan for the preparation of macro/meso/micro LHZ maps. The task force constituted by the DoM submitted its report to the government in September 2000 (NDMA. (2009 a)).

The GSI is the nodal agency for monitoring landslide activity and its mitigation. There are many government departments and organizations which are engaged in landslide hazard studies and hazard management in the country. These include the GSI, Central Road Research Institute (CRRI), Central Building Research Institute (CBRI), Indian Institute of Technology, Roorkee (IIT-R), Wadia Institute of Himalayan Geology (WIHG), Department of Space (DoS), National Remote Sensing Centre (NRSC), Defence Terrain Research Laboratory (DTRL), Bureau of Indian Standards (BIS), some academic institutions, and individual experts. In addition, the Border Roads Organization (BRO) is the principal agency responsible for the construction and maintenance of roads in almost all the hilly regions of the country and DST has been funding Research and Development (R&D) activities that include different types of landslide investigations. Landslide hazard management in India has till now been confined to ad-hoc solutions of site [63]specific problems and the implementation of immediate remedial measures including debris removal, and dumping of this debris either down slope or into a river. There is, however, a need to pre-empt disaster by making adequate information available in advance before it strikes, this is what NDMA Guidelines also emphasize.

Snow Avalanches Studies

Snow avalanches are not landslides in a strict sense of the term, but when snow and ice slides it causes devastation, especially during inclement weather at high altitudes. Snow avalanches lead to disasters in the snow covered mountainous terrain of the Himalayas. Avalanches are transient, three dimensional, gravity driven, free surface, rapidly moving shear flows that contain a dense granular core surrounded by a cloud of airborne turbulent powder. Snow and Avalanche Study Establishment (SASE), is the agency, under the Ministry of Defence dedicated to the study of avalanche related hazards and their mitigation in India. Avalanche forecasting, artificial triggering, structural control and generating awareness are prominent part of SASE's activities. Snow and Avalanche Study Establishment (SASE) was set up in 1969 near Manali under the aegis of DRDO in 1969 to combat the hazards of snow and avalanches not only to help the Armed Forces to fight and live in the mountains but also to accelerate the pace of socio-economic growth of the inaccessible snowbound hill regions. During the course of its growth over the last 40 years, SASE has created an impressive scientific and technical infrastructure which inter alia includes :Experimental Research Station for Structural control of Avalanches; Snow Chute for Avalanche Dynamics Modeling; Low Temperature Concrete Testing Facility; X-ray Micro tomography Facility; Radio based Remote Telemetry Systems; Remote Sensing Lab; and Earth Receiving Station. The SASE is engaged in carrying out research activities in the field of snow avalanche forecasting and avalanche control structures. The SASE is also responsible for conducting search and rescue operations as post-disaster activities in the case of snow avalanches.[64]

SASE collects snow and meteorological data at various[65] stations in J&K, Uttarakhand and Himachal Pradesh; records weather, temperature, sunshine and seasonal snow cover; and studies synoptic situation leading to precipitation and snow distribution effects. Albedo studies[66] are also conducted under simulated light. Snow and Avalanche Study Establishment has established a network of Automatic Weather Stations (AWS) in Himalayan region to collect various atmospheric and snow parameter data like temperature, humidity, albedo, wind speed and direction, snow depth, precipitation *etc.* using various sensors. This data is used at RDC Chandigarh to issue mountain weather and avalanche forecast.

Mountain weather forecasting is an important activity being undertaken by SASE. Functional role of SASE includes:

- ☆ Provision of precise avalanche and weather forecast;
- ☆ Creation of avalanche awareness among users;
- ☆ Controlled release of avalanches;
- ☆ Design and development of avalanche control structures;
- ☆ Remote sensing and instruments for mountain terrain/snow study;

- Provide advice to BRO, J&K, HP, and Uttarakhand governments regarding snow avalanches;
- Basic research on Snow Science;
- Cold region data bank for Himalayan region;
- Operational avalanche prediction;
- Computational snow material science;
- Snow cover information system and mountain hazard mapping using remote sensing;
- Instrumentation and sensors for snow-avalanche research;
- Snow climate research and mountain meteorology;
- Snow geo-intelligence using unmanned air borne systems;
- Cold region data bank;
- Advise to users on road alignments and avalanche control measures, and safe camping sites;
- Training on avalanche safety and rescue;
- Study of snow climate and its effect on mobility of troops;
- Avalanche dynamics study;
- Slope stabilization and afforesration;
- Research on glacier dynamics to study crevasse formation and glacier related hazards;
- Snow harvesting;
- Artificial triggering of avalanches; and
- Climate change study.

Heat Wave

Extreme positive departures from the normal maximum temperature result in a heat wave during the summer season. The rising maximum temperature during the pre-monsoon months often continues till June, in rare cases till July, over the north-western parts of the country. Slow and silent, a heat wave does not descend upon a city with the fury of a tornado, hurricane, or a winter storm. It moves over an area as a large, deep air mass with descending air, retarding the development of any significant precipitation that would provide relief to the ground surface's rising temperatures. As this air mass moves slowly or just sits over one area for days or even weeks, its rising surface temperatures begin to take their toll on the people who are trapped in it.

Despite the presence of improving technology (*e.g.*, air conditioning, architectural design, and improved accuracy in weather forecasting), heat waves continue to take many lives. Decrease in the Diurnal Temperature Range (DTR) due to urbanisation is a new factor leading to human mortality and discomfort. Increased minimum temperatures in summer do not allow the necessary nocturnal cooling to neutralise the high maximum temperature during a heat wave epoch. Aside from being a major health hazard, heat waves also put heavy stress on our energy resources.

Heat wave warnings are issued by IMD on regular basis and State Emergency Operation Centres shall be in contact with IMD to receive such information and disseminate it to the public. City of Ahmedabad launched (Vishal Dutta, Economic Times Bureau Apr 16, 2013,) a first-of-its-kind Heat Action Plan, making it the first city in South Asia to create a comprehensive early warning system and preparedness plan for extreme heat events fuelled by climate change. The project created by the Ahmedabad Municipal Corporation (AMC) in partnership with an international coalition of health and academic groups, that will help in reducing the deadly impact of extreme heat by initiating an early warning system for residents, providing preparation training to medical and community workers, building public awareness of the heat-related risks, and coordinating an inter-agency emergency response efforts when heat waves hit.

Cold Waves/Frost

Cold wave/frost is a localised seasonal phenomenon prevalent in the country except in Southern India. Frost damage to crops occurs when moisture within the plant is frozen, gets crystallised and expands. This causes cells to rupture and fluid to leak out, thus the watery appearance of plant tissue or seed after a damaging frost. Different parts of plant, different stages of development of plant, and different types of plants can have varying levels of these 'antifreeze' compounds that result in a range of susceptibility to frost.The extent of damage caused by cold wave depends on temperature, length of exposure, humidity levels, and the speed at which freezing temperature is reached. It is difficult to predict a definite temperature level up to which crops can tolerate cold wave/frost because many other factors also affect it. For example, when air temperature reaches Zero (0) degree centigrade, crop itself can be 4 or 5 degree cooler, because plants lose heat faster than the surrounding air temperature. Department of Agriculture and Cooperation, Ministry of Agriculture closely monitors cold wave situation in consultation with India Meteorological Department (IMD) and State Governments.

Mitigation Measures for Crops

As Cold Wave/Frost is a localised disaster event, location specific mitigation plans are to be drawn up by the concerned State Governments instead of a National level plan. In case of cold wave/frost situation, States needs to initiate location specific measures as outlined in District Crop Contingency Plans and in consultation with respective State Agricultural Universities to minimise its impact.

Farmers are to provide light irrigation as per need, immediately prune damaged tips of branches or shoot, burn leave/waste material in the orchard to create smoke and manage rejuvenation of damaged crops through pruning of dead material, application of extra doses of fertiliser through foliar sprays. [67]

Table 4.3: Mitigation Measures during different Vegetative Stages.

Sl.No.	*Stages of plant growth*	*Measures to be taken by farmers*
1	Seedling/Nursery Stage	Change of micro climate by smoking around the field especially during night.
2	Vegetative Stage	Irrigating the field, smoking the field during night.
3	Reproductive Stage	
4	Harvesting Stage	Harvest the crop at physiological maturity stage

Crops: Soybean, Maise, Jowar, Arhar, Cotton, Chick Pea, Wheat *etc.*.

Source: Contingency Plan developed by Central Research Institute for Dry land Agriculture (CRIDA), available at http://www.crida.in/

Disease Epidemics

Epidemics are disasters by themselves but these can emerge in the aftermath of other disasters as well. Several factors related to microbes, environment and host susceptibility contribute to the occurrence of epidemics. Because of prevalence of these factors, India is frequently affected by epidemics/outbreaks which result in high morbidity and mortality and affect the public health and economy adversely.

For epidemics Ministry of Health and Family Welfare, Government of India is nodal agency to play a vital role in management of such emergency situation. In case It has to prepare an effective prevention and mitigation strategies at the national level anticipating the magnitude of the impact of an epidemic, link surveillance and response measures, and provide operational procedures that involved a broad range of sectors in preparing for and responding to disease outbreaks.

Epidemic Management Measures

Epidemic management activities include taking appropriate control measures, such as treating those who are ill to reduce the reservoir of infection and providing health education to limit the transmission of the disease to others. Health professionals at higher levels will require peoples help in putting into taking any measures needed to control the spread of the disease, such as giving drugs to people in the community and providing health education. The community may be involved in the management of an epidemic once it is confirmed by the health authorities. The type of control measures for implementation will depend on the type of infectious agent, how the disease is transmitted, and any other factors contributing to the disease. Generally control measures should target the infectious agent, the source of any infection, and the treatment of those who became ill. Remember, the source of infection could be humans or animals, or non-living things in the environment.

Preventive Measures

i. Availability of good quality data to frame better policies and to make better decisions and the monitoring implementation of health care programme and quality of medical research.

ii. Availability of digitised data in unified format to help improving data quality and developing standards for knowledge generation.

iii. Efficiently handling health aspects of disaster by using data for quickly finding information about available resources around disaster site and mobilising these resources.

iv. Enhancing research to policy through collaborations and exchange of information between researchers and health professionals; between researchers and other stake holders involved in policy making.

v. Enhancing public private partnership in health and preparing national information system of research funding by different funding agencies.

vi. For developing and implementing a comprehensive plan for utilisation of knowledge network for health service delivery, medical education and research, an Appropriate Authority under the Department of Health Research in the Ministry of Health and Family Welfare, Government of India is to be setup. Such an authority with committed resources will be responsible for strategic planning, for implementation and review of the progress from time to time. This Authority/Cell may establish units in the States for action at the local level.

vii. To provide correct information and knowledge to end-users which include health professionals, research, patient and students and dissemination of quality information and knowledge through agencies like DAVP and appropriate media such as internet, information kiosks *etc.*

viii. Public needs to be educated about the signs and symptoms of likely diseases so as to enable the early detection and preventive measures of such disease. Local health authorities need to keep a constant vigil on the epidemic prone areas.

ix. A check on the spread of epidemics by means of adequate control and monitoring measures before and after the occurrence of epidemics has to be ensured.

x. Extensive training programme to update their knowledge and skills in this area is essential for health workers.

xi. To prepare a National Health and Emergency Management System to respond to major epidemic diseases.

xii. To identify, notify, verify, assess and respond to major epidemics.

xiii. To improve capacity at regional, sub-regional and country levels to support the detection, assessment, confirmation and response to major epidemics.

xiv. To strengthen technical and operational support for epidemic diseases control during humanitarian emergencies caused by conflict or natural disasters.

xv. To create an effective mosquito control program. Steps required to be taken to avoid uncontrolled urbanization, excessive population growth and urban decay characterised by substandard housing an inadequate water and water disposal system to control water borne epidemics.

xvi. House to house survey for detection of vector breeding sources, intensification of anti-larval operation/health education activities and focal spray with pyrethrum extract.

xvii. To prepare an effective management plans for each epidemic that goes beyond generic World Health Organization recommendations and also specific operational plans to respond to different phases of an epidemic.

xviii. To have a detailed vaccination strategies, logistics for drug provisions, or continuity plans for essential public services during epidemic outbreaks.

xix. To prepare guidelines for long term capacity development.

xx. In addition, governments and businesses need to cooperate more closely on long-term pandemic disease control. Some key measures could be to support regional and national pandemic preparedness planning, expand the role of pharmaceutical companies in Government preparedness and manufacturing of own vaccines that are best suited to national priorities and needs.

Forest Fire Risk Management

Nearly 55 per cent of the total forest cover in India is prone to fires every year.[68] Forest fires in India have environmental significance in terms of tropical biomass burning, which produces large amounts of trace gases, aerosol particles, and play a pivotal role in tropospheric chemistry and climate. Forests fires pose a threat not only to the forest wealth but also to the entire regime of fauna and flora seriously disturbing the bio-diversity and the ecology and environment of a region. During summer, when there is no rain for months, the forests become littered with dry senescent leaves and twinges, which could burst into flames ignited by the slightest spark.

Forest fires are caused by natural causes as well as manmade causes. Many forest fires start from natural causes such as lightning which set trees on fire. However, rain extinguishes such fires without causing much damage. Massive forest fire can be caused when a source of fire like naked flame, cigarette or *bidi*, electric spark or any source of ignition comes into contact with inflammable material. The youngest mountain ranges of Himalayas are the most vulnerable stretches of the world susceptible to forest fires. The forests of Western Himalayas are more frequently vulnerable to forest fires as compared to those in Eastern Himalayas.

This is because forests of Eastern Himalayas grow in high rain density. With large scale expansion of chirr (Pine) forests in many areas of the Himalayas the frequency and intensity of forest fires has increased. [69]

Fire prevention, detection and suppression activities are local governance subjects. The central governments has been formulating policy, planning and financing the states from time to time. Forest Protection Division of Ministry of Forest headed by DIG of Forests is responsible for forest fire management at central level. The Joint Forest Management Committees, more than 30000 in numbers across the country, have been given the responsibility to protect the forest from fire. The central government has issued the National Fire, Prevention Control guidelines and has also worked on a National Master Plan on Forest Fire Control. Active forest fires are detected from the satellite images and the information is uploaded daily to the Indian Forest Fire Response and Assessment System (INFFRAS) website during the forest fire season – February to June (www.inffras.gov.in). Current satellite based forest fire detection system use the data gathered by two satellites; Advanced Very Resolution Radiometer (AVHRR) which was launched in 1998 and Moderate Resolution Imaging Spectroradiometer (MODIS) which was launched in 1999.

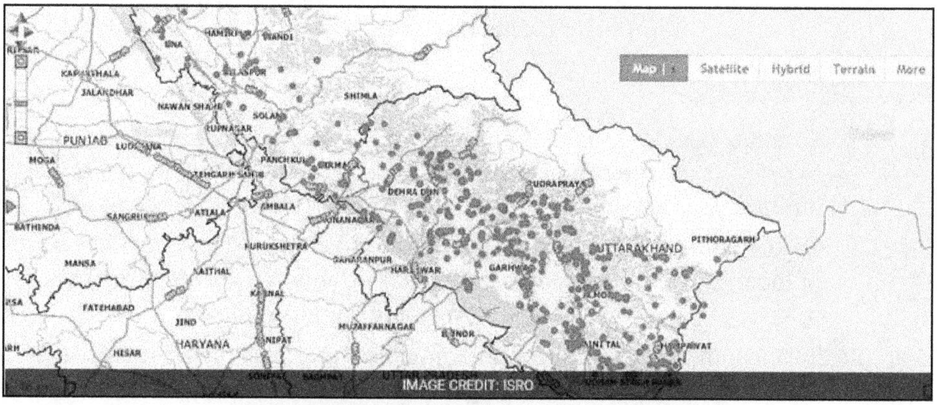

Figure 4.8: ISRO's Forest Fire Rating System – Map Showing Fire Spots in Uttarakhand.

Wireless sensor network have a broad range of applications in the category of environmental monitoring. The wireless sensor network unlike that of traditional approaches, can easily forecast forest fires before the fire is spread uncontrollable. This framework includes designing of sensor node (Figure 4.9) and data transmission process. Some important technologies and systems that are currently used towards this goal are: employing charged coupled device (CCD) cameras and Infrared (IR) detectors, satellite based monitoring systems and wireless sensor networks. In case of charged coupled device (CCD) cameras and IR detectors based system, the accuracy of the system is affected by weather conditions such as clouds, light reflection and smoke from various activities. Three factors compose the basis of a forest fire: the fire source, environmental elements and combustible material.

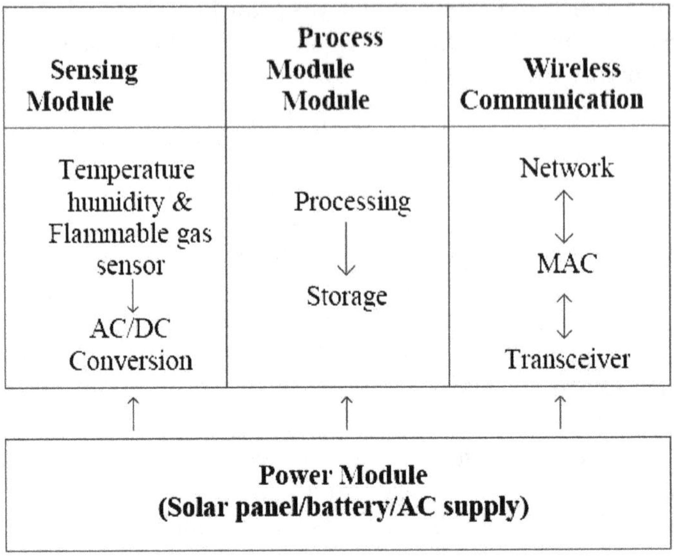

Figure 4.9: Wireless Sensor Node for Fore Fire Detection (*Source*: Chauhan *et al.*, 2013).

Preventive Measures[70]

i. To create firebreaks in the shape of small clearings of ditches in the forests to prevent the fire from spreading. To keep the source of fire or source of ignition separated from combustible and inflammable material.

ii. To keep the source of fire under watch and control. The volunteer teams of local communities may be created not only for fire- fighting but also to keep watch on the start of forest and sound an alert.

iii. To not allow combustible or inflammable material to pile up unnecessarily and to stock the same as per procedure recommended for safe storage of such combustible or inflammable material.

iv. To adopt safe practices in areas near forests namely. Factories, coalmines, oil stores, chemical plants and even in household kitchens.

v. To incorporate fire-reducing and fire-fighting techniques and equipments while planning a building or coal mining operation in forest region.

vi. To arrange fire fighting drills frequently.

Railway Accidents

Railway Accidents as an emergency situation is responsibility of Railways. The core values of the Indian Railways (IR) include safety apart from security, punctuality and reliability.

To avoid Accidents Indian Railways take various measures such as:

i. To stimulate the implementation of modern, proactive and systematic safety measures
ii. To bring about both qualitative and quantitative improvements in safety performance
iii. To encourage safety research and development
iv. To reduce consequential train accidents
v. To develop techniques, models and expertise for risk assessment and Mitigation methods.
vi. To bring in a culture of Safety as Safety is an ethos that should pervade all activities of railway operations and maintenance. The safety organisation has essentially to be a coordinating service department helping the concerned departments to discharge their safety functions effectively
vii. Timely sanction and execution of works for replacement of signalling installations to maintain the required level of safety and Reliability.
viii. Long-term policy on mandatory provision of proven On -Board Train Protection and Warning System and Anti-Collision Device of approved design

There are many areas requiring decisions and directives for long-term policies for the prevention and Mitigation of Railway Accident. They primarily relate to maintenance of assets, quality of material related to safety components, train operations and industrial relations *etc.* Review of existing Maintenance schedules and issue of yardsticks for deployment of skilled manpower to ensure proper upkeep of all signalling gears and to optimise Mean Time To Repair (MTTR). Indian railways has come out with railways disaster management plans as available at http://www.iritm.indianrailways.gov.in/.

Aviation Accidents

An aircraft crash may result in a very large number of casualties and considerable disruption to both air traffic and essential services. There would also be a considerable risk of fire, and resultant consequences, if the accident takes place either at the airport or in the populated area. The consequences of the crash and the measures needed to deal with it would vary according to the circumstances. The plan dealing with the worst conceivable case of an aircraft crash involves coordination among various government authorities, operator involved and the nearest airport.

Section 7 of the Aircraft Act 1934[71] empowers the Central Government to make rules providing for the investigation of any accident or incident arising out of or in the course of the navigation in or over India of any aircraft, or anywhere of aircraft registered in India. International Civil Aviation Organization (ICAO) had continuously placed responsibility on Contracting States for safe air operations. The recent one is assimilation of all the safety aspects in a comprehensive programme for example State Safety Programme (SSP). The Programme is an integrated set of Regulations and activities aimed at improving safety. Directorate General of Civil Aviation (DGCA) has regulatory responsibility for aviation safety, and has released its SSP document recently.

Conclusion

Early warning is considered to be a very important component for Disaster Risk Reduction. EW facilitates risk predictions and risk communication and plays major role in prevention, mitigation, and prepared for Disasters. In slower onset disasters that can be pre-empted days or months in advance, early warning systems can provide enough time for risk reduction measures to be put in place. Early warning alone cannot save lives and structures. It requires a proper response mechanism to be in place. Besides, multiple stakeholders need to be on common ground for effective disaster management.

Early warning Organizations such as IMD, CWC, ISRO, GSI, NRSA, and SASE *etc.* in India are all playing their respective roles for early warning; however they all require cooperation, and long-term planning for an integrated EW system. *Integration of the academia with the government and corporate sector is very crucial for better data integration and data organization. Participatory Approach* (*Early Warning Practices can Save Lives: Selected Examples Good Practices and lessons learned, Bon2010, United Nations 2010*) on early warnings bank upon the coping capacity and capability of local populations and empowers people to play a pro-active role in ensuring the security of their families and livelihoods.There is a need for local knowledge and practices to be integrated with those of the science community, to improve forecasts and increase acceptance, ownership and sustainability of early warning systems. *The UNISDR's Hyogo Framework for Action emphasised the importance of encouraging the use of traditional knowledge.*[72] *New Sendai Frame Work on DRR also indicate dependency on efficient early warning systems as part of preparedness and risk reduction.* Impacts of EW systems can be maximised only when all necessary steps are taken to enhance the effectiveness of technological tools and scientific forecasts that governments and communities rely on, it may require some more time for appropriate action.

Lastly, it is pertinent to understand about the vulnerability of the population whom early warning needs to be directed. Ability of locals to comprehend the message rightly and taking apt actions are essential for effectiveness of an EW system.

Reference

Chauhan, A., R. Chauhan, S. Semwal (2013). Early Detection of Forest Fire Using Wireless Sensor Network. Int. J. Engineering Research and Applications. www.ijera.com 3(4):163-168.

Notes

1. National Disaster Management Plan
2. http://www.un.org/esa/population/publications/pop_challenges/Population_Challenges.pdf
3. WDR
4. http://unchronicle.un.org/article/global-early-warning-systems-needed-creating-partnerships-cope-natural-disasters/
5. http://www.unisdr.org/wcdr/
6. Yokohama Strategy and Plan of Action for a Safer World (1994), http://www.unisdr.org/eng/about_isdr/bd-yokohama-strat-eng.htm

   http://www.un.org/esa/sustdev/documents/agenda21/english/agenda21toc.htm
7. Barbados Plan of Action for Small Island Developing States (1994), http://www.un.org/documents/ga/conf167/aconf167-9.htm
8. World Summit on Sustainable Development (Johannesburg, South Africa, 2002) http://www.un.org/esa/sustdev/documents/WSSD_POI_PD/English/WSSD_PlanImp.
9. The Gleneagles G8 meeting (2005). http ://www.g8.gov.uk/Files/KFile/PostG8_Gleneagles_Tsunami.pdf
10. The International Early Warning Program (IEWP), was first proposed at the Second International Early Warning Conference (EWCII) in 2003 in Bonn, Germany. It developed increasing importance in the wake of the 2004 Indian Ocean tsunami, which claimed over 200,000 lives and injured over half a million people.
11. Southasiadisasters.net special issue October 2013, http://reliefweb.int/sites/reliefweb.int/files/resources/99per cent20Cycloneper cent20Phailin.pdf
12. www.scidev.net/.../early-warning-of-disasters-facts-and-figures-
13. http://www.scidev.net/global/communication/feature/early-warning-of-disasters-facts-and-figures-1.htm
14. Global Assessment Report for Disaster Risk Reduction. (Report, UNISDR, 2011)
15. Early Warning Systems: State-of-Art Analysis and Future Directions by Veronica F. Grasso Ashbindu Singh United Nations Environment Programme (UNEP)
16. Checklist prepared in Third International conference on early warning at Bonn Germany 2006
17. Early Warning Systems: State-of-Art Analysis and Future Directions by Veronica F. Grasso Ashbindu Singh United Nations Environment Programme (UNEP)

18. http://www.scidev.net/global/communication/feature/early-warning-of-disasters-facts-and-figures-1.html
19. Ogallo, L., *et al.,* Adapting to climate variability and change: the Climate Outlook Forum process [837kB]. BAMS 57, 93–102 (2008)
20. http://www.ptwc.weather.gov
21. http://www.scidev.net/global/communication/feature/early-warning-of-disasters-facts-and-figures-1.html
22. www.ushahidi.com
23. Ibid
24. Veronica F. G. and Singh A. (2012)
25. Asian Disaster Management News January-April 2007
26. Developing Early Warning Systems: A checklist. Report from the Third International Conference on Early Warning (Report, UNISDR, 2006) and Global Assessment Report for Disaster Risk Reduction. (Report, UNISDR, 2011)
27. Kettlewell, J. *Early Warning Technology* – is it enough? (BBC, 2008)
28. http://www.imd.gov.in/doc/mandate.htm
29. www.imd.gov.in/section/nhac/dynamic/RSMC-2011.pdf
30. http://www.imd.gov.in/services/agri_new/forecast.htm
31. http/aviation/services-to-aviation.htm
32. SIGMET or Significant Meteorological Information is a weather advisory that contain meteorological information concerning the safety of all aircraft. https://en.wikipedia.org/wiki/SIGMET
33. http://www.imd.gov.in/services
34. SIGMET, or Significant Meteorological Information, is a weather advisory that contains meteorological information concerning the safety of all aircrafts. https://en.wikipedia.org/wiki/SIGMET
35. A description of present weather conditions and forecast of those immediately expected.
36. http://www.imd.gov.in/section/rth/static/country_net.htm
37. The Global Telecommunication System (GTS) is defined as: "The co-ordinated global system of telecommunication facilities and arrangements for the rapid collection, exchange and distribution of observations and processed information within the framework of the World Weather Watch." As on https://www.wmo.int/pages/prog/www/TEM/GTS/index_en.html
38. http://imd.gov.in/doc/Nowcast.pdf
39. www.imd.gov.in/section/hydro/static/welcome.htm

40. www.imd.gov.in/section/seismo/dynamic/eq.doc
41. http://www.imd.gov.in/section/seismo/static/welcome.htm
42. http://www.imd.gov.in/section/seismo/static/welcome.htm
43. http://www.imd.gov.in/section/seismo/static/welcome.htm
44. http://moes.gov.in/programmes/observational-networks-and-data-centre
45. http://www.imd.gov.in/op.cit
46. http://www.incois.gov.in/Incois/tsunamicontents
47. http://www.tsunami.incois.gov.in/ITEWS/tsunamimodeling.jsp
48. www.imd.gov.in/section/seismo/static/welcome.htm
49. http://www.incois.gov.in/Incois/tsunamicontents
50. www.rmcchennaieatlas.tn.nic.in
51. http://www.rsmcnewdelhi.imd.gov.in/index.php?lang=en
52. http://www.nrsc.gov.in/Earth_Observation_Applications_Agriculture_Drought_Assessment.html
53. http://www.nrsc.gov.in/Drought
54. http://india-wris.nrsc.gov.in/wrpinfo/index.php.
55. http://www.cwc.nic.in/newsite/FloodForecasting.html
56. http://india-wris.nrsc.gov.in/wrpinfo/index.php/title=CWC_National_Flood_Forecasting_Network
57. http://www.india-water.gov.in/
58. http://www.india-water.gov.in/eSWIS-MapViewer/
59. http://www.india-wris.nrsc.gov.in/wris.html
60. Malpa landslide. On 11 and 17 August 1998,. wiped away the entire villiage of Malpa in the Pithoragarh district of Uttarakhand (Then in Uttar Pradesh) in Kali Valley of Higher Kumaun Himalaya. About 220 people were reported to be died.
61. National Disaster Management Guideline Management of Landslide and Snow Avalanches by NDMA (2009)
62. Landslide Hazard Zonation Map can be seen at http://www.bmtpc.org/topics.aspx?mid=56 and Mid1=186 and http://www.ndma.gov.in/en/landslides-zone-map.html
63. Ibid
64. http://drdo.gov.in/drdo/labs/SASE
65. Ibid From: http://www.eoearth.org/view/article/149954/
67. http://www.crida.in/

68. www.isro.gov.in/applications/disaster-management-support-programme
69. www.fire.uni-freiburg.de/iffn/iffn_34/09-IFFN-34-India.pdf
70. nidm.gov.in/PDF/pubs/Forestper cent.20Fireper cent.202013.pdf
71. http://www.dgca.nic.in/rules/act-ind.htm
72. http://www.scidev.net/global/communication/feature/early-warning-of-disasters-facts-and-figures-1.htm

Chapter 5

COMPONENTS OF DISASTER COMMUNICATION

Emergency Communication

An Emergency communication system (ECS) is "any system (typically, computer-based) that is organized for the primary purpose of supporting one-way and two-way communication of emergency messages between both individuals and groups of individuals. These systems are commonly designed to integrate the cross-communication of messages between varieties of communication technologies, forming a unified communication system intended to optimize communications during emergencies". ECS provides for information, message, voice, data, and video required to be shared during an emergency situation. As situation demands for resiliency such systems need to be disaster resilient, robust, and have sufficient redundancy (WIKIPEDIA).

All disaster events create physical, emotional, and social disarray. Communications become very important during such crisis to bring back things in order. Communications during emergencies employ a wide range of measures to manage the situation. Telecommunication sources such as telecommunication satellite, radar, telemetry, meteorology and remote sensing make early warning possible and facilitate early actions before disasters strike. Telecommunications is a conduit for disseminating information about the impending danger, and thus, making it possible for people to take the necessary precautions to mitigate the impact of hazards. Other telecommunication applications, including remote sensing and global positioning system (GPS), have critical roles to play in tracking approaching hazards, alerting authorities, warning affected populations, coordinating relief operations, assessing damages and mobilising support for reconstruction; thus it is evident that telecommunications play a pivotal role in disaster prevention, mitigation, and management (Fakaruddin, 2007).

Communication facilities are essential for pre-disaster alerts and during disaster information sharing. Emergency telecommunications play a critical role in the immediate aftermath of disasters for ensuring timely flow of vital information which is much needed by government agencies, and other humanitarian actors involved in rescue operations and providing medical assistance to the injured.[1] International Telecommunications Union (UN-ITU) Recommendation ITU-R F.1105 on "Transportable Fixed Radio Communication Equipment for Relief Operation" deals with various types of transportable wireless equipment and their broad characteristics.

Communication Definitions and Types

Communication broadly means exchange of information by speaking, writing or using some other medium. Information can be voice, data (in any form may be text as well) or video signals and can be in two forms, analog or digital. Analog signal is considered continuous, on the contrary, digital signal is non-continuous characterised by two states "1" or "0" (indicating presence or absence of a voltage for example). Natural voice is analog in nature and data is digital in nature. However, now voice, by various sampling and coding techniques, is converted in to digital signal. Since modern telecommunication networks are digital, voice is invariably coded as '1' and '0' for its transmission in the network. Therefore voice and data are no more segregated for transmission networks. With new technologies of next generation networks, voice is interchangeable to data for communication purpose.

The term 'communication' originates from the Latin word communicare, which means to share or impart. When used as per its function, it means a common ground of understanding.[2] Communication is the process of exchanging facts, ideas and opinions and a means that individuals or organisations use for sharing meaning and understanding with one another. In other words, it is the transmission and reception of facts, ideas, opinions, feelings or attitudes. Communication is an interdisciplinary concept as theoretically it is approached from various disciplines such as mathematics, accounting, psychology, ecology, linguistics, systems analysis, etymology, cybernetics, auditing etc. Communication enables us to do important things: to grow, to learn, to be aware of ourselves and to adjust to our environment.

- ☆ As per a generic definition, 'communication' is "*any act by which one person gives to or receives from person information about that person's needs, desires, perceptions, knowledge, or affective states. Communication may be intentional or unintentional, may involve conventional or unconventional signals, may take linguistic or nonlinguistic forms, and may occur through spoken or other modes*".
- ☆ Communication is the process of transferring information from a sender to a receiver with the use of a medium wherein the communicated information is understood by both sender and receiver.

- In its simplest form communication is the use of words to describe and convey a message or give information to another person. We communicate using language as a code to share information, ideas and feelings.
- Telecommunication system consists of three main components: namely (i) the information, (also called baseband), (ii) the medium and (iii) the carrier. The medium is the entity the signal travels through. It can be air, space, copper wire or optic fiber. Each of these media offers its own unique set of advantages and disadvantages (noise and signal distortions). Carrier, as the name suggests, carries the information through the medium to be delivered at the other end. Selection of a transmission medium decides the frequency range used for carrying information for example if optical fiber is used as communication medium the transmission signal is light rays, thus electromagnetic frequency range is in the optical spectrum. The communication information which is transmitted over a medium, for many scientific reasons, (which are beyond scope of this chapter and are not explained here) are modulated, multiplexed, coded and then transmitted. At the receiver side same process in reverse order takes place to retrieve original information.
- *A communication model developed by Claude Shannon and Warren Weaver is depicted in the Figure 5.1 (1949 The mathematical theory of communication) to explain how communication works:*

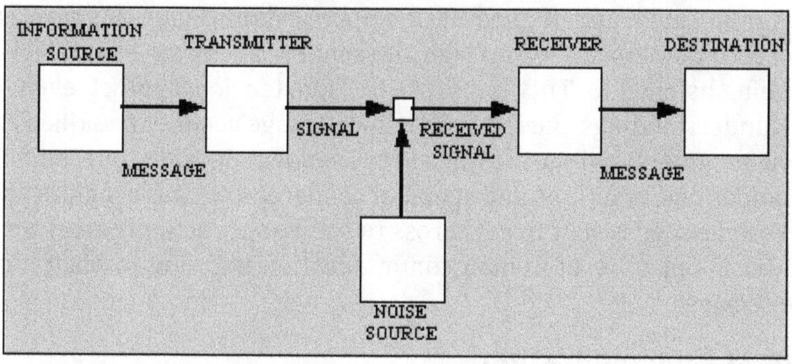

Figure 5.1: Shannon's (1948) Model of the Communication Process.[3]

- Yet another definition says that 'communication' is giving, receiving or exchanging ideas, information, signals or messages through appropriate media, enabling individuals or groups to persuade, to seek information, to give information or to express emotions. This broad definition includes body-language, skills of speaking and writing. It outlines the objectives of communication. It emphasises listening as an important aspect of communication.[4]

Therefore, communication is a continuous method of exchanging messages which are verbal and non verbal. Message is a precondition of communication and it must be expressed through some method to the receiver. It is very vital that the receiver fully understands this message completely in the same way as it was meant to be understood by the sender and after receiving a message, depending upon requirement of communication (point to point/point to multipoint; real time/store and forward message transfer), receiver may/may not reply back or acknowledge the reception of message to transmitter within a stipulated time. Communication is assumed to be the basic building block for social entities, such as personal relationships and organisations. Communication is thus a process by which meaning is assigned and conveyed in an attempt to create shared understanding. This process, which requires a vast range of skills in interpersonal information sharing, listening, observing, speaking, reading, processing, questioning, analysing, gestures, and evaluating enables collaboration and cooperation.[5]

Verbal Communication

Verbal communication occurs when a person puts across his/her message by speaking. The person sending the message is expected to be able to convey a message which clearly expresses intended expressions of a communication, namely feelings, needs, wants, hopes, messages, expectations, values, beliefs and thoughts using a common language between sender and receiver. The receiver has to be able to listen to the information, understand the message which was meant to be communicated, and, if need be, act upon/acknowledge the message. If the receiver misunderstands the message, the sender can get upset and feel like they are not being listened to. This can cause friction, tension, conflict, even anger. To avoid misunderstandings, the receiver of the message needs to be skilled enough to know how to understand the message of the sender.[6] Besides, tone, pitch, volume, voice modulations, emotions and speed of sender's voice has a significant impact on how the message is sent to get across to someone is actually received by him/her. The basic objective of human communication is trying to elicit a response from the receiver.

Non-verbal Communication

It is commonly said that only one third of a message that is sent in a person-to-person exchange is in words alone, the other two thirds of the message is made up of non-verbal communication. There are many types of non-verbal communication which inter alia include[7]:

- ☆ *Body language;*
- ☆ *Physical characteristics and appearance;*
- ☆ *Personal space; and*
- ☆ *Environment.*

Nonverbal communication is the process of communication through sending and receiving wordless (mostly visual) cues between people. It is sometimes mistakenly referred to as body language (kinesics), but nonverbal communication encompasses much more, such as use of voice (paralanguage), touch (haptics),[8] *distance and physical environments/appearance.*[9] *Typically overlooked in nonverbal communication is proxemics, or the informal space around the body and Chronemics* [10] *the use of time. Not only considered eye contact, actions of looking while talking and listening, frequency of glances, patterns of fixation, pupil dilation, and blink rate etc. (WIKIPEDIA).*

Even speech contains nonverbal elements known as paralanguage, including voice quality, rate, pitch, volume, and speaking style, as well as rhythm, intonation, and stress. Likewise, written texts have nonverbal elements such as handwriting style, selection of words, length of sentences, spatial arrangement of words, selection and highlighting of words, or the physical layout of a page. Only a small percentage of the brain processes verbal communication. As infants, nonverbal communication is learned from social-emotional communication, making the face rather than words the major organ of communication. As children become verbal communicators, they begin to look at facial expressions, vocal tones, and other nonverbal elements more subconsciously

Nonverbal communication involves the processes of encoding and decoding. Encoding is the act of generating the information such as facial expressions, gestures, and postures. Decoding is the interpretation of received information; it is based on receiver's abilities to comprehend, interrelate and assimilate nuances of non verbal communication, based on previous experiences.

Written Communication

Over a long period of time communication as a subject has evolved through the continuing progression of technology. Progression of written communication may be divided into three revolutionary stages called as Information Communication Revolutions.[11].[12] *During the first stage, written communication first emerged through the use of pictographs. The pictograms were made in stone; hence written communication was not yet mobile.*[13] During the second stage, writing began to appear on paper, papyrus, clay, wax, *etc.* The third stage is characterised by the transfer of information through controlled waves and electronic signals; telegram, telex, fax and now emails, chats are examples of transfer of written messages through various mode of communications.

Written communication involves any type of interaction that makes use of the written word.[14] Communicating through writing is essential in the modern world and is becoming ever more so as we are now living in information age. In fact, written communication is the most common form of business communication. One of the various forms of written communication that are used internally for business operations include internal memos, reports, duty sheets, duty charts, bulletins, job descriptions, employee manuals, written feedbacks, evaluation sheets, meeting minutes, notice, circulars, orders, e-mail, and Instant Messages.

Business Communication

As the name suggest, when communication is related to business purpose namely management addresses, internal memos, reports, duty sheets, duty charts, bulletins, job descriptions, employee manuals, written feedbacks, evaluation sheets, meeting minutes, notice, circulars, orders, e-mail, instant messages, business policies, business reports, bulletins *etc.* An efficient business communication should be complete, concise, clear, concrete, correct, and courteous. There should be ambiguity and scope for interpretations. Unclear, inaccurate, or inconsiderate business communication can waste valuable time, alienate employees or customers, and destroy goodwill toward management or the overall business. The importance of business communication has been greatly emphasised by all management experts and is considered to be an important subject being taught in management studies. The modern business setups rely heavily on communication for its survival and growth. Right information flow within and out of an organization is considered to be the lifeline of an organization.

Effective and Ineffective Communication

Communication is transfer of information from one entity to other through an intended medium. Communication is very crucial and unavoidable, and can be segregated broadly in an effective and an ineffective communication.

Effective Communication

Communication is effective if everything in the communication process goes as planned that is the receiver understands the message in the way it was intended. Effective communication occurs when a desired effect is the result of intentional or unintentional information sharing in a desired way. It means messages are not altered/distorted/lost during the communication process. Effective communication should elicit the desired response. Therefore, effective communication serves the purpose for which it was planned or designed. Possible purposes might be to elicit change, generate action, create understanding, inform or communicate a certain idea or point of view.[15] Professionally, this allows proper procedures to be executed correctly and on time, increasing productivity and saving money. Personally, it prevents misunderstandings and can enrich the overall process of information sharing. Understanding strategies for effective communication such as listening, focusing on the others in the process, and asking questions for clarification improve chances for effective communication.

Ineffective Communication

Communication becomes ineffective when it does not go as per plan, it means it is not received and acted upon in a manner it was intended to be. There may be numerous reasons for a communication to be ineffective, for example if a sender speaks/writes in a language the receiver doesn't understand, the communication fails. Writing a message to a person who has difficulties reading/understanding language in which communication is written, is also ineffective communication. Disruption or failure in the communication process, communication system, mode of communication *etc.* may cause ineffective communication. An ineffective communication may cost heavily on time, money, resources, and overall energy of the communication system.

Crisis and Emergency Risk Communication

Risk communication is a field wherein the communicator hopes to provide the receiver with information about an expected outcome (good or bad) and magnitude (weak or strong). Usually, it is a communication about an expected adverse outcome and the probability of that outcome occurring. Crisis and emergency risk communication encompasses the urgency of disaster communication with the need to communicate risks and benefits to stakeholders and the public. Crisis and emergency risk communication differs from crisis communication in that the communicator is not perceived as a participant in the crisis or disaster, except as an agent to resolve the crisis or emergency. Crisis and emergency risk communication is the effort by experts to provide information to allow an individual, stakeholder, or an entire community to make the best possible decisions about their well-being within very less time constraints. Since time taken for a decision is very less the decision may be irreversible, the outcome of the decision may be uncertain, and the decision may need to be made with imperfect or incomplete information (CDCP, 2002).[16]

Shannon's Model of the Communication Process

Shannon's (1948) model of the communication process may by many ways, may be considered as the beginning of the modern field. It provided, for the first time, a general model of the communication process that could be treated as the common ground for many diverse disciplines such as journalism, public speaking, linguistics etc. Part of its success is due to its structural manifestation of communication composed of basic constituents that not only explain how communication happens, but also why communication sometimes fails. The Shannon's Model of the Communication Process due to its wider applications has got specific mention in many introductory textbooks on communication. Shannon's Model of the Communication is found in introductory textbooks in both mass communication and interpersonal communication.[17]

Shannon's model, as shown in Figure 5.1, breaks the process of communication down into eight discrete components:

1. An information *source*. Presumably a person/machine/system that creates a message.
2. The *message,* which is both sent by the information source and received by the destination.

3. A *transmitter.* For Shannon's immediate purpose a telephone instrument that captures an audio signal, converts it into an electronic signal, and amplifies it for transmission through the telephone network. Transmission is readily generalised within Shannon's information theory to encompass a wide range of transmitters. The simplest transmission system, which associated with face-to-face communication, has at least two layers of transmission. The first, the mouth (sound) and body (gesture), create and modulate a signal. The second layer, which might also be described as a channel, a medium is built of the air (sound) and light (gesture) that enable the transmission of those signals from one person to another. However new generation of communication systems include many more layers, with the addition of cameras and microphones, digital systems and ICT tools.
4. The *signal,* which flows through a channel. There may be multiple parallel signals, as is the case in face-to-face interaction where sound and gesture involve different signal systems that depend on different channels and modes of transmission. There may be multiple serial signals, with sound and/or gesture turned into electronic signals, radio waves. Now in time of high bit rate digital transmission, original signal is modulated, multiplexed and coded as a bit streams of "1" and "0".
5. A carrier or *channel,* which is represented by the small unlabeled box in the middle of the model. The most commonly used channels include air, space, wire, fibre, light, electricity, radio waves, paper, and postal systems. There may be multiple channels associated with the multiple layers of transmission and combination of different channels and carriers are used in today's communication systems.
6. *Noise,* in the form of secondary signals that obscure or confuse the signal. Given Shannon's focus on telephone transmission, carriers, and reception, it should not be surprising that noise is restricted to noise that obscures or destroy or distort some portion of the signal within the channel.
7. A *receiver.* In Shannon's conception, the receiving telephone instrument. In face to face communication a set of ears (sound) and eyes (gesture). In modern systems several layers of receiver, including an antenna and a television set with set top box or DTH in case of television transmission; in Internet systems a laptop/desktop/phone and in mobile technologies a mobile handset.
8. A *destination.* Presumably a person/machine/system which consumes and processes the message.

Connotations of various Components of Communication Process, involved in personal communication with explanation of main components of the communication process are following:

Context

Communication is influenced by the situation, ambience and ecosystem in which it happens and this may be cultural, political, social, economical, or physical. Hence, any communication continues depending on the situation or context and the sender of the message selects it in reference with the context.

Sender/Encoder

The sender or the encoder is an individual who conveys the message and to do this he uses certain symbols or signs such as visual aids, graphics, words, texts, body language, gestures, gaps and silence in order to convey the message and to receive the needed response. The sender may be a person or he may belong to a group or to an organization. The impact of the message depends highly on the views, approach, skills and competencies, background and the experience or knowledge of the sender.

Message

The message is the main information which the sender wants to communicate the recipient. The process of communication starts around the subject matter of the message to be communicated and therefore, the main goal of transmission of the message is that it must be understandable.

Medium

Medium is a method that is utilised for transmitting or exchanging the message. Hence, it is important that the sender must select a proper medium for conveying the message or else it may be misinterpreted or not conveyed properly to the recipients. For this reason, it is important to appropriately select the medium in order to convey the message efficiently and the recipient to be able to interpret it correctly and respond at par with expectations. The mode and choice of communication differs and is dependent on its features. For example, when the message has to be conveyed to a tiny group of persons the written medium is selected, whereas when an instant feedback is needed from the recipient a vocal medium is selected to avoid any future misunderstandings.

Recipient/Decoder

It is a person or a group for whom the message is meant or intended or expected to be conveyed. The extent in which the recipient understands the message depends upon many factors like the recipient's knowledge, capability to understand, assimilation ability, correlation with past experiences, capacity of value addition *etc.* which culminates in appropriate response to the received message.

Feedback

It is the main element of communication as it gives permission to the sender to evaluate the effectiveness of the message. This aids the sender in being able to approve the proper understanding by the decoder. The feedback received may be either verbal/oral or non-verbal (in the way of gestures, sighs, smiles or facial expressions *etc.*). It may also be through written form using reports, memos, notes, emails *etc.*

Telecommunication

As the name indicates, "*telecommunication is communication at a distance with help of some means such as cables, wires, optical fibers, antennas, radio or a wide range of electronic mediums. Now we may say that in a broad sense communication refers to sending, receiving, processing of information by electronic means*" (Davis, 1999). Therefore "*A Telecommunication Network is a collection of elements necessary to provide telecommunications services. These elements vary depending on the type and Size of networks. Telecommunication network generally include switches, transmission equipments, operations systems, adjuncts, and databases. When a user wants to be served, the user indicates this desire to a network and receives an indication that this service can be provided by a telecommunication service provider. Access to facilities of telecommunication network occurs over an access transmission line which is called last mile connectivity.*" If this connectivity is a wire of any type (copper, fiber), the network access is wire line, or else if it consists of radio transmission equipment, the local loop is wireless".[18] A network is set of transmission links and switches/routers as well as control software meant to connect many geographically distributed users.

Shannon's model, as shown in Figure 5.1, breaks the process of communication down into eight discrete components and is very generic and simplest explanation of telecommunication process. As technology advanced, basic components of ibid simple telephony model became more complex and efficient, its message carrying capacities enhanced by manifolds, however its basic type and purpose remained same. Based on; types of technologies used, medium of communication, mode of communication and types of sender (Communication Transmitter) and receiver (Telecommunication Receiver) changes.

Initial telecommunication started with manual trunk board services, wherein two users were connected via human operator intervention and all telephone lines were distributed and terminated in a board. These manual boards used to be of capacities 5-100 lines. Subsequently capacities of such boards were enhanced through multiplications of operators and extension of boards. Long distance calls also started through these boards. Size of such boards became very long with extension of number of subscribers and then came automatic dialing switches such as strowger telephone exchanges and X-bar telephone exchanges. These were first generation of automatic telephone switching systems. These were also called analog switching systems. Subsequently as technology advanced, analog systems got converted in to software based digital communications networks. A digital telephone network communicates information between users in fractions of a second using digital transmission technology.

Similarly roots of data (initially non verbal messages only) transmission may be traced back to telegrams, which subsequently advanced to telefax, ISDN calls, video conferencing, internet, emails, chats, now real-time multimedia streaming. Now a day there is no segregation between voice calls and data transfers every mode of telecommunication has now been converted into data transmission. This technology convergence is due to end to end digitization of communication technologies.

Telecommunication network based on coverage may again broadly be classified as Personal Area Network, Local Area Network or Wide Area Network *etc.* Networks may also be classified based on communication technologies used such as satellite, landline, or undersea, fiber, copper, or wireless, it provides connectivity among the community of users and host computers.

Wired and Wireless Communication

Traditionally the telecommunication was through a telephone network catering to the needs of voice telephony. Network was providing interconnection between various customers for speaking to each other. This collection of interconnected systems, or network in broader sense, was called Public Switched Telephone Network (PSTN). Public switched as it was meant to serve public's requirement of communication through this network. PSTN normally refers to voice connectivity. The PSTN started as human-operated analogue circuit switching systems (plug boards), progressed through electromechanical switches to software based modern electronic switches. Now the pure telephone networks do not exist, as technological advancement has enabled telecommunication networks to offer many services other than voice. With rapid technological advancement, digital switches came in to being. Digital switches came in early 90s in India; the French technology based E10 B system from Alcatel was first digital software switching system introduced in India. Very soon came, other digital technologies, which initiated the first step towards the convergence of voice and data by introducing Integrated Services Digital Network (ISDN).[19]

Based on communication channel used for transfer of signal/message/ information, from sender point to receiver point, a telephone network can be divided in to wire line and wireless communication network. There are various technological options available for both kinds of communications. The main difference between a wired and wireless data communication infrastructure is the existence of physical cabling. The same or similar techniques are employed for both types of communication infrastructure in terms of the core elements of essential network services.

Wired Communication Network

A wired network uses copper cables/wires/fibers as medium of communication. Wire line networks; serve users through a series of network elements and interconnecting lines. A user is connected to a network via a hard wire over which all necessary communications between the user and the network take place.

Therefore wired communication refers to the transmission of data over a wire-based communication technology such as telephone networks, cable television or internet access through Ethernet cables[20], Local Area Networks connected through wires and fiber-optic communication *etc.* Wired networks have been around for quite some time now. A wired network takes a lot of time for physical installation of communication network, as it involves laying of cables/wires/fibers. Right of way (permission to lay cables from civic authorities) is a major bottleneck for expansion of a wired line network. Wired networks do not offer users, facility of mobility while they communicate, because of the wiring, except for cordless phones which offer mobility within the premise to an extent of some meters only.

Wired networks offer better transmission speeds than wireless ones. In a wired network, a user does not have to share access to network with other users and thus gets dedicated speed of transfer of information through wired communication channel, while in wireless networks, the same connection may be shared by multiple users. A wired network provides high performance and better security than wireless networks.

Wireless Communication Network

Wireless networks do not use any form of cable. The transmission of data occurs over radio waves[21]. In wireless network connectivity of user device with core network that is access to network facility is not connected with physical cable/wire/fiber. Therefore susceptibility to physical disconnection, which leads to interruption of communication, is not a problem in wireless communication. A wireless network saves time and efforts in installing the lot of cables. The major advantage of having a wireless device is the mobility and freedom that comes with it. Moreover, there is less clutter and fewer wires to worry about. However now a day, with increase in public awareness to EMF radiations, installation of telecom towers in residential areas are getting difficult.

Wireless network, as compared to wire line network offer lesser transmission speed and security. However, with advent of new coding mechanism, wireless networks are capable of offering higher speed, mobile GPRS, 3 G, 4G, 5G systems are examples of high speed and secure wireless communications.

Core Network and Access Network

Telecommunication network can be divided into core network and access network. Access network is the part of a telecommunications network which connects subscribers to their immediate telecom service provider. In other words access network is the part of the network through which users get access to the network facilities/services. It is contrasted with the core network, which connects core network elements required for service provisioning. The access network may be further divided between distribution network, and edge network.

Core Network

A core network, or network core, is the central part of a telecommunication service-provisioning network. Typically the term refers to the high capacity communication facilities that connect primary nodes. The term Core/Backbone networks are interchangeably used. The devices and facilities in the core/backbone networks are switches and routers. Core/backbone network provides paths for the exchange of information between different sub-networks. Many main service providers would have their own core/backbone networks, which are interconnected. Some large enterprises have their own core/backbone network, which are typically connected to the public networks. One of the main functions of the core network is to route information from one end to the destination namely telephone calls across the PSTN or data communication from one machine (may be computer) to another device. Normally backbone network is used for central network components that ties various edge network components, for example part of a Wide Area Network (WAN)connecting server clusters needs to be accessed by different departments, people located at different geographical locations and connected through the WAN. The latest trend is to push the intelligence and decision making into access and edge devices and keep the core devices dumb and fast. As a result, switches are more and more often used in the core/backbone network facilities. Technologies used in the core and backbone facilities are data link layer and network layer technologies such as SONET, DWDM, ATM, IP, *etc.* For enterprise backbone network, Gigabit Ethernet or 10 Gigabit Ethernet technologies are also often used.

Core networks typically providing the following functionality:[22]

i. *Aggregation:* The core network is the highest level of aggregation in a service provider network. The next level in the network hierarchy level are the distribution networks and then the edge networks. Customer-premises equipment (CPE), do not normally connect to the core networks of a large service provider. CPE are connected to core network through access network.

ii. *Authentication:* The function to decide whether the user requesting a service from the telecom network is authorized to do so within this network or not.

iii. *Call Control/Switching:* call control or switching functionality decides the future course of call based on the call signaling processing. *e.g.* switching functionality may decide based on the "called number" that the call be routed towards subscriber within this operator's network or more prevalent to another operator's network.

iv. *Charging:* This functionality handles the collation and processing of charging data generated by various network nodes. Two common types of charging mechanisms found in present day networks are prepaid charging and postpaid charging.

v. *Service Invocation:* Core network performs the task of service invocation for its subscribers. Service invocation may happen based on some explicit action (*e.g.,* call transfer) by user or implicitly (call waiting).

vi. *Gateways:* Gateways shall be present in the core network to access other networks. Gateway functionality is dependent on the type of network it interfaces with.

Access Network

Access network is the part of the network through which users get access to the network. Access network, based on type of connectivity and services provided to the subscriber, can further be divided in to wired and wireless network. In fixed/wired network, access network or outside plant refers to the series of wires, cables and equipment lying between a consumer/business telephone termination point (the point at which a telephone connection reaches the customer) and the local telephone exchange. Access network connects directly to the end user or customer. It connects subscribers to their immediate service provider.

Wired Access Network

Wired access networks consist largely of pairs of copper wires, each traveling in a direct path between the exchange and the customer. The access network is perhaps one of the oldest assets a telecoms operator owns, and is constantly evolving, growing as new customers are connected, and as new services are introduced. Access networks around the world for wired line communication, has now evolved to include more and more optical fibre. Optical fibre already makes up the majority of core networks and has started coming closer and closer to the customer premises. Some contemporary Optical fiber access network components are:

(i) Passive Optical Networks

Providing the triple play services of voice, video, and high-speed data access is an important way for carriers to increase their revenue and compete with other access providers. Passive Optical Network (PON) is a single, shared optical fiber that uses inexpensive optical splitters to divide the single fiber into separate strands feeding individual subscribers. PONS are called "Passive" because, other than at central office and subscriber endpoints, there are no active electronics within the access network

(ii) EPON (Ethernet Passive Optical Network)

Ethernet in the First mile PON builds a point-to-multi-point fiber topology that supports a speed of 1 Gbps (speed of Giga bits per second) for a distance up to 20 km. While subscribers are connected via dedicated distribution fibers to the site, they share the Optical Distribution Network (ODN) trunk fiber back to the Central Office.

(iii) GPON (Gigabit Passive Optical Networks)

GPON provides a capacity boosts in both the total bandwidth and bandwidth efficiency through the use of larger, variable-length packets in PON technology. GPON is standardised in ITU-T G.984 (GPON) that permits several choices of bit rate. GPON Encapsulation Method (GEM) allows very efficient packaging of user traffic, with frame segmentation to allow for higher QoS(Quality of Service) for delay-sensitive traffic such as voice and video communications. GPON not only provides substantially higher efficiency as a transport network, but also delivers simplicity and superb scalability for future expansion in supporting additional services. GPON, through the Generic Framing Procedure (GFP)-based adaptation method, offers a clear migration path for adding services onto the PON without disrupting existing equipment or altering the transport layer in any way.

Wireless Access Network

A radio access network (RAN)[23] is part of a mobile telecommunication system. It implements a radio access technology. Conceptually, it resides between a devices such as a mobile phone, a computer, or any remotely controlled machine and provides connection with its core network. Wireless access points (APs or WAPs), working on wireless protocols are specially configured devices on wireless local area networks (WLANs). Access points act as a central transmitter and receiver of wireless radio signals including Wi-Fi. APs are most commonly used to support public Internet hotspots and also on internal business networks to extend their Wi-Fi signal range.[24]

Wireless technologies enable one or more devices to communicate without an actual wired connection. Radio frequency is used to transmit the data. Such technologies are rapidly evolving to meet a variety of communications needs, from simple to complex. Wireless communications networks may be classified based on the distance covered/area served: wireless personal area networks (WPAN), wireless local area networks (WLAN), and wireless wide area networks (WWAN). WPANs are limited to distances under about 10 meters and include technologies such as infrared (IR), Bluetooth technology, and ultra-wideband (UWB). WLANs cover a local area with distances of individual access points reaching to about 100 meters, and include technologies such as Wi-Fi. WWANs cover even larger areas, using cellular data networks.[25]

WPANs typically provide dynamic network connections between devices within close range of each other. These connections are different because they do not generally need to connect to any network infrastructure to operate. They can simply connect to each other and perform necessary communications without the need of any access network devices, such as access points or base stations. Bluetooth is most widely used WPAN network standard. Bluetooth network applications include wireless headsets, hands-free operation, wireless synchronization, wireless printing, and many types of file transfer.

WLANs provide connections designed to connect devices to wired networks. Unlike a wired LAN, a WLAN does not require cabling to connect the device to a switch or router. Devices connect wirelessly to nearby wireless access points that are attached to the local network using an Ethernet connection. A single access point communicates with nearby WLAN devices in a coverage area of about 100 meters. This coverage area allows users to move freely within range of an access point with their notebook computers, handhelds, or other network devices. Multiple access points can be coordinated together by a network WLAN switch to allow users to hand off between access points. Wi-Fi is the set of standards established to define wireless LANs. The ITU 802.11 workgroup currently documents use in five distinct frequency ranges: 2.4GHz, 3.6 GHz, 4.9 GHz, 5 GHz, and 5.9 GHz bands. These frequencies are free and spectrum allocation/user charges are not required for using these frequencies for Wi-Fi. With increasing demand of Wi-Fi, Telecom Regulatory Authority of India (TRAI) is now recommending allocation of higher order frequencies wideband for utilization in Wi-Fi.

WWANs provide broadband data networks with a far greater range, using cellular technologies such as GPRS, HSPA, UMTS, 1xRTT, 1xEV-DO, and LTE. Wireless data devices connect to a wireless broadband network through a Telecom Service Provider (TSP) data network, allowing broadband performance without the need for a cabled connection to a network infrastructure (much like a WLAN), while providing end users with far greater mobility.

(iv) GSM

The Global System for Mobile communication (GSM), is the most widely used cellular technology in use in the world today. The letters GSM originally stood for the words *Groupe Speciale Mobile*, but as it became clear that the cellular technology was being used worldwide the meaning of GSM was changed to Global System for Mobile Communications. It has been very successful cellular phone technology for a variety of reasons including the ability to roam worldwide with the certainty of being able to operate on GSM networks. Since this cellular technology was first deployed in 1991, the use of GSM has grown steadily, and it is now the most widely used cell phone system in the world. GSM use digital cellular technology and time division multiple access (TDMA) transmission methods. GSM offers data transmission rates of up to 9.6 Kbps, while enabling such services as short messaging service (SMS) or text messaging, as it is more commonly known, and international roaming.

Subsequently 2G GSM evolved to 3G. Wideband Code Division Multiple Access (W-CDMA) brings GSM into 3G. W-CDMA is a type of 3G cellular network and is a high-speed transmission protocol used in Universal Mobile Telecommunications System (UMTS). UMTS offers packet-based transmission for text, digitized voice, video, and multimedia content.

High-Speed Packet Access (HSPA) is mobile telephony protocols that further enhance data communication performance of UMTS. HSPA uses improved modulation schemes, and protocols that mobile devices and base stations use to communicate. These processes improve radio bandwidth utilization provided by UMTS. High-Speed Downlink Packet Access (HSDPA) and High-Speed Uplink Packet Access (HSUPA) are 3G mobile telecommunications protocols from the HSPA mobile protocol family. HSDPA and HSUPA enables higher data transfer speeds and capacity in UMTS-based networks. Evolved HSPA (HSPA+) is a wireless broadband standard that provides peak speeds of up to 42 Mbps on the downlink

(v) *CDMA*

CDMA uses a "spread-spectrum" technique which makes electromagnetic energy to spread to allow for a signal with a wider bandwidth. This entails multiple people on multiple cell phones to be "multiplexed" over the same channel to share a bandwidth of frequencies. With CDMA technology, data and voice packets are separated using codes and then transmitted using a wide frequency range. Since more space is often allocated for data with CDMA, this standard became attractive for 3G high-speed mobile Internet use.

First-generation (1G) radio networks were analog-based and limited to voice services and capabilities only. 1G technology was vastly inferior to today's technology.Then came 2G; Second-generation (2G) CDMA-based wireless networks, known as cdmaOne, proved their effectiveness in delivering high-quality voice traffic to subscribers. 2G networks made the transition from analog signals to all-digital signals, expanding network capabilities to include both voice and data services1xEV-DO is the broadband wireless network standard developed by the Third-Generation Partnership Project 2 (3GPP2) as part of the CDMA2000 family of standards. As the use and number of wireless devices increased, more and more demands were placed on the underlying technologies to deliver enhanced capabilities and services. This section discusses the evolution of WWAN technologies and their capabilities. CDMA2000 1X EV-DO Rel. 0 EV-DO Rev. A CDMA IS-95-A LTE.

(vi) *4G Mobile Broadband Technology*

Basic GSM and CDMA mobile communication technologies have evolved and are continuously evolving from their original prototypes. Mobile technologies are indexed wide different technological generation such as 2G, 3G, 4G and 5G. In GSM technology there is 2.5 G also which is known for GPRS/EDGE systems for enhanced data rate. With advancement of technologies segregation between basic GSM/CDMA are blurred and current technologies LTE (Long Term Evolution) or the E-UTRAN (Evolved Universal Terrestrial Access Network), introduced in 3GPP R8, are the access part of the Evolved Packet System (EPS) being employed in mobile access networks. The main requirements for the new access network are high spectral efficiency, high peak data rates, short round trip time as well as flexibility in frequency and bandwidth.

4G technology standards offer greatly improved data rates over previous generations of wireless technology. Faster wireless broadband connections enable higher-level data services, including business applications, streamed audio and video, video messaging, video telephony, mobile TV, and gaming. LTE is a global 4G standard. The LTE standard is integrated with GPRS/UMTS networks and represents an evolution of radio access technologies and networks for UMTS. The 4G channel offers four times more bandwidth than current 3G systems and is scalable. It is very efficient in spectrum utilization. Improved spectrum efficiency allows more information to be transmitted in a given bandwidth, while increasing the number of users and services the network can support. Better spectral efficiency of LTE has also increased its coverage area, which is most desirable aspect in wireless access technology. This is pure IP technology and offers better data security.

LTE network uses all IP network architecture. Due to this fact, it is dedicated for packet switched operations. It supports data as well voice. The voice can be transported using voice over LTE protocols (VoLTE) and fall-back to legacy networks (*i.e.* 2G/3G GSM/W-CDMA). In LTE mobile terminals can have low power during transmissions and hence battery life can be enhanced on user side. LTE works faster therefore average network connection time is reduced and results in decrease traffic load on the LTE network. Enhanced data speed ensures live data streaming. Higher versions of LTE such as LTE advanced will further improve the performance of existing LTE standard based products.

Local Area Network (LAN)

A LAN is a type of communication network that is usually owned and operated by such consumers whose requirement is to share and process information amongst various user terminals belonging to similar institute/organization/business. A LAN enables many independent peripheral devices, such as computer terminals, Printers, servers, workstations, other personal computing devices and processing units to be linked to a network through which they can share expensive central processing units. These devices communicate over short distances may be from 10s of meters to maximum 1-2 Km. Local networks can be categorized according to their transmission bandwidth and configurations. In wired broadband networks, the capacity of the cable is divided into many channels, which can transmit higher speed signals. In the past, wired LAN used to operate over coaxial cables now Ethernet over optical fiber is increasingly being preferred. Fiber optic cables initially cost more than coaxial cable; however, the fiber optic cable allows faster communication and now with economy of scale of catching up; making fibers more cost effective. As explained above, WLANs provide connections designed to connect devices to wired networks. Unlike a wired LAN, a WLAN does not require cabling to connect the device to a switch or router. Devices connect wirelessly to nearby wireless access points that are attached to the local network using an Ethernet connection

Telecom Infrastructure

Telecommunication started with transmission of short written text messages (Telegraphy) and transmission of voice over a distance (Manual Telephony). It required a series of keying equipments, boards and wires to carry out the task.

Thereafter the automatic telephone exchanges followed, where with the help of movement of mechanical relays, user could dial and could get connected to the telephone of intended receiver. It involved a lot of technological acumen to convert manual telephony to automatic electro-mechanical telephone exchanges. Subsequently software based electronic telephone exchanges were designed. Electronic exchanges paved way for data communication along with voice communication. Transmission of information initially started with copper cables, subsequently with technological advancements new technologies such as microwave, optical fibers, satellite, DSL, WiMax, Wi-Fi, Bluetooth *etc.* took charge as transmission medium for voice and data (NDMICS (2011).

Succeeding part of this chapter describes about some important components of telecom infrastructure. Telecom infrastructure may further be divided amongst active and passive infrastructure based on method of power supply or energizing of components. Physical sites, telecom towers which do not require power/energy to operate and act as support to other network components are considered as passive infrastructure, whereas antenna systems, cables, transmission system and core infrastructure involving switches and networking, are called active infrastructure.

Figure 5.2 shows a telecom tower structure. Telecom tower structure is meant to hoist antenna, support cables, feeders, and waveguides. These equipments are connected to telecom site at ground. A robust tower base with standard foundation is designed for ground towers whereas in case of rooftop towers roofs are strengthen to take load of towers and equipments. Heights of rooftop towers are lesser than ground base towers. Power failure to sites are a major problem in India, thus Diesel Generators (DG Sets) are normally used to supply power to these equipments along with regular power supply to avoid interruption of services due to power failure.

Types of Towers

Several types of towers are used in mobile networks depending on the applicability, antenna load and terrain of the installation site. Some of the popular types of telecom tower structures are given below:

i. *Steel Lattice structure:* Steel lattice structures are modular in construction and are made up of latticed steel elements. Such structures are usually triangular or square in cross-section. Steel lattice constructions are very widely used since they offer the advantage of minimum wind resistance and easy maintenance. These towers have good strength, low cost and can be erected very quickly. Most of the power transmission, telecommunication and broadcasting towers are of this type.

ii. *Tubular Steel structure (Monopole):* Tubular steel structures offer high load bearing capacity. These towers are made up of elements with tubular section joined together to get great strength. The sections of tubular towers have their own advantages, but offer low durability especially at

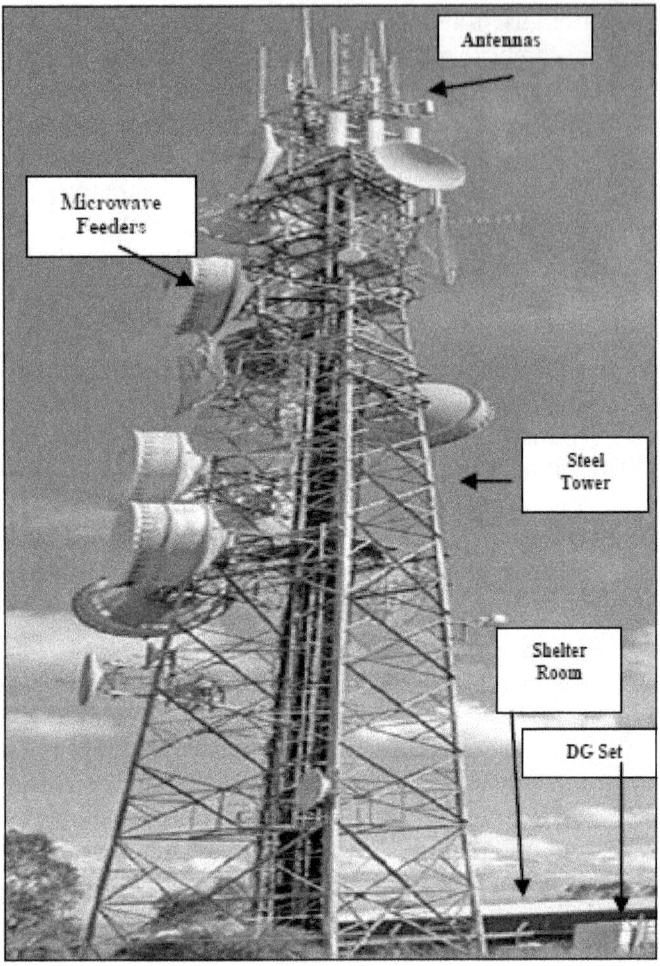

Figure 5.2: Telecom Tower Structure[26,27].

places with high corrosive environment, like a coastal area or an industrial zone.

iii. *Reinforced concrete towers:* These towers are among the most expensive to construct. However, reinforced concrete towers provide maximum rigidity against high wind-loads. These towers can house within it, a control centre to monitor and control operations of the tower and associated equipment. These towers due to its high strength can hoist many antennae, are therefore, considered ideal for telecom tower sharing amongst service providers.

iv. *Fibre glass towers:* These towers are used for very niche and limited applications.

Based on the installation type, mobile towers may be further classified as Ground Based Towers (GBT) or Roof Top Towers (RTT).

(i) Ground-Based Tower: Erected on the ground, ground-based towers (GBTs) are taller (typically 30mts to 110mts) and are mostly used in rural and semi-urban areas because of the easy availability of real-estate space there.

(ii) Roof-Top Tower (RTT): Roof-top towers (RTTs), which are generally placed on the roofs of high-rise buildings, are shorter (than GBTs) and more common in urban and highly populated areas, where there is paucity of real-estate space.

It is the structure of a telecom tower that defines strength of a tower to hold antennae. Therefore structure of a telecom tower determines the number of antennas that can be accommodated, which in turn determines the capacity of the towers, apart from factors such as location and geographical conditions (wind speeds, type of terrain, *etc.*). Hence, typically, while Ground-Based Towers can accommodate up to six tenants, Roof-Top Towers, in order to limit the tower load on the building, can accommodate two to three tenants (NDMICS, 2011).

Antenna and Power Supply System

Antenna System

An antenna is a device used to transmit and receive signals. It does not offer any added power to the signal. Instead, an antenna simply redirects the energy it receives from the transmitter. The antennae are designed in a manner that redirection of this energy has the effect of providing more energy in one direction, and less energy in all other directions (Directional antenna is designed for this purpose). There are three main categories of antennae:-

(i) Omni-directional: Omni-directional antenna radiates Radio Frequency (RF) signals in all directions. These antennae radiate RF in a fashion similar to the way a table or floor lamp radiates light. They are designed to provide general coverage in all directions. The small rubber dipole antenna, often referred to as a "rubber duck" antenna, is the classic example of an Omni-directional antenna.

(ii) Semi-directional: Unlike Omni-directional antennae, semi-directional antennae are designed to direct signals in a specific direction. Semi-directional antennas are used for short- to medium-distance communications. There are three types of antennae that fit into the semi-directional category namely patch antenna, panel antenna, yagi antenna.

(iii) Highly-directional: Highly-directional antennae are strictly used for point-to-point communications, typically to provide network bridging between two buildings. They provide the most focused, narrow beam width of any of the antenna types. Highly-directional antennas radiate RF in a fashion similar to the way a spotlight is designed to focus light on a flag or a sign. There are two types of highly-directional antennas namely parabolic dish and grid antennas.

Power Supply System

i) Conventional

The DC power supply equipment is powered by commercial three-phase A.C. Mains supply. In India and many other under developed/developing countries do face problem of interrupted power supplies. Therefore an on-site Diesel Generator (DG Set) of optimum capacity is installed to provide captive A.C. power when commercial AC supply fails. The Storage Batteries provide DC supply to equipment for 2 - 4 hours in absence of AC input to AC/DC rectifier.

ii) Non-Conventional

Considering the fact that in many under developed parts of the world, people do not have access to reliable power supplies or no access to electricity at all. In order for them to be able to take advantage of mobile communications, base stations have to be powered by such means as diesel-driven generators. However, as diesel prices rise and network infrastructure spreads to more remote areas, other alternatives are required, not only in order to save money, but also to help combat climate change and ensure utilization of green energy sources. GSM Association (GSMA) launched in September 2008 a programme called Green Power for Mobile to promote the use of renewable energy sources by the mobile phone industry. It would give a boost to the provision of mobile phone services in places that cannot yet use them.

a) Solar Power

There is often an abundance of sunlight in rural areas of developing countries, and this, together with the increasing availability of solar equipment and its relatively low running cost, makes solar power a popular choice for sites that need up to 2 kW of power. Solar solutions are less economically attractive for larger sites, the report says, but it foresees that the price of installing solar power is likely to fall in the coming years.

b) Wind

The equipment to trap wind energy is cheaper than for solar-powered stations that have standard load requirements. However, wind power is only viable in such areas as coastal and mountainous regions, where wind blows sufficiently strongly and frequently. In other places, hybrid solutions may be used that combine wind and solar power.

c) Pico-hydro

Pico-hydro refers to very small hydroelectric power generators that typically produce up to 10kW from the energy of streams and rivers. It is a very good alternative energy technology for other applications such as rural electrification and has one of the lowest capital investment requirements of all. But it is usable for a limited number of places.

d) *Biodiesel*

Biodiesel fuel (derived from vegetable oils or animal fats) can be used as a direct replacement for conventional diesel in base station generators, but it is not necessarily a universal solution. Among the factors that must be considered are local access to supplies of biodiesel, and how its production could affect agriculture.

e) *Fuel Cells*

Fuel cells, or batteries, are mainly used as back-up electricity supplies for base stations that have limited power requirements. So far, the commercial viability of using fuel cells as the prime power source has not been greatly tested. However, research and development is being undertaken on the technology.

Optical Fiber Cable Infrastructure for Backhaul[28]

One of the most significant recent developments in communications is fiber-optics technology. Fiber-optic communication is a method of transmitting information from one place to another by sending message in form of pulses of light through an optical fiber. The light forms an electromagnetic carrier wave that is modulated to carry information. Compact light source, optical fiber and photo detectors form basic components of optical fiber communication. Depending upon requirement of transmission, Light Emitting Diodes (LED) or Laser Diodes are selected as light source. Optical fibers have become omnipresent in core and access networks as optical communication permit transmission over longer distances and at higher bandwidths (data rates) than wire cables. Fibers are used instead of metal wires because optical signals travel with lesser loss, better noise immunity, data security and zero electromagnetic interference. When light traveling in an optically dense medium hits a boundary at a steep angle (larger than the critical angle for the boundary), the light is completely reflected. This is called total internal reflection. This effect is used in optical fibers to confine light in the core as is shown in Figure 5.3[29]

Optical fibers typically include a transparent core surrounded by a cladding material also transparent,with a lower index of refraction. Light is kept in the core by total internal reflection. This causes the fiber to act as a waveguide. Fibers that support many propagation paths or transverse modes are called multi-mode fibers (MMF), while those that only support a single mode are called single-mode fibers (SMF). Multi-mode fibers generally have a wider core diameter, and are used for short-distance communication links and for applications where high power must be transmitted. Single-mode fibers are used for most communication links longer than 1,000 meters (3,300 ft).

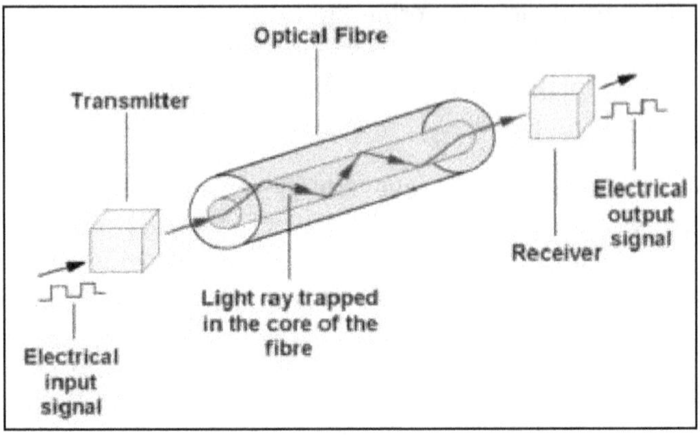

Figure 5.3: Optical Fiber Communications.

Joining lengths of optical fiber is more complex than joining electrical wire or cable. The ends of the fibers must be carefully cleaved, and then carefully spliced together with the cores perfectly aligned. A mechanical splice holds the ends of the fibers together mechanically, while fusion splicing uses heat to fuse the ends of the fibers together. Special optical fiber connectors for temporary or semi-permanent connections are also available. Fiber optic equipment include a wide range of the products, most commonly seen fiber optic equipment in the market are the connectivity fiber optic equipment, media conversion fiber optic equipment, test fiber optic equipment and cable management fiber optic equipment.

Fiber optic equipment for connectivity includes the cables and cable assemblies like fiber patch cables, pigtail cable, and fiber loop back testers. Cables and cable assemblies are used to link between the devices and form the fiber optic equipment network, as we know fiber optic communication use optical fiber as the media to transmit the signal, it is not wireless (NDMICS, 2011).

Telecommunication Network Technologies

Telephone System

The Telephone system was designed to communicate voice message from one place to another. It was considered as a major technological breakthrough, as conversion of voice to electrical signal, transmitting this to a distant point through a medium and again successfully converting electrical signal back to voice signal at receiving end involved considerably high knowledge of applied science. The telephone system since has evolved from a manual board system, to mechanical switches to analog network to digital network to now all IP digital network which can simultaneously carry multiple voice, data and video signals in packet forms over IP.

Analog transmission services used continuous waves or signals (such as the human voice) for which conventional telephone lines were designed, whereas the digital transmission is binary (off/on) in nature. Modems[30] convert the digital signals into analog waves for transmission over conventional analog telephone lines.

Microwave Communication

Microwave radio has been the backbone transmission system of the long distance telephone network for a very long time. Until the mid-1970s, nearly all radio systems were analog, transmitting analog multiplexed signals originating from analog telephone systems. Although such networks provide good voice telephone service, they are currently less than satisfactory for data transmission at high speed and low error rates. Therefore digital microwave transmission systems have replaced all analog systems. These systems are able to detect and reshape the transmitted pulses. The latest development in microwave radio systems expands the bandwidth to match the land-based optical fiber systems. Microwave transmission system as the name suggest, operate in frequency range of 6-7 GHz for intercity transmission of signals and therefore these band is used for microwave backbone network. For access network frequency band of 13GHz, 15 GHz, 17 GHz and 21 GHz are used in India.[31,32]

VHF Radios

Very high frequency (VHF) is the radio frequency range from 30 Mhz to 300 Mhz. Frequencies immediately below VHF are denoted High frequency (HF), and the next higher frequencies are known as Ultra high frequency (UHF). VHF propagation characteristics are ideal for short-distance terrestrial communication, with a range somewhat farther than line-of-sight from the transmitter. Unlike high frequencies (HF), the ionosphere does not usually reflect VHF radio and thus transmissions are restricted to the local area (and do not interfere with transmissions thousands of kilometers away). VHF based communication technologies have been widely used for Disaster communications due to their simplicity of operation and fast deployment and particularly where the towers for mobile network do not exist or disrupted. Defence also widely uses VHF for its ease of deployability.

Portable AM/FM Broadcasting Radio Station

Radio broadcasting is an efficient method for disseminating information at the disaster site. Now completely portable radio station in a box are commercially available, which can be set-up at the disaster site within few hours. Tailor-made programs as per the requirement of disaster site can be transmitted to the community over a large area. The packaged portable radio station typically contains three portable packs namely broadcast studio, transmitter rack and antenna system. The broadcast studio provides features such as audio mixer, CD players, audio recorder, cassette player, monitoring headsets, which are essential for creating the information content.

There are two types of broadcast technologies namely FM (Frequency Modulation) broadcast and AM (Amplitude Modulation) broadcast. AM radio operates in the frequency range 535 to 1705 KHz whereas FM radio coverage is the frequency range of 88 to 108 MHz In AM transmission, coverage distance depends on various factors like soil conductivity, time of day (day/night) and RF propagation condition as the transmission is through reflection of waves by the ionosphere layer (which shifts between day and night). Thus, portable AM equipment range for quality reception is typically around 15 km radius. FM radio broadcast on the other hand depends on line-of-sight transmission and can easily cover a radius of 25km. The power output of a typical transmitter is 500 watts.

Optical Fiber Communication (OFC) Networks

The advent of optical fiber technology has changed the dynamics of telecommunication networks. It has provided solution to the bandwidth limitations imposed by other modes of communication, namely copper cable, microwave and satellite, and it offers large bandwidth capacity at an economical price. Optical fiber communication has wide ranging applications in all types of communications. The economics of optical networking has already resulted in its total dominance of long-distance segment of the global terrestrial and undersea communication network. With exponential advancement of optical technology, its value proposition has equally become significant even for shorter distance applications and the optical network has steadily advanced moving to the customer premises for high bandwidth applications. In OFC communication system, information to be communicated travels in form of light rays for example optical signal. Different voice/data/video signals are multiplexed and converted in to different colors of light (varying in their wavelength/frequency) and are transmitted (WDM). The baseband is formed in OFC networks using SDH (Synchronous Digital Hierarchy) standard, which provides high bit rate up to 10 gigabits. The baseband (electrical signal) is converted into optical signal using lasers and transmitted over the optical fiber. Since transmitted signals are light waves they are free from noises which are introduced due to presence of electromagnetic fields. This is considered to be the most efficient transport system for information and thus now a days all backbone networks are OFC based. Moreover, nowadays, access networks are also being built over various optical technologies, PON/GPON/EPON have already been explained in previous section of the chapter.

Rapid developments in optical technology has brought a revolution to modern day networks with the capability to carry terabits of data employing Dense Wave Division Multiplexing (DWDM) technology, which has increased the capacity of a single strand of fiber by several fold. DWDM makes it possible to carry the entire global traffic (including telephone calls, emails, Web page down load *etc.*) using optic fiber communication. With the deployment of Optical Cross Connects and Optical Add-Drop Multiplexers, wavelengths of light can be added to or dropped from the optical fiber en route between the main hubs. It allows the end user to access wavelengths of light from the optical network directly thus providing enormous bandwidth for various services such as internet, IPTV, voice *etc.*[33]

Propagation Modes through Optical Fibers

Optical fiber, as optical waveguide, can propagate in several modes depending upon their physical structure. Depending upon the propagation properties, optical fibers are classified as single mode and multimode fibers. As the name suggests, single mode fibers support only one mode of propagation, while multimode supports several modes.

The multimode propagation result in modal dispersion and modal noise as result of interaction between the modes and it limits the transmission carrying capacity of the fiber. Hence, single mode fibers are used for telecommunication purposes as it avoids the deficiencies due to multimode and it can carry signal at a much higher speed than multimode fibers. Multi-mode fibers generally have a wider core diameter, and are used for short-distance communication links and for applications where high power must be transmitted. Single-mode fibers are used for most communication links longer than 1,000 meters (3,300 ft).

As the light travels through silicon fibers, it suffers loss (attenuation) of signal strength from two controlling factors namely Raleigh scattering[34] and IR absorption. For all telecommunication requirements, single mode fibers operating in 1330nm and 1550 nm are used, as the attenuation at these wavelengths are lowest.

Dense Wave Division Multiplex (DWDM)

DWDM is a technology where incoming signals wavelengths are multiplexed by assigning different optical wavelength to each incoming signal and simultaneously, transmitting all of them over a single fiber. With the rapid advancement in optical fiber technology, DWDM has provided terabits of bandwidth capability as well as bit transparent medium over optical fiber links. DWDM provides bit rate and format independent (transparent) paths.

Broadband Services

Broadband is often called "high-speed" access to the Internet, because it usually has a high rate of data transmission. In India, any connection to the customer of 256 kbit/s or greater has traditionally been considered as broadband Internet access. However TRAI has suggested that 512 Kbps or more to be considered as broadband. In forthcoming National Telecom Policy (NIP) 2 Mbps is being targetted to be achieved as basic broadband bandwidth rate.

The standard wired network broadband technologies in most areas are various variance of Digital Subscriber Loop or DSL technologies.Newer wired broadband technologies in use are pushing optical fibre connections closer to the subscriber. Fibre-optic communication, while only recently being used in fibre to the premises and fibre to the curb schemes, has played a crucial role in enabling broadband internet access by making transmission of information over larger distances much more cost-effective than copper wire technology.

An internet leased line is a premium internet connectivity product, delivered over fiber normally, which is dedicated and provides uncontained, symmetrical speeds, Full Duplex. It is also known as an Ethernet leased line, DIA line, data circuit or private circuit.

Satellite Internet is usually among the most expensive ways of gaining broadband Internet access, but in rural areas, tough and hilly terrain, isolated Islands (Andaman and Nicobaars, Lakshya Deeps) it may be the only choice other than cellular broadband. However, costs have been coming down in recent years to the point that it is becoming more competitive with other broadband options.

In a few areas not served by cable or ADSL, community organisations have begun to install Wi-Fi networks. The newest wireless technologies being deployed for broadband services for both mobile and stationary devices are WiMax/LTE (Long term evolution). Newer Access technologies like Broadband Wireless Access (BWA) and 3G and 4G can significantly transform the character of internet/ broadband scenario in India.

As a result of these measures, the broadband subscribers grew from 0.18 million in 2005 to 8.8 million as on 31st March 2010, 100.76 million at the end of Apr- 2015 to a broadband subscriber base of 291.61 million in May 2017. This growth pattern is going to be may fold in near future as fiber reach to rural India will futher increase.[35]

Broadband have created profound impact on the society. It is the basic tool of bridging the digital divide and financial inclusion. As stated by the Broadband Commission, " *Broadband is playing a role in opening up young minds to new horizons through educational technologies; in empowering women to expand their opportunities through genuine choices; in improving awareness of hygiene and healthcare; and in helping family breadwinners find work, a better salary or return on their goods. Through broadband, the provision of public services is transformed to make them global public goods for the global good. Greater access to the Internet and broadband applications and services help accelerate achievement of internationally-agreed development goals, including the Millennium Development Goals (MDGs).*"

Mobile Communication

Mobile Communication means, ideally, communicating from anywhere, anytime, anyplace. Mobile communication, as name suggests, offers a lot of convenience and flexibility to end users. It took immense technical advancement to shift from fixed telephony to mobile telephony. Introduction of mobile communication services revolutionized the telecommunication service scenario in India, providing an on-demand telephone at affordable cost. The Indian telecom sector was having subscriber base of about 28.53 million in year 2000 which has phenomenally grown to 1,194.58 Million at the end of 31st March 2017. The major contribution has been from wireless segment. Subscription in Urban Areas is 692.97 Million with a tele-density of 171.80, whereas rural area subscription is 501.61 Million with a tele-density of 56.91at the end of March 2017.[36] India has become amongst world's fastest growing telecom market. Cellular mobile usage has become very popular among the masses, and has almost become kind of right to communication.

Basically, there are two standards, namely, GSM (Global System for Mobile) and CDMA (Code Division Multiple Access) for operating the mobile system. In India GSM operates in 900 and 1800 MHz band and CDMA operates in 800 MHz band. Mobile systems are evolving over the years and the evolution is denoted by their Generation (1G,2G,3G). Presently the mobile systems in India belong to 2G 2.5 G and 3G Categories and evolving towards 4G technology. Bharti Airtel has already launched 4G services and other service providers are about join the stream. Presently all social media sites access/conversation, e-mail, chat, location based services, M-commerce are available through 3G; 4G mobile systems are expected to further the speed of services such as broadband, interactive, multi-media services, video streaming *etc.*

Signal Transmission and Mobile Communication Technologies

For the purpose of efficient communication, transmitted signals (voice, video and data) are to be carried through a carrier radio frequency (with minimum possible bandwidth of the frequency spectrum, bandwidth being a scarce resource) to the destination. One of the basic objectives in data communication is to allow several transmitters/users to send voice/data simultaneously over a single communication channel of given bandwidth without causing interference. In other words, all communication technologies have to solve the same fundamental problem of dividing the given RF spectrum amongst multiple with best possible spectrum-utilized efficiency. This task of channel access in radio-communication is realized through the concept of channel access/multiple access, using various modulation/multiplexing techniques like TDMA (Time Division Multiple Access), FDMA (Frequency Division Multiple Access) and CDMA (Code Division Multiple Access).In theory, TDMA, FDMA and CDMA have the same spectral efficiency but in ground reality each has its own challenges and resulting deviation from theoretical efficiency.

- ☆ Time Division Multiple Access (TDMA) divides the access by time, and hence the data communicated is not continuous, but the time gap between two bursts of communication is too small to affect quality. In fact, TDMA uses different time slots of transmission and reception of data. TDMA systems must carefully synchronize the transmission times of all the users to ensure that they are received in the correct timeslot and do not cause interference. Since this cannot be perfectly controlled in a mobile environment, or user mobility, each timeslot must have a guard-time, to reduce the probability of users interfere, at the cost of decreases in the spectral efficiency. An analogy to the problem of multiple access is a room (channel) in which people wish to talk to each other simultaneously. However, to avoid confusion, people are asked to take turns while speaking (time division) or speak at different pitches (frequency division), or speak in different languages (code division).

- ☆ Frequency-Division Multiple Access (FDMA) divides the access by frequency. FDMA systems must use a guard-band between adjacent channels, due to the unpredictable Doppler shift of the signal spectrum, in case of user mobility. The guard-bands will reduce the probability adjacent channels interfere, at the cost of spectral efficiency.
- ☆ Code Division Multiple Access (CDMA) is adopted for the access where the signal generated in a particular bandwidth is deliberately spread over the entire frequency band. Each group of users is allotted a shared code in the spread-spectrum and many codes occupy the same channel, each channel (code) being over laid on the top of one another. Only the users assigned with a particular code can communicate amongst themselves resulting in uses of the transmission bandwidth several orders of magnitudes greater than the minimum signal bandwidth required. Spread spectrum techniques were initially used for military applications because of its security and resistance to jamming, to prevent detection (because jamming signal has only a finite amount of power available that can either be spread over the entire bandwidth of the signal to be jammed or can jam only part of the entire signal).

Global System for Mobile (GSM)

GSM is a cellular (In a cellular system, the geographical area is divided into adjacent, non-overlapping, hexagonal shaped cells) standard developed to cater voice services and data delivery using digital modulation. In addition to providing a range of supplementary services, GSM differentiates itself from previous and competing technologies in two key ways. First, it enables a user to roam between networks - a user can use the services of their home network and in other, interconnected networks. These services include both the original voice-based service and the supplementary and text-based services. Second, it offers in-built security, authentication and encryption. GSM devices use a smartcard, known as a Subscriber Identity Module (SIM), that contains operator and subscriber specific data that enable users to be authenticated and sessions (voice or data) to be encrypted over the air.

GSM Mobile Network Architecture

In a cellular system, the geographical area is divided into adjacent, non-overlapping, hexagonal shaped cells. Each adjacent cell operates at different frequencies from one another to maintain distinctness and to avoid confusion and interference amongst communication information. Since cells operate at different frequencies each cell has its own Transmitter and Receiver (called Base Stations) to communicate with the Mobile units in that cell; whenever a mobile unit crosses cell boundaries, a Mobile Switching Centre (MSC) in the network coordinates the handoff of mobile units to maintain the communication. The system hands over calls in almost seamless manner from transmitter to transmitter, as customers move around from one cell site to other.

Cellular systems are based on the concept of frequency reuse, for example the same frequency is used by several sites, which are physically separated far enough from one another, thus conserving the frequency spectrum. One of the most important concepts for any cellular telephone system is that of multiple accesses, meaning that multiple, simultaneous users can be supported through frequency reuse. In other words, a large number of users share a common pool of radio channels (spectrum) and any user can gain access to any channel (each user is not always assigned to the same channel), which is temporarily allocated for the communication. A schematic overview of the GSM system is shown in the Figure 5.4 given below. The system is composed of three main components:

(i) Base Station Subsystem

(ii) Mobile Switching Subsystem and;

(iii) Mobile handset.

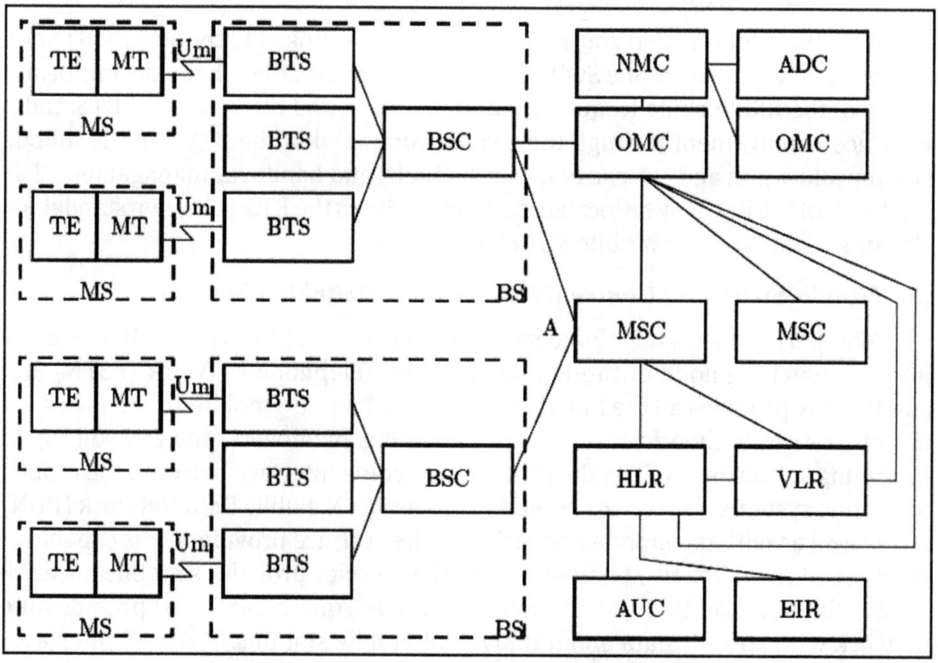

Figure 5.4: Simplified Structure of GSM PLMN.
(*Source*: Hanzo and Steele (1994)[37])

Base station part controls the communication across the air (radio) interface. The Switching Subsystem enables the connection between the two mobile users. The Mobile handsets act as the transmitter/receiver for the user for communicating with others. The functions of various components in GSM mobile network are described in brief in the following paragraphs.

(i) Base Station Subsystem (BSS)

Base Station Subsystem provides the direct contact with the mobile handset through air interface and links it to the Switching centre. Base Transceiver Station (BTS) and Base Station Controller (BSC) together, form a functional entity of the Base Station Subsystem

Base Transceiver Station (BTS)

BTS provides the mobile's interface to the network. Each cell site is equipped with a BTS, which is usually located in the center of a cell to provide coverage within a cell, which may cover an area of 30 - 40 sq kms. However, in a congested, urban location, the BTS coverage area is much smaller. The transmitting power of the BTS and number of customers in the cell determine the absolute cell size. BTS houses the radio transceivers that define a cell and handles the radio-link protocols with the Mobile handsets.

Base Station Controller (BSC)

One BSC monitors and controls several BTS. It is linked to the several BTSs on one side and with the Mobile Switching Centre (MSC) on the other side. Important works of the BSC include frequency administration and the control of BTS; radio interface management through the remote command of the BTS and the mobile station; allocation and release of radio channels; and handover management. The hardware of the BSC may be located at the same site as the BTS, at its own standalone site, or at the site of the Mobile Switching Centre.

(ii) Mobile Switching Centre (MSC)/Mobile Switching System

MSC performs the telephone switching functions of the system. It acts like a normal switching node of the Public Switched Telephone Network (PSTN), but additionally provides all the functions needed to handle a mobile subscriber, such as registration, authentication, location updating, handovers, and call routing to a roaming subscriber. MSC is the primary switching interface between the mobile telephone systems and other networks such as PSTN, Public Data Network (PDN) *etc.* as well as with the mobile networks of other service providers. It is capable of routing calls from the fixed network and other service provider's mobile network via the BSC and the BTS to an individual mobile station. MSC also provides the network with specific data about individual mobile stations. The MSC interfaces with BSS on one side (through which it is in contact with GSM users) and with the external networks on the other.

Home Location Register (HLR)

HLR is a database about subscribers; it stores the identity and user data of all the subscribers belonging to the area of related MSC. These are permanent data, such as the International Mobile Subscriber Identity (IMSI) of an individual user, authentication key, including a subscriber's service profile, location information, activity status and some temporary data.

Temporary data on the SIM include such entries as (i) the address of the current Visitor Location Register (VLR), which currently administers the mobile stations, (ii) the number to which the calls must be forwarded (if the subscriber selects call forwarding), and (3) some transient parameters for authentication and ciphering.

Visitor Location Register (VLR)

The VLR contains the relevant data of all mobiles currently located in a serving MSC. It is the database that contains temporary storing subscription data for those subscribers currently situated in the service area of the corresponding MSC as well as holding data on their location at a more precise level than the HLR. The VLR is always integrated with MSC. The permanent data are the same as data found in the HLR; the temporary data differ slightly. For example, the VLR contains the Temporary Mobile Subscriber Identity (TMSI), which is used for limited periods to prevent the transmission of the IMSI via the air interface.

Code-Division Multiple Access (CDMA) System

CDMA is a digital cellular technology that uses spread-spectrum techniques. CDMA does not assign a specific frequency to each user, instead, every channel uses the full available spectrum. Individual conversations are encoded with a pseudo-random digital sequence. CDMA consistently provides better capacity for voice and data communications than other commercial mobile technologies, allowing more subscribers to connect at any given time. It operates in 800 MHz band

Value Added Services (VAS)

The mobile value added services include, text or SMS, menu based services, downloading of music or ring tones, mobile TV, video, streaming, M- banking and other sophisticated M-commerce applications *etc.* Prior to 2008, a majority of VAS revenues were attributable to SMS's, today cloud computing, remote surveillance, fleet management, telemetric,retail supply chain, mobile TV, mobile banking and mobile governance are already available on mobiles of data hungry, smart phone user generation.

Satellite Communications

Satellite communication was initiated in the early sixties; with this the problem of connectivity to remote and far flung areas has been resolved. Satellite communication provides the facility of covering areas (national and international) using one or several satellites. Broadly there are two types of satellites, used for satellite communications; geostationary satellites and medium and low orbit satellites.

Satellites are used for communications, when two Stations on Earth want to communicate through radio spectrum but are unable to do so by conventional terrestrial communication methods. A communication satellite acts as a repeater/

relay station. The two stations can use a satellite as a relay station for their communication. One Earth Station sends a transmission to the satellite. This is called an Uplink. The satellite Transponder converts the signal and sends it down to the second earth station/mobile user. This is called a Downlink.

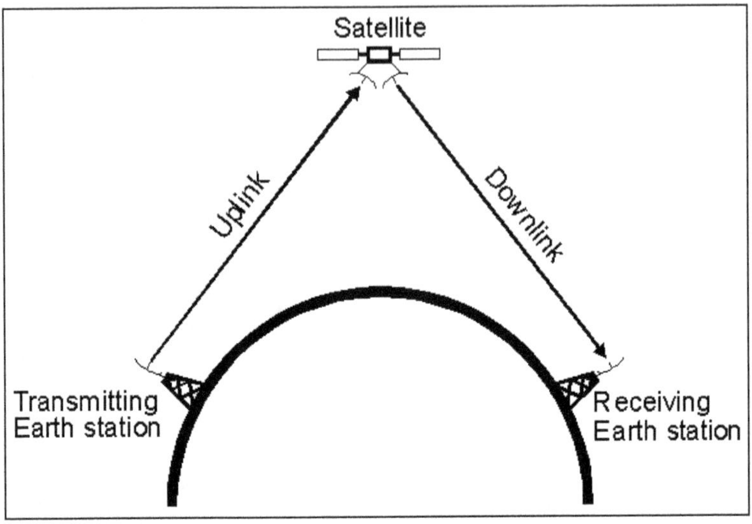

Figure 5.5: Satellite Communication[38]

Some inherent characteristic of Satellite communications are:

Propagation delay or latency: In view of the very altitude of satellites there is significant propagation delays involved in satellite communication.

Limited bandwidth: All satellites are designed for specific purposes. A very complex technology is involved in satellite design and launch. Almost all satellites have limited transponder capacity thus limited bandwidth for commercial communication purpose. Accordingly many systems will require using the available bandwidth very effectively.

Noise: The high altitude or longer communication path length and limited power levels are main reasons communication noise in satellites. To overcome this, directive antennas are normally employed. However in addition to this robust error correction techniques are normally required for data transmission.[39]

Satellites, based on design, purpose of utilization, adopted technology, power utization *etc.* are launched to three types of orbits. Geostationary orbit (GEO), low earth orbit (LEO) and medium earth orbit (MEO). A geostationary satellite is placed at an altitude of approximately 35,800 kilometers (22,300 miles) directly over the equator that revolves in the same direction the earth rotates. Due to height of orbit and speed of rotation this satellite appears as a fixed, static point to any observer or antenna tracking it on the ground. While used for applications such as broadcast communications and weather satellites, communication through a

GEO system becomes more difficult as the ground observer's latitude increases the further north or south it travels due to occurrences such as atmospheric refraction, thermal emission, line-of-sight obstructions and signal reflections from structures on the ground. Since GEO Satellite remains approximately fixed relative to Earth, a few satellites can provide global coverage, making this most economical option. However there is a transmission delay of 250 ms to complete up/down link.

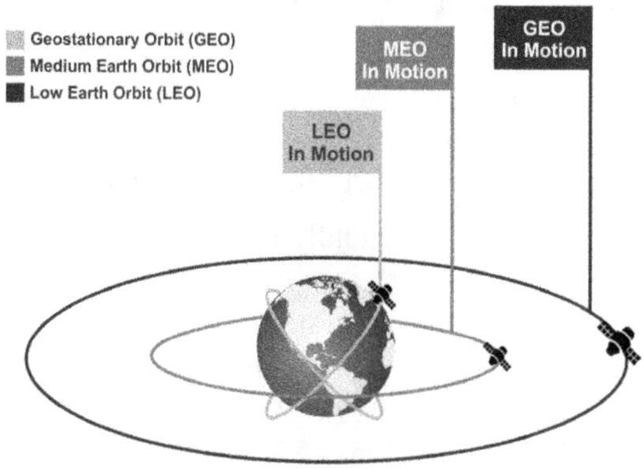

Figure 5.6: Satellite Orbits.
(*Source*: http://www.harriscaprock.com/blog/high-throughput-satellite-communications-systems-meo-vs-leo-vs-geo/)

As the names imply, Low Earth Orbit (LEO) is relatively low in altitude; the LEO altitude range is between 200 and 1200 km above the Earth's surface. A variety of different types of satellite use the LEO orbit levels. Some communications satellites including the Iridium phone system use LEO. Earth monitoring satellites use LEO as they are able to see the surface of the Earth more clearly as they are not so far away. They are also able to traverse the surface of the Earth.[40] LEO provides constant coverage through a constellation of several satellites that are closer to Earth, offering a significant performance advantage with low latency (transmission delay).

Medium Earth orbit (MEO), sometimes called intermediate circular orbit (ICO), is the region of space around the Earth above low Earth orbit and below geostationary orbit. The most common use for satellites in this region is for navigation, communication, and geodetic/space environment science. Reduced latency, improved look angles to ground stations at higher altitudes are characteristics of MEO satellites. However it requires more expensive and complex ground stations

The primary difference between MEO and LEO systems as compared to the majority of communications satellites in geostationary orbit is altitude. The position of the satellite in this orbit is continually changing with respect to earth; they require several satellites in their constellation for linking to ground earth station and providing continuous communication services. Similarly, since each satellite serves a smaller area of earth's surface, several earth stations need to be established. They have, therefore, been not popular for communication purposes. In geostationary systems, any single satellite loss or failure is catastrophic to the system. Low Earth Orbit (LEO) satellites orbit around earth at a distance of few hundred kilometers providing the facilities of remote sensing *etc.* Location is everything. MEO and LEO satellites travel overhead at all times, instead of tracking with a fixed point on Earth as GEO satellites do. This type of orbit allows them to operate with lower free space losses, lesser latency and better frequency reuse.

Satellite Communication has an inherent advantage over terrestrial communication of better disaster resilience. When terrestrial communication infrastructure gets damaged due to natural calamities satellite communication offers an alternate method to connect to disaster affected areas.

Communication satellites are usually designed to have a typical operating lifetime of 10 to 15 years. They operate on different frequencies for uplink (earth station to satellite) and downlink (satellite to earth station). Since higher frequencies suffer higher propagation loss and there are limitations on the power output from the satellite (due to constraint on overall weight), the earth station transmits at higher frequency (uplink) as the earth station can compensate the higher propagation loss with higher power output. Hence, uplink frequency is higher than the downlink frequency. Uplink and Downlink are also referred as Transmit and Receive frequency respectively. Satellite communication systems are dominated by the need to receive very weak signals. In the early days, very large receiving antennas at the earth stations with diameters upto 30 metres were required. As more heavy and powerful satellites are launched nowadays, much smaller earth station antennas have become feasible. Networks for satellite communication system on the ground have evolved from centralised systems built around a single mainframe computer to distributed networks of interconnected systems. Space-based satellite networks are evolving from centralised networks relying on a single geostationary satellite to distributed networks of interconnected low Earth orbit satellites. At the height of 36,000 Km, communications through a GEO travels at the speed of light, entailing a round trip transmission latency of about one-half second. This means that GEO technology does not provide fiber-like low latency services. This is the source of the delay in many intercontinental phone calls, impeding understanding and distorting the personal nuances of speech.

Satellite systems operate in the microwave and millimeter wave frequency bands, using frequencies between 1 and 50 GHz. Above 10 GHz, rain causes significant attenuation of the signal and the probability that rain will occur in the path between the satellite and an earth station must be suitably factored into the system design. The frequency bands of operation are denoted by uplink frequency/ downlink frequency and bands currently used are: predominantly C-band (6/4 GHz) and Ku-band (14/11 GHz) with a few services in Ka-band (30/20 GHz). Ku and Ka-band enable use of smaller diameter antenna (typically less than 1.5m), narrower antenna beams and better control of coverage patterns compared to

C-band. Maritime services use C-band for shore station and L-band (1.5 GHz) for ship terminals.

Each satellite contains several transponders which basically do the function of frequency conversion (uplink to downlink frequency) and amplifying the weak signal received at the satellite after travelling about 36000 km. Typically, each transponder supports 36 MHz, 54 MHz or 72 MHz (depending on the design) and the available total bandwidth is typically 500 MHz for each of frequency band (C, Ku and Ka bands) of operation. Large GEO satellites also use both the C-band (6/4 GHz) and Ku-band (14/11 GHz) for providing more bandwidth. Further increase in bandwidth is also possible by frequency reuse and polarization for example polarizing the same carrier in LHCP (Left Hand Circular Polarization) or RHCP (Right Hand Circular Polarization) in order to nearly double the bandwidth.

Three basic multiple access techniques namely; FDMA (Frequency Division Multiple Access), TDMA (Time Division Multiple Access), CDMA(Code Division Multiple Access) where the baseband signal (information) from various users are differentiated by frequency, time and code respectively are used. Multiple accesses enables users to share the same transponder bandwidth by directly accessing the satellite more efficiently by transmitting the signals (voice, video, data) from the earth station using one of the multiple access techniques. The baseband can be based on one of the multiplexing techniques and it is modulated using a modulation technique. However, for digital baseband, digital modulation techniques need to be used. Two modulation techniques are in use in satellite communication system: namely Frequency Modulation (FM) and Phase Shift Keying (PSK). In FM, the frequency of the carrier is changed in accordance with the amplitude of the information signal. FM is widely used in analog system. Presently, satellite systems widely use digital transmission techniques and baseband consist of digital bit stream, PSK technique is used for modulating the digital signal.

Satellite communication, being essentially broadcast in nature (in addition to its use in interactive mode), provides direct access to many terminals (users) simultaneously in its coverage area with the same quality (which is referred to as "multiple access" feature). This is the distinct advantage of satellite communication when compared with terrestrial and landline communication, in addition to its high reliability with the absence of problems related to last mile connectivity.

Very Small Aperture Terminals (VSATs)

Geo stationary and lower orbit Satellites (MEO and LEO) have been supporting telecommunication applications for providing voice and data communications. Satellite links are the primary means of connectivity to remote and far-flung regions of the country and they are the backup links for large number of terrestrial connectivity in the mainland. Now a days, satellite networks to a large extent, use Very Small Aperture Terminals (VSATs) to cater to the traffic and application requirements of varied users. The term VSATs stand for Very Small Aperture Terminal, these are fixed satellite terminals that are used to provide interactive or receive -only communications and are used for a wide variety of telecommunications applications such as Corporate WAN, Rural Telecom, Distance Learning, Telemedicine, video conferencing, defense operations, defense establishments, stock exchanges, manufacturing and FMCG companies Disaster Recovery, Ship - Board communications (communication on large ships) *etc.*

A VSAT network comprises of a central hub and hundreds of terminals, which are further interfaced to computers and other peripheral devices. The hub acts as a gateway with interface to external connectivity and several application servers. Rapid technological advancements and reduction in the cost of user equipment are increasing the popularity of VSAT network. A VSAT network works out to be a cheaper option while establishing a network to cover a wide geographical area, state wide or nationwide. VSAT networks presently in India are operating in C, Extended C and Ku-bands[41] Internationally Ku-band is a popular frequency band in used for VSAT, due to its smaller antenna size.

VSATs have become increasingly popular, because VSAT networks are designed to support all kinds of applications supporting video, voice and data, with a wide range of data rates from few kilobits per second (kbps) to few megabits per seconds (mbps). VSAT networks are a flexible communication platform that can be installed quickly and cost effectively to provide telecom solutions to consumers, governments and corporations[42]. Some of the advantages of using VSATs highly reliability higher uptimes (as high as 99.5 per cent), wider area coverage, centrally managed network, good Mean Time to Repair, collocated last Miles connectivity. Due to aforementioned properties VSAT are considered to be very effective for emergency communication and PPDR (Public Protection and Disaster recovery) networks.

A typical satellite network consists of a Hub and VSATs (Very Small Aperture Terminals). Hub is a central station where all the signals from VSATs are integrated and their connectivity with other networks is established. VSATs are located at the customer premises and these are invariably fixed terminals providing multiple connections for voice, video and data including LAN for a particular location. The primary advantage of satellite communication is the ease with which it can be used for emergency communications, disaster management *etc.* Other applications are: remote area communication, weather forecasting, TV broadcasting, mobile communication *etc.*

Handheld satellite phones/briefcase terminals (like INMARSAT sat phones) provide direct connectivity to the satellite along with mobility (as it is easy to carry extending voice and data and, in some cases, video facility for individual users). However, use of mobile satellite phones services involve very high recurring expensive call charges (though the capital charges are much lower) compared to call charges of satellite services provided by fixed VSAT terminals (which incur high investment cost compared to satellite phones).[43]

Single Channel Per Carrier (SCPC)

SCPC is a method employed to connect remote places with low density of traffic using satellite communication. In SCPC system, a carrier is permanently assigned for each satellite terminal and it is connected to other terminal on a point-to-point basis. A pair of carrier frequency assignment is essential for each terminal for transmit and receive. In order to conserve the frequency spectrum, DAMA (Demand Assigned Multiple Access) mode of working for SCPC terminals became popular wherein, there is no permanent carrier assignment to each satellite terminal. In DAMA, a pair of carrier frequency to a terminal is assigned from Hub on demand basis. When a terminal intends to communicate, a request message from the terminal is automatically sent in random transmission mode on a common carrier shared by all the terminals. The request message contains the originating terminal number and type of service. DAMA server at Hub receives the message through satellite, allots free satellite channel (a frequency pair) if available and sends allocated satellite channel number in the return common broadcast channel. When the calling terminal receives the allocation message meant for it, it tunes it's transmit and receives frequency corresponding to the allocated channel pair and begins the communication. SCPC is employed for providing access to remote stations with low traffic in a cost effective manner, as terrestrial systems may not be economically viable due to remoteness.[44]

Satellite Phones

Satellite phones are the mobile handsets accessing the Satellite or accessing through a portable terminal. Among the various satellite technologies, INMARSAT BGAN (Broadband Global Area Network) service is accessed via a small, lightweight satellite terminal that can be quickly and easily connected to the laptop PC and voice handset. They are designed to suit single user whose primary concern is portability. All terminals are robust and able to withstand challenging environments and extremes of temperature.

Iridium and INMARSAT have introduced satellite phones (which are similar in Size to a GSM/CDMA mobile phones) with flip antenna that are to be projected pointing upwards while talking. Primarily these phones offer voice service with limited data rates up to 2.4kbps which can be mainly used for file transfers. Iridium service is based on Low Earth Orbit (LEO) constellation of 66 satellites orbiting from pole to pole while INMARSAT employs 3 Geo-stationary satellites for worldwide coverage. Both are portable just like a cellular mobile phone and cost of a handset ranges between USD 700 to USD 1200. These satellite phones enable communication to be established from a remote site, very much similar to a mobile phone.

INSAT MSS Type-D System

Space Applications Centre (SAC), of ISRO (with active participation of Indian industries) has developed a small portable satellite terminal that works with INSAT Satellite for voice/data communication. The INSAT satellite contains specialised Mobile Satellite Transponder by which communication through a small terminal is possible. The terminal was developed keeping its primary use for voice communication during disaster when other communication means are likely to have broken down. It can be used from any location in India for emergency communication. The portable Satellite terminal at a remote place can be connected through the satellite to an EPABX at the Hub. In this way, a remote place can be linked as one of the extensions of EPABX and telephone calls can be made between portable satellite terminal at the remote site and local phones connected to the EPABX.

The Central Hub station uses a 6.3m antenna system. It transmits and receives frequencies in C band. The portable satellite terminals transmit and receive frequencies of the terminal is in S band. The satellite channel works in DAMA (Demand Assigned Multiple Access), where, a pair of satellite channels is assigned from Hub to a terminal on demand basis. When a person lifts the telephone handset to make a call, a request message from the terminal is automatically sent in random transmission mode on a common carrier shared by all the terminals. The request message contains the originating terminal number and type of service requested for. DAMA server at Hub receives the message through satellite, allots free satellite channel (a frequency pair) if available and sends the allocated satellite channel number in the return common broadcast channel. When the calling terminal receives the allocation message meant for it, it tunes it's transmitter and receives a frequency corresponding to the allocated channel pair and gets a dial tone from the exchange, subsequently.[45]

DTH (Direct-To-Home)

Direct Broadcast Satellite (DBS), is a term used to refer to satellite television broadcasts intended for home reception, also referred to more broadly, as Direct-To-Home (DTH) signals. The expression direct-to-home or DTH was, initially, meant to distinguish the transmissions directly intended for home viewers from cable television distribution services that sometimes carried on the same satellite. The direct-to-home (DTH) market in India has evolved considerably over the past four years with the entry of private players. DTH can apply to several services transmitted over a wider range of frequencies (including standard Ku-band and Ka-band).[46,47]

Satellite Based Earth Observation

Earth observation is the gathering of information about planet Earth's physical, chemical and biological systems. It is used to monitor and assess the status of, and changes in, the natural environment and the built environment. In recent years, Earth observation has become technologically more and more sophisticated. It has also become more important due to the dramatic impact that modern human civilization is having on the Earth, the need to minimize the negative impacts and the opportunities Earth observation provides to improve social and economic well-being" (WIKIPEDIA).[48]

With the increasing prevalence of Earth Observation (EO) technologies being employed in development work, the World Bank and the European Space Agency

(ESA) have partnered for the purpose of mainstreaming the use of EO into the World Bank's lending operations, across all of the Bank's Sustainable Development Network's sectors. July 2010, the collaboration was scaled up. Twelve World Bank Teams from across a variety of sectors and representing all regions of Sustainable Development Network were selected and received the pro-bono assistance of ESA to apply Earth Observation services in selected operations. Sectors benefitting include:

- ✰ *Disaster Risk Management;*
- ✰ *Urban Development;*
- ✰ *Agriculture and Forest Management;*
- ✰ *Water Resources Management;*
- ✰ *Coastal Zones Management;*
- ✰ *Marine Environment; and*
- ✰ *Climate Change Adaptation.*

This initiative aims to establish a stable connection between the specific information needs of World Bank projects and the new developments in Earth Observation programmes and services. In particular, *eoworld* gives the World Bank the opportunity to benefit from ESA's 30 years of experience in collecting and distributing Earth Observation (EO) data to users around the world, and in developing innovative information services for various applications to be used by public organisations, national authorities and private sector.

EU-ESA has "GMES (Global Monitoring for Environment and Security)" a space program for new national space missions in Europe and Canada. Earth Observation technology will be a key enabler for long-term sustainable data flow, for objective and continuous views of our planet supplied by satellite images and data provide scientists and decision makers with the desired information for understanding climate change and environment protection. EO *inter-alia* has applications for monitoring air, seas and land; providing the basis for accurate weather reports; and supplying national and international relief agencies with data when disasters strike

Future Trends for Disaster Communication

The technologies of cellular mobile network are evolving to provide progressively higher data rates and improved spectral efficiency with 3G and 4G systems. Ultimately, it will enable to achieve diversification in the deployment scenarios for each technology so that they operate in an integrated network to optimize network resources. Some of latest promising communication technologies, which have been widely promulgated, are described in succeeding part of the chapter.[49,50]

Wi-Fi Technology

Wi-Fi stands for Wireless Fidelity and it is a way to get Internet access using microwave wireless medium. The Wi-Fi Alliance is the body responsible for promoting the term Wi-Fi and its association with various wireless technology standards. A Wi-Fi hotspot is defined as any location in which coverage to internet is available through this technology and is available for common use by consumers. Generally, the most common usage of Wi-Fi technology is for laptop users to gain Internet access in locations such as airports, coffee shops, and so on, where Wi-Fi technology can be used to help consumers in their pursuit of work-based or recreational Internet usage. It also useful in small offices and house where multiple users want to use internet through common Wi-Fi router. Nowadays there are many small devices are available which convert Ethernet to Wi-Fi, to enable users to be mobile while using internet. These devices are very useful for mobile internet users. Wi-Fi has many advantages. Wi-Fi networks are easy to set up and inexpensive. Wi-Fi most commonly uses the 2.4 gigahertz and 5 gigahertz radio bands. Wi-Fi has become mostly used for connectivity of personal devices such as personal computers, video-game consoles, smartphones, digital cameras, tablet computers, smart TVs, digital audio players and modern printers and selfie sticks.

WiMAX

WiMAX (Worldwide Interoperability for Microwave Access) is a family of wireless communications standards initially designed to provide 30 to 40 megabit-per-second data rates, with the 2011 update providing up to 1 Gbit/s for fixed stations (WIKIPEDIA). The name "WiMAX" was created by the WiMAX Forum, which was formed in June 2001 to promote conformity and interoperability of the standard. The forum describes WiMAX as "a standards-based technology enabling the delivery of last mile wireless broadband access as an alternative to cable and DSL". IEEE 802.16m or Wireless MAN-Advanced is a candidate for the 4G, in competition with the LTE Advanced standard".[51] WiMAX Technology works very much like cell phone technology because a reasonable proximity to a base station is required to establish a data link. Users within 10 kms of the base station will be able to establish a link using non-line-of-sight (NLOS) technology with data rates as high as 75 Mbps. However, users up to 50 kms away from the base station need an antenna, mounted for line-of-sight (LOS) to the base station, to provide data rates approaching 280 Mbps. WiMAX is a wireless access service designed to cover wide geographical areas serving large numbers of users at low cost providing IP connectivity. WiMAX is considered one of the best solutions for "last mile" distribution. WiMAX uses microwave radio technology to connect computers to the (IP) data network in place of wired connections such as DSL or cable modems. Although, primarily a wireless broadband data network, the advent of Voice over Internet Protocol (VoIP) and native quality of service capabilities enable WiMAX to offer quality voice services at relatively low cost (being an all-IP network).

Heavy storms and manmade disasters can knock out DSL and cable broadband, satellite, dial-up and Wi-Fi. The city of Brownsville, TX lost Internet access for 18 hours when Hurricane Emily made landfall in 2005. To ensure residents weren't cut off again from weather reports and Web communications tools, the city installed security-enhanced world interoperability for microwave access (WiMax) system that provides wireless broadband. The constant wireless Internet connection ensures continuous Internet telephone and weather reports during storms, while everyday safety benefits include local police's wireless access to criminal databases. The WiMax base towers can withstand hurricane winds and provide a safer solution because they require no above-ground utility poles and lines that can fall in hurricane winds.[52]

Public Mobile Radio Trunking System (PMRTS)

Public Mobile Radio (PMR) is a two way radio, employing one-to-many, point- to multipoint type of calling system. Two-way radio operates in frequencies ranging from 66-88 MHz, 136-174 MHz, 403-520 MHz and 800 MHz frequency. This results in faster communication of urgent messages or vital information within any organization. Conventional VHF system is the foundation of two way radios. In conventional two-way radio communications, users have a dedicated frequency channel for their own use. However, a shortcoming of the two-way radio communications is that there can only be one person speaking at a given point of time and as the numbers of users grow, frequency congestion occurs since there are only a finite number of frequencies. This short coming is overcome by another variant of two-way radio communication known as Radio Trunking, (namely Public Mobile Radio Trunking system (PMRTS)) in which a set of frequencies are used for enabling simultaneous multiparty communication within the group.

When radio channels are trunked, it means that radios automatically get a free 'path', at the press of a button, to communicate with each other. The handsets can be installed in vehicles or small enough to be carried around in hand. In radio trunking, a large number of users share a relatively small number of communication paths or trunks. The service may be very useful for emergency response and relief communication. In India, the main users consist of government departments, securities, defense establishments, hotels, municipal corporations, schools, cash vans, radio-taxis, courier and other fleet management systems. Of the existing technologies, "TETRA" is the most popular technology in India, which offers open standard.

IP Multimedia Subsystem (IMS)

IP(Internet Protocol) Multimedia Subsystem (IMS) is a generic architecture for offering multimedia and voice over IP services, defined by 3rd Generation Partnership Project (3GPP).[53] IMS is access independent, as it supports multiple access technologies including GSM, WCDMA, CDMA2000, WLAN, Wireline broadband and other packet data applications. IMS will make Internet technologies, such as web browsing, e-mail, instant messaging and video conferencing available to everyone from any location (NDMICS, 2011).

RF Identification Systems (RFID)

Another technology that will help disaster management in a big way is RF identification systems, simply called RFID. They are small, low-cost tags that can be attached to objects or persons to identify and track their position. RFID provides both proximity and dynamic identification, so that tagged objects do not need to be physically contacted or stationary in order to be identified. RFID creates an automatic way to collect information about a product, place, time or transaction quickly, easily and without human error. It provides a contactless data link, without the need for line of sight or concerns about harsh or dirty environments that restrict other auto ID technologies such as bar codes. In addition, RFID is more than just an ID code; it can be used as a data carrier, with information being written to and updated on the tag on the fly.

RFID systems consist of a number of components including tags, handheld or stationary readers, data input units and system software. The tags are the backbone of the technology and come in all shapes, Sizes and read ranges including thin and flexible "smart labels", which can be laminated between paper and plastic. RF tags are either active or passive. Active tags are usually more expensive because they can transmit a low-powered signal. These tags come with batteries and permit farther distances for RF identification. Passive tags do not come with batteries and are detected solely by the RF receiver. It will be possible to create a network to cover a large area with RF that will support reading of RF tag and reporting to control station about the movement of consignments mounted on moving vehicles. This will result in timely assessment of material movement and planning during disaster relief.[54]

Telepresence

Telepresence refers to a set of technologies which allow a person to feel as if they were present, to give the appearance of being present, or to have an effect, via telerobotics, at a place other than their true location. Telepresence requires that the users' senses be provided with such stimuli as to give the feeling of being in that other location. Additionally, users may be given the ability to affect the remote location. In this case, the user's position, movements, actions, voice, etc. may be sensed, transmitted and duplicated in the remote location to bring about this effect. Therefore information may be traveling in both directions between the user and the remote location. A popular application is found in telepresence videoconferencing, the highest possible level of video telephony. Telepresence via video deploys greater technical sophistication and improved fidelity of both sight and sound than in traditional videoconferencing. Technical advancements in mobile collaboration have also extended the capabilities of videoconferencing beyond the boardroom for use with hand-held mobile devices, enabling collaboration independent of location" (WIKIPEDIA).[55] High definition audio and video are the key features of telepresence that are used for a new generation of high 'Video Conferencing System' which facilitates virtual meeting that gives the feeling of "Almost like being there". With life like crystal clear imagery, high fidelity sound and high reliability, Telepresence takes video communications to a new level of virtual reality with associated economies as regards time saving and related economic benefits. The emerging trend is proliferation of Telepresence for individuals like senior executives, top-end managers and also for trainings and awareness creation.

Multiple Input Multiple Output (MIMO) Wireless Technology

MIMO is effectively a radio antenna technology. It uses multiple antennas at the transmitter and receiver end to enable a variety of signal paths to carry the data, choosing separate paths for each antenna to enable multiple signal paths to be used. Wi-Fi, LTE and many other radio, wireless and RF technologies are using the new MIMO wireless technology to provide increased link capacity and spectral efficiency combined with improved link reliability using what were previously seen as interference paths.

Software Defined Radio (SDR)

A Software Defined Radio (SDR) is a transmitter in which operating parameters, including transmission frequency, modulation type, and maximum radiated or conducted output power can be altered without making any hardware changes. SDR is a technology that enables "reconfigurable system" for wireless networks. The goal of SDR is to produce a radio that can receive and transmit, whenever there is a change in the radio protocol, simply by running new software. In the long run, antenna will remain the only band-specific item and SDR is expected to become the dominant technology in radio communications which is also likely be the enabler of cognitive radio.

Cognitive Radio

Studies have shown that spectral utilization is relatively low when examined not just by frequency domain, but also across the spatial and temporal domains. Thus, an intelligent device, aware of its surroundings and able to adapt to the existing RF environment in consideration of all three domains, may be able to utilize spectrum more efficiently by dynamically sharing spectral resources. While evolutionary technological developments are required to improve spectral efficiency by increasing the number of users in each frequency band, revolutionary technological advancements are to be introduced in terms of cognitive radios, which promise more efficient spectrum utilization through environmental awareness. This technology builds on the accomplishments of Software Defined Radio, which replaces the classical heterodyne radio (with a digital signal processor).[56]

Cognitive radio is the "intelligence" that sets above the SDR and let SDR determines which mode of operation and parameters to use. By using advanced computing power to sense the existing radio frequency (RF) environment, cognitive radios become aware of and can respond to that environment through adaptation to statistical variations in the input stimuli, with two objectives in mind namely highly reliable communication whenever and wherever needed and efficient utilization of radio spectrum further.

Unified Messaging System

Unified Messaging is the integration of several different communications media, such that users will be able to retrieve and send voice, fax and email messages from a single interface. This evolutionary vision envisages the integration and inter-operability of fixed, mobile, and mobile-satellite networks through a common terminal. Such networks, called hybrid networks, use a common spectrum to be shared between satellite space segment and the terrestrial network, making frequency planning and network coverage much easier. In addition, messages can be managed from various terminal types including phones, web-browsers and standard e-mail clients. Different service levels for different end-user groups can be formulated to achieve various service definitions with multiple classes. The products of Unified Messages are highly scalable and platform agnostic thus, making existing resources reusable.

IPV6

Most of today's internet uses IPv4, (Internet Protocol Version 4) which is now nearly twenty years old. There is a growing shortage of IPv4 addresses, which are needed by all new machines added to the Internet. The limited address range forces organisations to use Network Address Translation (NAT) firewalls to map multiple private addresses to a single public IP address and NATs does not support standards-based network-layer security. The Internet Engineering Task Force (IETF) has designated IPv6 as the successor to version 4 for general use on the Internet. It significantly increases IP address size. IPv6 has 128-bit (16 byte) source and destination IP addresses while IPv4 supports only has 32-bit (4byte) IP address. The large address space of IPv6 has been designed for multiple levels of subnetting and address allocation from the Internet backbone to the individual subnets within an organization.

The government of India has decided to speed up deployment of IPV6 compliant equipments in India. Department of Telecommunication is driving implementation of IPv6 for all devices across various sectors/organisations/institutions in India.

Next Generation Network (NGN)

ITU-T's Definition of NGN: "A Next Generation Networks (NGN) is a packet-based network able to provide Telecommunication Services to users and able to make use of multiple broadband, QoS-enabled transport technologies and in which service-related functions are independent of the underlying transport-related technologies"[57]

The present telecom networks are service dependent. Separate networks exist for voice, data, mobile, Internet, *etc.* Next Generation Networks (NGN) are envisaged to facilitate the convergence of voice and data networks into a single unified multi-service network capable of providing futuristic services including multimedia.

Machine to Machine (M2M)

Broadly, Machine to Machine (M2M), [58] refers to technologies that allow both wireless and wired systems to communicate with other devices (machines) of similar abilities. M2M uses a device (such as a sensor/meter/actuator[59]) to capture an event, which is relayed through a network (wireless, wired or hybrid) to an application, that translates the captured event into meaningful information. There are various definitions given by Global Standardization organizations for M2M. European Telecommunications Standards Institute (ETSI) [60] has defined M2M2 as 'Physical telecommunication based interconnection for data exchange between two ETSI M2M compliant entities, like: device, gateways and network infrastructure.' Accordingly M2M may be described as machines are connected to the Internet, using a variety of fixed and wireless networks and communicate with each other and with the wider world.

M2M is also being inter-changeably used with various other terms - Internet of Things (IoT), Internet of Everything (IoE), Smart systems (Homes, Cities, Meters, Grids *etc.*). IoT refers to addition of communications and sensing capabilities to a wide range of physical objects. In the coming decade it is expected that millions of IoT devices will be deployed in electric meters, parking meters, thermostats, car components, roads *etc.* IoT devices will send data directly using protocols such as Wi-Fi and Bluetooth and via mobile networks, specialized networks and over the global network.[61]

Advancement M2M technology will add entirely a new dimension to disaster early warnings. Technologically empowered sensors, gauges, meters will take measurements *viz.* hydrological measurements will talk to their respective devices at back end which in turn will inputs to weather prediction models and more accurate forecast/nowcast/early warnings will be generated. This machine trails would be extended to provide necessary triggers to mitigation and risk aversion mechanical models as well.

International Telecommunication Union (ITU) had established various Groups (Foucus Groups, study Group) with the objective of developing recommendations relevant to M2M from telecom/ICT perspectives. Study Group 20 (SG 20) is currently working to address the standardization requirements of Internet of Things (IoT) technologies, with an initial focus on IoT applications in smart cities and communities (SC and C). Mandate of SG20 is to develop international standards to enable the coordinated development of IoT technologies, including machine-to-machine communications and ubiquitous sensor networks. SG 20 core objectives inter alia include standardization of end-to-end architectures for IoT, and mechanisms for the interoperability of IoT applications and datasets employed by various vertically oriented industry sectors.[62]

Role of Communication in Disasters Management

It is said that for keeping people safe it is necessary to keep people connected. Communication is a part of everyday life. It has long been recognised that if a community is capable of receiving or communicating advance information on hazards, their associated adverse effects can be minimized. Effective early warning and proper evacuation orders empower individuals and communities to respond appropriately to a threat in order to reduce the risk of death, injury, and property loss and damage. The communication of warning message should inform what is happening, what it is relevant to that person, and what that person can do. It should be communicated in a clear, simple language and delivered with enough lead-time for the recipient to take any necessary action. In disaster prevention, the avoidance of hazards, telecommunications have a key role in the distribution of knowledge and in raising awareness. They are vital tools for day-to-day prevention and early warning. Telecommunications under these circumstances are a great challenge. One of the important aspects related to disaster prevention is a well-designed telecommunication network and information system that is suitable in coping with chaotic situations during and after disaster emergencies. Telecommunication and IT facilities are the two expected aspects in dealing with disaster and emergency situations (Fakaruddin, 2007).

The Tampere Convention

When disaster strikes immediate requirement is to communicate for response. The mode of communication may vary according to level of preparation of administration of the locality and local community. Rapid and accurate information exchange is vital for effective response. In many cases, it happens that when disaster strikes, due to chaos situation, wired and mobile communication traffic increase beyond traffic handling capacity of the system, and results in breakdown of communication systems. Sometimes due to disaster conditions such as cyclone, high wind, earthquake, flash flood *etc.* communication link disrupts and make the situation worse, as for rescue and relief workers who arrive on the scene these links are essential. Considering non-availability of essential communication during emergency situation, The Tampere Convention, fully entitled as "*The Tampere Convention on the Provision of Telecommunication Resources for Disaster Mitigation and Relief Operations*" *is a multilateral treaty,governing the provision and availability of communications equipment during disaster relief operations, particularly as regards the transport of radio and related equipment over international boundaries by radio amateurs was proposed at the First Intergovernmental Conference on Emergency Telecommunications (ICET-98) in Tampere, Finland, in 1998, and went into effect on 8 January 2005. Before this, the trans-border use of telecommunication equipment by humanitarian organisations was often impeded by regulatory barriers that make it extremely difficult to import and rapidly deploy telecommunications equipment for emergency without prior consent of the local authorities. The treaty simplifies the use of life-saving telecommunication equipment. The Tampere Convention calls on States to facilitate the provision of prompt telecommunication assistance to mitigate the impact of a disaster, and covers both the installation and operation of reliable, flexible telecommunication services. Regulatory barriers that impede the use of telecommunication resources for disasters are waived.* These barriers include the licensing requirements to use allocated frequencies, restrictions on the import of telecommunication equipment, as well as limitations on the movement of humanitarian teams. The Convention describes the procedures for request and provision of telecommunication assistance, recognising the right of a state to direct control and coordinate assistance provided under the convention within its territory. It calls on states to provide prompt telecommunication assistance to mitigate the impact of a disaster, and to ensure the installation and operation of reliable telecommunication resources. The treaty sets out procedures for requesting and providing telecommunication assistance and recognises the right of a State Party to direct, coordinate and supervise telecommunication assistance provided within its territory under the terms of this treaty.[63]

Public Protection and Disaster Relief (PPDR)

Public Protection and Disaster Relief (PPDR) radiocommunication systems, as name suggest are meant for maintenance of law and order, response to emergency situations, protection of life and property and response to disaster and for disaster relief facilitation. The new age broadband technologies have immense potential of enhanced capability and capacity to facilitate the achievements of both public protection operations and responding to major emergencies and catastrophic disasters.[64]

PPDR network are specifically designed to suit operational requirements and matching technologies to fulfill desirable operational requirements. Public protection functionality primarily requires:

- Wider Coverage;
- Easy Accessibility;
- Faster Service Continuity;
- High Reliability;
- Sufficient Failure Tolerance and Resilience;
- Multi-Agency Technical Interoperability;
- Fast Call Set-Up;
- One-Touch Broadcasting;
- One-Touch Emergency Alert (For Life Threatening Situations).

For disaster response and relief operations, communication network requires to have:

- Multimedia capability or integration of voice, data, video and image communications;
- User authentication (*e.g.* public key cryptography) among PPDR agencies
- Capability of working in extreme weather conditions;

- Robust equipment support;
- Local communication operations;
- Localized Communication Services;
- Direct Mode Operation (DMO) (In absence of availability of trunks);
- Push to Talk (Walki Talki type of communication);
- Scalability of Network infrastructure;
- Mobility of Network infrastructure;
- Quick deployment of temporary infrastructure;
- Network Resilience;
- Security which includes optional end-to-end encryption;
- Wide and reliable coverage;
- Useable and easily configurable user and system devices;
- Support to many types of PPDR applications;
- Prioritization of service access (Giving priority to PPDR agencies);
- Interoperability and interworking capability with national and international networks.

Abovementioned requirements of PPDR communication network demands for very high order of standardization of telecom equipment devices, technologies, network protocols, and frequency spectrum allocation for wireless devices. ITU and many telecom technology fora, consortium, technology invention and standardization bodies (ITU-R, European conference on Postal and Telecommunication, ETSI, 3GPP *etc.*) are working technological requirements of PPDR networks. LTE is being evolved as a future technology for PPDR networks.

ITU document Report ITU-R M.2377-0 (07/2015) on “Radiocommunication objectives and requirements for Public Protection and Disaster Relief (PPDR)” is very detailed document on PPDR. In India Telecom Regulatory Authority of India (TRAI) is to provide detailed recommendations on PPDR and regulatory framework for its implementation in India.

Mass Communication and Disaster Risk Reduction

The Sendai Framework for Disaster Risk Reduction 2015-2030 in clause 35(d) says that “ *Media to take an active and inclusive role at the local, national, regional and global levels in contributing to the raising of public awareness and understanding and disseminate accurate and non-sensitive disaster risk, hazard and disaster information, including on small-scale disasters, in a simple, transparent, easy-to-understand and accessible manner, n close cooperation with national authorities; adopt specific disaster risk reduction communications policies; support, as appropriate, early warning systems and life-saving protective measures; and stimulate a culture of prevention and strong community involvement in sustained public education campaigns and public consultations at all levels of society, in accordance with national practices".*

Mass communication, as its name suggests, is en masse communication to a group using mass media as medium of communication. It usually relates to print media such as Newspaper, Magazines, Books, and audiovisual media such as Radio, Television, Films *etc.* Study of mass communication specifies how the content of mass communication persuades or otherwise affects the behavior, attitude, opinion, or emotion of the person or people receiving the information (Bhatti and Ariyabandhu, 2002).

Mass communication plays a very important role in dissemination of early warning messages. There are two types of communication methods available for warning dissemination: mass notification methods; and addressable notification methods. Mass notification methods are not individually addressable and generally provide the same alert or message to everyone within a particular geographic area, regardless of level of individual risk. These include: outdoor systems; sirens; loud speakers, mobile electronic signs; and mass broadcasting systems such as conventional radio, television, cable television and low power radio. Addressable notification methods are tailored and targets alerts and messages only to those at risk or to specific groups (such as emergency responders). Some latest addressable technologies are flexible enough to support many of the same functions of traditional mass notification systems (Fakaruddin, 2007). These include:

- ☆ Broadcasting systems – provincial radio broadcasting, amateur radio, VHF/HF radios, weather radio, micro phone, mosques, madarsas,[65] temples, chuch, gurudwaras *etc.*
- ☆ Telecommunication systems – telephone, fax, cellular mobile, Short Message Service, paging and tone-alert radio, internet, multimedia broadcast, point to multipoint communications, VoIP, and satellite.
- ☆ Inter personal communication – door-to-door, peer to peer *etc.*

Prevention starts with information. It has been proved in many cases that where information was adequately transmitted to at-risk communities through the media, communities were able to take timely actions to avert considerable damage to lives and livelihoods.

Disasters and Communication Media

The media throughout the world play a vital role in educating the public about disasters, warning of hazards, gathering and transmitting information about disaster affected areas, alerting government officials, relief organisations and public to specific needs and facilitating discussions about disaster preparedness and response. Timely reporting media may result in saving of lives and reducing damages.

In the year 1979 Australian Counter-Disaster College published a report of a media (Broadcasters) Study Group. The group was invited to study the broadcast media in disasters to formulate guidelines for commercial radio broadcasters. A member of the study group (Mr A. Thompson) laid down the following guidelines for communicators to in relation to disasters" (Bhatti and Ariyabandhu, 2002)[66]

i. Inform the public on various matters to raise their level of preparedness
ii. Inform the public about the current situation.
iii. Inform the public about what has happened.
iv. Inform the public about the prognosis or the diagnosis.
v. Advise the public on course of action appropriate for the event.
vi. Advise the public on the actions being taken by authorities and aid groups.
vii. Relay messages concerning the welfare of isolated groups within the community.
viii. Maintain a reassuring presence.

Therefore the role of media, both print and electronic, in informing the people and the authorities during emergencies is very critical, as media can play a vital role in public awareness and preparedness through educating the public about disasters; warning of hazards; gathering and transmitting information about affected areas; alerting government officials, helping relief organisations and the public towards specific needs; and even in facilitating discussions about disaster preparedness and response. During any emergency, people are in distress they need to be given reliable, accurate and important information. The main principle of information provision, therefore, should be an ethical one: and so, during an emergency, the media should be sensitive to the needs of the public in affected areas and should avoid broadcasting unconfirmed reports that may lead to despair and panic. Reliable and timely information provided through the media can help people overcome their fears and insecurities in aftermath of an emergency. Media today has arguably penetrated every household in the world, in one form or another. This puts more and more responsibilities over them for fair and just reporting of an event.[67]

Conclusion

Communications is the key to the success of disaster mitigation, preparedness, response and recovery. Accurate information disseminated to the people at risk, first responders, concerned officials, community leaders, media, and all important stake holders reduces risk, saves lives and property, and speeds up recovery process. The ability to communicate is no longer an afterthought or a luxury; timely communication is now as important as logistics or the pre-deployment of materials. Planning and controlling the flow of information before, during and after a disaster defines credibility, trustworthiness, authority, and effectiveness of disaster management.The emergence of new tools and communication resources such as internet, e-mail, blogs, text messaging, cell phone photos, social media sites, instant messaging and chat, satellite phones, new RF technology instruments,voice over internet and multimedia connectivity are playing ever-increasing role in disaster communication and emergency management. In addition roles now played by "first informers" witnesses who now have the ability to transmit information immediately from the event are redefining the roles of government and media. The tools and rules of communications are evolving and disaster communications must evolve to capitalize on these changes and exploit the opportunities they provide.[68]

References

Bhatti, A. and Ariyabandhu, M. M. (2002). Disaster Communication : A Resource Kit for Media.

Davis, G. K. B. (1999). Electronic Communication Systems.

Fakaruddin, S H M (2007). Emergency Communications for Disaster Management. ADPC. Asian Disaster Management News. Jan-April 2007.

Notes

1. http://www.itu.int/ITU-D/emergencytelecoms
2. http://www.academia.edu/4358901/managerial-skill-development.pdf
3. Retrieved from: http://worrydream.com/refs/Shannonpercent.20-per cent.20Aper cent.20Mathematicalper cent.20Theoryper cent.20ofper cent.20Communication.pdf
4. http://communicationtheory.org/definitions-of-communication/
5. Encyclopedia of communication Theory: available at https://books.google.co.in/
6. Reflective listening is a communication strategy involving two key steps: seeking to understand a speaker's idea, then offering the idea back to the speaker, to confirm the idea has been understood correctly.
7. Nonverbal Communication Theories. (2009). In the Encyclopedia of Communication Theory available at https://books.google.co.in/
8. Haptics is any form of nonverbal communication involving touch.
9. Nonverbal Communication Theories. (2009). In the Encyclopedia of Communication Theory available at https://books.google.co.in/
10. Chronemics is the study of the use of time in nonverbal communication. The way that one perceives and values time, structures time and reacts to time frames communication. Across cultures, time perception plays a large role in the nonverbal communication process.
11. http://www.primidi.com/written_communication
12. en.wikipedia.org/wiki/Communication
13. Recently in Kanker a Tahsil in Jagdaplur distt. of Chhattisgarh state of India about 33000 years old pictures on walls of caves similar to Bhimbetka near Bhopal in Madhya Pradeh, India are found by archaeologists. These pictures

carved on stones are depicting stories of that time and are important historical written communication.

14. http://education-portal.com/academy/lesson/what-is-written-communication-in-business-definition-types-examples.html
15. https://books.google.co.in
16. "Crisis and Emergency Risk Communication" (2002) Centre for Disease Control and Prevention. Available at http://emergency.cdc.gov/cerc/resources/pdf
17. Retrieved from: http://worrydream.com/refs/Shannonper cent20-per cent20Aper cent20Mathematicalper cent20Theoryper cent20ofper cent20Communication.pdf and http://communicationtheory.org/shannon-and-weaver-model-of-communication/
18. training.fema.gov/EMIWeb/edu/docs/techem/
19. Draft Disaster Risk reduction guidelines for banking sector as lending institutions for Telecom Sector, prepared by Telecommunications Consultants India Limited (TCIL)
20. A local-area network (LAN) architecture developed by Xerox Corporation in cooperation with DEC and Intel in 1976. Ethernet uses a bus or star topology and supports data transfer rates of 10 Mbps. The Ethernet specification served as the basis for the IEEE 802.3 standard, which specifies the physical and lower software layers. Ethernet uses the CSMA/CD access method to handle simultaneous demands. It is one of the most widely implemented LAN standards. A newer version of Ethernet, called *100Base-T* (or *Fast Ethernet*), supports data transfer rates of 100 Mbps. And the newest version, Gigabit Ethernet supports data rates of 1 gigabit (1,000 megabits) per second. (http://www.webopedia.com/TERM/E/Ethernet.html)
21. *Radio waves* are a type of electromagnetic radiation with wavelengths in the electromagnetic spectrum longer than infrared light. Radio waves have frequencies from 300 GHz to as low as 3 kHz, and corresponding wavelengths ranging from 1 millimeter. (0.039 in) to 100 kilometers. (62 mi). Like all other electromagnetic waves, they travel at the speed of light. Radio waves are used for fixed and mobile radio communication, broadcasting, radar and other navigation systems, communications satellites, computer networks and innumerable other applications. Different frequencies of radio waves have different propagation characteristics.
22. http://en.wikipedia.org/wiki/Core_network
23. http://en.wikipedia.org/wiki/Radio_access_network
24. http://innovation.verizon.com/content/dam/vic/PDF/LTEper cent20Theper cent20Futureper cent20ofper cent20Mobileper cent20 Broadbandper cent20Technology.pdf

25. http://innovation.verizon.com/content/dam/vic/PDF/LTEper cent20Theper cent20Futureper cent20ofper cent20Mobileper cent20Broadbandper cent20Technology.pdf
26. Draft Disaster Risk reduction guidelines for banking sector as lending institutions for Telecom Sector, prepared by Telecommunications Consultants India Limited (TCIL) for NDMA.
27. NDMICS (2011) ibid
28. In a hierarchical telecommunications network the backhaul portion of the network comprises the intermediate links between the core network, or backbone network and the small sub-networks at the "edge" of the entire hierarchical network. As at https://en.wikipedia.org/wiki/Backhaul_(telecommunications)
29. https://www.online-sciences.com/the-waves/the-importance-of-lasers-in-communication-and-advantages-of-fiber-optics-cables/
30. Modem is a combined device for modulation and demodulation. Short for modulator demodulator. A modem is a device or program that enables a computer to transmit data over, for example, telephone or cable lines. Computer information is stored digitally, whereas information transmitted over telephone lines is transmitted in the form of analog waves. A modem converts between these two forms.
31. In India, spectrum for various telecommunication services, broadcast, cable TV, amateur radio *etc.* are allocated by Wireless Planning Coordination(WPC) wing of Department of telecommunication, Ministry of Telecommunication and IT.
32. NDMICS (2011)
33. Draft Disaster Risk reduction guidelines for banking sector as lending institutions for Telecom Sector, prepared by Telecommunications Consultants India Limited (TCIL).
34. Rayleigh scattering is the (dominantly) elastic scattering of light or other electromagnetic radiation by particles much smaller than the wavelength of the light. After the Rayleigh scattering the state of material remains unchanged, hence Rayleigh scattering is also said to be a parametric process. The particles may be individual atoms or molecules. It can occur when light travels through transparent solids and liquids, but is most prominently seen in gases. Rayleigh scattering results from the electricpolar is ability of the particles. The oscillating electric field of a light wave acts on the charges within a particle, causing them to move at the same frequency. The particle therefore becomes a small radiating dipole whose radiation we see as scattered light. Rayleigh scattering of sunlight in the atmosphere causes diffuse sky radiation, which is the reason for the blue color of the sky and the yellow tone of the sun itself.

35. Telecom subscription data as on 31st March 2017 as released by TRAI. http://trai.gov.in/
36. Telecom subscription data op.cit
37. Retrieved from: http://www.scholarpedia.org/article/Global_system_for_mobile_communications_(GSM)
38. Picture source (Source: http://www.radio- electronics.com/info/satellite/communications _satellite/)
39. http://www.radio-electronics.com/info/satellite/communications_satellite/communications-satellite-technology.php
40. http://www.radio-electronics.com/info/satellite/satellite-orbits/low-earth-orbit-leo.php
41. http://www.isro.gov.in/applications/telecommunication
42. https://www.cse.iitb.ac.in/synerg/lib
43. Draft Disaster Risk reduction guidelines for banking sector as lending institutions for Telecom Sector, prepared by Telecommunications Consultants India Limited (TCIL).
44. Ibid.
45. http://www.sac.gov.in/SACSITE/TTIDWebsite/gallery/MSSDAT.pdf
46. https://en.wikipedia.org/wiki/Broadcasting.
47. www.dth.co.in/dth-satellite.html
48. (http://en.wikipedia.org/wiki/Earth_observation).
49. Draft Disaster Risk reduction guidelines for banking sector as lending institutions for Telecom Sector, prepared by Telecommunications Consultants India Limited (TCIL).
50. NDMICS(2011) ibid
51. https://en.wikipedia.org/wiki/WiMAX
52. http://www.ehow.com/about_6690422_disaster-management-communication.html
53. The 3rd Generation Partnership Project (3GPP) unites six telecommunications standard development organisations (ARIB, ATIS, CCSA, ETSI, TTA, TTC), known as "Organizational Partners" and provides their members with a stable environment to produce the reports and specifications that define 3GPP technologies. The project covers cellular telecommunications network technologies, including radio access, the core transport network, and service capabilities including work on codec, security, quality of service and thus provides complete system specifications. The specifications also provide hooks for non-radio access to the core network, and for interworking with Wi-Fi networks.

54 Draft Disaster Risk reduction guidelines for banking sector as lending institutions for Telecom Sector, prepared by Telecommunications Consultants India Limited (TCIL).

55. https://en.wikipedia.org/wiki/telepresence

56. https://en.wikipedia.org/wiki/Cognitive_radio

57. www.itu.int/en/ITU-T/gsi/ngn/Pages/definition.aspx

58. http://www.dot.gov.in/sites/default/files/Nationalper cent20Telecomper cent20M2Mper cent20Roadmap.pdf

59. An actuator requires a control signal and a source of energy and is responsible for moving or controlling a mechanism or system. The control signal may be low energy electric voltage or current, pneumatic or hydraulic pressure, or even human power. (https://en.wikipedia.org/wiki/Actuator)

60. TR 102 725 V1.1.1 (2013-06)

61. http://trai.gov.in/sites/default/files/Consultation_Paper_M2Mper cent20_18_October_2016.pdf

62. Taken From http://www.itu.int/en/ITU-T/about/groups/Pages/sg20.aspx

63. http://www.itu.int/ITU-D/emergencytelecoms/tampere.html

64. Report ITU-R M.2377-0 (07/2015)

65. A place, where followers of Islam go for formal and religious study.

66. Disaster Communication : A Resource Kit for Media by Amjad Bhatti and Madhvi Malagoda Ariyabandhu, 2002.

67. Retrieved from: http://nidm.gov.in/idmc2/pdf/concept/Media.pdf

68. A book on Disaster Communications in a Changing Media World By George Haddow and Kim Haddow.

Chapter 6

ICT ENABLED DISASTER COMMUNICATION SYSTEMS IN INDIA

Introduction

ICT (information and communications technology or technologies) is an all-inclusive term for communication devices or applications, such as radio, television, cellular phones, satellite systems, computer, laptops, memory storage and network hardware and software, and so on, as well as the various services and applications associated with them. Chapter 3 of this book "ICT: An enable for effective disaster management" establishes important link between ICT and efficacy of disaster management. Chapter 4 of the book "Early Warning Systems" emphasized on role of early warning systems as an important contributor for disaster-risk reduction, which could save lives and help protect livelihoods and national development gains. Early warning systems have been recognized as an effective tool to reduce vulnerabilities and improve preparedness and response to natural hazards. Communication systems are needed for delivering warning messages to alert local and regional governmental agencies of the likely affected locations. Warning messages need to be reliable, consistent and simple to be understood by authorities and public.

ICT consist of range of tools, applications, systems, used to input, store, edit, retrieve, analyze, synthesize and process information and share data in all its forms. ICT encompasses: radio, television, broadband, satellite and cellular mobile phones, computer and network hardware and software, websites, portals, remote sensing, satellite systems and so on, as well as the various services and applications associated with them, such as data storage, analysis and integration, videoconferencing and distance learning[1] (Nonita, 2011). For ICT to play its role communication medium or channel is required through which interconnection of various equipment and transmission of signals/messages could take place. Therefore, communication infrastructure is the backbone for the so called "magic wand" of ICT to act.

The telecom sector of India has a big success story and shown spectacular growth in last 10-15 years. It has grown multifold in size from both in terms of increase in tele-density[2] as well as service offerings through technological advancements. The Indian telecom sector in the year 2000 had subscriber base of about 28.53 million which had phenomenally grown to 1,204.98 million consists of wireless segment of 1,180.82 million and wireline subscriber base of 24.16 million as on 31 May 2017. Subscription in urban areas was 697.1 million with a teledensity of 172.3, whereas rural area subscription was 508 million with a teledensity of 57.6 at the end of May 2017. Total number of broadband subscribers was 291.61million.[3]

National Telecom Policy

There are multiple factors which contributed to the phenomenal growth of telecommunication sector in India. In 1994, the Government announced the National Telecom Policy (NTP 1994), which defined important objectives, including availability of telephone on demand, provision of world class services at reasonable prices, ensuring India's emergence as major manufacturing/export base of telecom equipment and universal availability of basic telecom services to all villages. It also announced a series of specific targets to be achieved by 1997.

NTP 1994 also recognized that requisite resources for achieving the targets aimed in NTP may not be met from Government sources and concluded that private investment and involvement of the private sector to bridge the resource gap is critically important. Government invited private sector participation in a phased manner. In the early nineties initially it was for value added services such as Paging Services and Cellular Mobile Telephone Services (CMTS) and thereafter for Fixed Telephone Services (FTS). After a competitive bidding process, licenses were awarded to 8 CMTS operators in the four metros, 14 CMTS operators in 18 state circles, 6 BTS operators in 6 state circles and paging operators in 27 cities and 18 state circles. VSAT services were liberalized for providing data services to closed user groups.[4]

Subsequently in the year 1999, a new Telecom Policy (NTP 99) [5] was adapted with broader objectives, such as, to:

- ✰ Provide a balance between the provisions of universal service to all uncovered areas, including the rural areas, and the provision of high-level services capable of meeting the needs of the country's economy.
- ✰ Encourage development of telecommunication facilities in remote, hilly and tribal areas of the country.

- Create a modern and efficient telecommunications infrastructure taking into account the convergence of IT, media, and telecom and consumer electronics and thereby propel India into becoming an IT superpower.
- Convert PCO's, wherever justified, into Public Tele-info centers having multimedia capability like ISDN services, remote database access, government and community information systems *etc.*
- Transform in a time bound manner, the telecommunications sector to a greater competitive environment in both urban and rural areas providing equal opportunities and level playing field for all players.
- Strengthen research and development efforts in the country and provide an impetus to build world-class manufacturing capabilities.
- Achieve efficiency and transparency in spectrum management.
- Protect defense and security interests of the country.
- Enable Indian telecom companies to become truly global players.

Therefore, broader vision of NTP 1999 *inter-alia*, were to:

- Make available telephone on demand;
- Encourage development of telecom in rural areas making it more affordable by suitable tariff structure and making communication mandatory for all fixed service providers;
- Increase rural tele-density and provide reliable transmission media in all rural areas;
- Achieve telecom coverage of all villages in the country and provide reliable media to all exchanges by the year 2002;
- Provide Internet access to all district headquarters by the year 2000;
- Provide high speed data and multimedia capability using technologies including ISDN to all towns with a population greater than 0.2 million by the year 2002;

It is evident that most of the objectives as envisioned by NTP 99 were not only achieved but surpassed by huge margins well before the timelines. Policy framework as envisioned in NTP 99 created an environment, which enabled continued attraction of investment in the sector and allowed creation of communication infrastructure by leveraging on technological advancements. Meanwhile government also allowed Foreign Direct Investment (FDI) in telecom sector, facilitated growth of the sector. The preamble of NTP 99 says 'The Government of India (Government) recognizes that provision of world class telecommunications infrastructure and information is the key to rapid economic and social development of the country. It is critical not only for the development of the Information Technology industry, but also has widespread ramifications on the entire economy of the country. It is also anticipated that going forward, a major part of the GDP of the country would be contributed by this sector. Accordingly, it is of vital importance to the country that there be a comprehensive and forward looking telecommunications policy which creates an enabling framework for development of this industry'.

Subsequently, NTP 2012 has also been formed with a vision of "Broadband on Demand". It envisages leveraging telecom infrastructure to enable all citizens and business both in rural and urban areas to participate in internet and web based economy, thereby ensuring equitable and inclusive development across the nation. Various majors taken under policy framework of NTP 99, proved to be very successful and it resulted in spectacular growth of telecom sector in India. Based on this, importance of ICT, as an enabler for overall development of the country, was further reinforced; as a consequence, telecommunication has been identified as driver for economic and social development in current telecom policy (NTP 2012).

Broadband Policy

Government of India embarked on multiple actions focused on ICT for inclusive development of people from all strata of the society. Department of Telecommunication brought out Broadband Policy in 2004. Its preamble says "*Recognizing the potential of ubiquitous broadband service in growth of GDP and enhancement in quality of life through societal applications including tele-education, tele-medicine, e-governance, entertainment as well as employment generation by way of high speed access to information and web-based communication, Government have finalized a policy to accelerate the growth of broadband services.*"

Universal Service Obligation Fund (USOF)

Universal Services refers to the idea that basic infrastructure such as electricity, transportation, water and telecommunication *etc.* should be accessible to every citizen in the country. Apart from the higher capital cost of providing, telecom services in rural and remote areas, these areas also generate lower revenue due to lower population density, low income and lack of commercial activity. Thus normal market forces alone would not direct the telecom sector to adequately serve backward and rural areas. Keeping in mind the inadequacy of the market mechanism to serve rural and inaccessible areas on one hand and the importance of providing vital telecom connectivity on the other, most countries of the world have put in place policies to provide Universal Service and Universal Access to ICT.[6,7]

NTP'99 provided that the resources for meeting the Universal Service Obligation (USO) would be raised through a 'Universal Access Levy (UAL)', which would be a percentage of the revenue earned by the operators under various licenses. The Universal Service Support Policy came into effect from 01.04.2002. The Indian Telegraph (Amendment) Act, 2003 giving statutory status to the Universal Service Obligation Fund (USOF) was passed by both Houses of Parliament in December 2003. The Rules for administration of the USOF known as Indian Telegraph (Amendment) Rules, 2004 were notified on 26.03.2004. As per the Indian Telegraph Act 1885 (as amended in 2003, 2006 and 2008), the USOF is to be utilized exclusively for meeting the requirements for universal service provisioning.

As per various recommendations of TRAI (TRAI, Telecom Infrastructure policy Recommendations, 2011, p. 102), the ubiquitous service was aimed to be achieved through roll out obligation under licensing conditions for telecom service providers and through use of USOF. Various schemes are run by USOF to provide voice and data connectivity, through fixed and wireless mode of communication to rural and far flung areas of India. The TRAI, in its recommendations on "Spectrum Management and Licensing Network" dated May 2010 and "National Broadband Plan" dated December 2010, revisited the issue of penetration of mobile services and broadband in the rural areas. It spelt out about need of creation of a robust telecom infrastructure which can meet the backhaul bandwidth requirement and create competition. TRAI, recommended that optical fiber should be laid up to villages so that broadband requirement of end user can be met by different services providers such as Internet Service providers (ISP), cable TV operators, wireless broadband providers. Accordingly it suggested various conditions to be imposed to telecom service providers under roll out obligations and for USOF to implement new schemes.

Other Important Initiatives

Over the years, a large number of initiatives have been taken by various state governments and central ministries to usher in an era of e-Government[8]. Sustained efforts have been made at multiple levels to improve the delivery of public services and efforts are made to simplify the process of accessing such services. e-Governance in India has steadily evolved from computerization of government departments to initiatives such as citizen centricity, service orientation and transparency. Lessons from previous e-Governance initiatives have played an important role in shaping the progressive e-Governance strategy of the country. Due cognizance has been taken of the notion that to speed up e-Governance implementation across the various arms of government at national, state, and local levels, a programme approach needs to be adopted, guided by common vision and strategy. This approach has the potential of enabling huge savings in costs through sharing of core and support infrastructure, enabling interoperability through standards, and of presenting a seamless view of government to citizens.

The 'National e-Governance Plan (NeGP)', takes a holistic view of e-Governance initiatives across the country, integrating them into a collective vision, a shared cause. Around this idea, a massive ICT infrastructure is reaching down to the remotest of villages in the country. Creation of ICT infrastructure have far reaching benefits such as easy access to Internet which in turn have many benefits for education, telemedicine, healthcare, common service delivery to citizens *etc.* NeGP envisions bringing all public services closer to home for citizens of India.

The government approved the National e-Governance Plan (NeGP), comprising of 27 Mission Mode Projects (MMPs) and 8 components, on May 18, 2006. In the year 2011, 4 projects - Health, Education, PDS and Posts were introduced to make the list of 27 MMPs to 31MMPs. The Government has accorded approval to the vision, approach, strategy, key components, implementation methodology, and management structure for NeGP. However, the approval of NeGP does not constitute financial approval(s) for all the MMPs and components under it.

In order to promote e-Governance in a holistic manner, various policy initiatives and projects have been undertaken to develop core and support infrastructure. The major core infrastructure components are State Data Centers (SDCs), State Wide Area Networks (SWAN), Common Services Centers (CSCs) and middleware gateways that is National e-Governance Service Delivery Gateway (NSDG), State e-Governance Service Delivery Gateway (SSDG), and Mobile e-Governance Service Delivery Gateway (MSDG). The important support components include core policies and guidelines on security, HR, citizen engagement, social media as well as standards related to metadata, interoperability, enterprise architecture, information security *etc.*

"National Information Technology Policy 2012" has been brought out by the government with a vision "*To strengthen and enhance India's position as the Global IT hub and to use IT and cyber space as an engine for rapid, inclusive and substantial growth in the national economy*". Mission and objectives of the policy inter alia include leveraging ICT for enhanced competitiveness and productivity of key economic and strategic sectors and provisioning of ubiquitous affordable access to information and public services for enhancing efficiency, transparency, accountability and reliability using ICT as a driver for social inclusion of everyone.

Telecom connectivity is being assigned a very high priority as it is the vehicle for launch of all e-governance initiatives, and achievement of targets of central and state government social sector schemes. Therefore, upgradation of rural telecom infrastructure is being assigned utmost priority. Broadband coverage of all 2,50,000 village Panchayats and setting up of Common Service Centers (CSCs) at all Panchayats is the basis of "Bharat Nirman" programme. Now the latest government initiative is "**Digital India**" programme. This ambitious programme is centered on 3 visions: digital infrastructure as a utility to every citizen; governance and services on demand; and digital empowerment of citizens. The "**Digital India**" initiative has true potential to completely transform India to a developed nation. It is intended to develop ICT infrastructure so that it becomes all pervasive and has an impact in every aspect of lives of Indian citizens. The Digital India programme was launched in 2015. The program has now moved from the planning phase towards execution phase. There are many schemes/programmes such as Development of smart cities, e-governance initiatives, Optical fiber connectivity of rural areas through Bharat Net, financial inclusion of people from all strata of society through M- banking initiatives *etc.* which are being run along with **Digital India.**

Current Telecom Scenario

At present in India multiple telecom service providers are operating under licenses as issued by Department of Telecommunication (DOT), Ministry of

Communication. These are public and private operators who have created their telecom networks, based on various terrestrial and wireless technologies. Core infrastructures of almost all networks are primarily built up using Optical Fibre (OFC) network. In areas where it was not feasible to lay OFC, wireless and satellite communication technologies are used. OFC based communication networks have been extensively installed throughout the country by government Public Sector Undertakings such as Bharat Sanchar Nigam Limited, Power Grid Corporation of India Limited, Railtel and others as well as private telecom service providers (TSP). Hybrid Fibre Coaxial (HFC), Fibre to the Curb (FTTC) and Fibre to the Home (FTTH) networks make use of fiber cabling into the last mile. The fiber based models are future proof as they are able to provide huge amounts of bandwidth in the last mile[9] and deliver high quality voice, data and video. A fairly large satellite based communication networks also exist in the country namely NICNET, POLNET, ISRODMS and EDUSAT. In addition, there are multiple Internet service providers and Virtual Network Operators (VNO) operating in the country. Similarly, there are networks specially established by the concerned agencies in the country disseminating early forecasting of different natural hazards. International long distance calls are covered with help of submarine cables as well as communication satellites.

Holistic management of disasters call for assigning priority to prevention, mitigation and preparedness activities in the pre-disaster scenario of the DM continuum, while strengthening the efforts for faster and more efficient response through properly trained personnel equipped with advanced equipment/ instruments and better recovery during rehabilitation and reconstruction period. Timely deployment and use of telecommunication resources play a crucial role in disaster mitigation and relief operations. It is extremely important to establish telecommunication resources at the national level and cooperation amongst various states and non-state entities to facilitate the use of the same. Identifying importance of communication during various phases of DM, DoT in its draft Telecom Policy (NTP)-2011 has emphasized upon the need to build resilient telecom networks in India. Some references of draft NTP 2011, which specifically deal with disaster communication needs, are mentioned in succeeding part of this paragraph:

"Para 13 of Preamble - NTP-2011 recognizes the importance of creation of the robust and resilient telecom networks for adequately addressing the need for proactive support for mitigating disasters, natural and manmade."

"Para 5.13 of Strategies - To prescribe sectoral standard operating procedures for effective and early mitigation during disasters and emergencies. To mandate telecom service providers to provide alternative reliable means of communication at the time of disaster by creating appropriate regulatory framework".

"Para 7.7 of Strategies - To promote creation of robust, reliable and resilient communication networks"

These provisions of the draft telecom policy have far reaching implications for a country like India which is prone to natural disasters like floods, earth-quakes, coastal cyclones *etc.* on one hand and manmade disasters like accidents, terrorist attacks *etc.* on other. Hence, putting in place processes and structures to ensure reliable means of communications during disasters/emergencies in India will definitely help in better response and recovery (Telecom Regulatory Authority of India Consultation paper On Telecom network failures during Emergencies/ Disasters Priority routing of calls of persons engaged in 'response and recovery' released on 10th May, 2012).

NTP-2012 also "*recognizes the importance of creation of the robust and resilient telecom networks for adequately addressing the need for proactive support for mitigating disasters, natural and manmade*". Besides NTP 2012 also emphasizes importance of Disaster Management and various provisions it mentions about requirement "*To create appropriate regulatory framework for provision of reliable means of public communication by Telecom Service Providers during disasters (Para 5.13 of Strategies)*"; *and "To encourage use of ICTs in prediction, monitoring and early warning of disasters and dissemination of information* (*Para 5.14 of Strategies*)". NTP 2012 also provides "*To facilitate an institutional framework to establish nationwide Unified Emergency Response Mechanism by providing nationwide single access number for emergency services*".

Accordingly Department of Telecommunication, under Ministry of Communication of Government of India, has prepared a Standard Operating Procedure (SOP-2017) for responding to disasters and to be adopted by all telecom service providers in India. SOP lays down procedures for:

- ☆ Co-ordination of national actions to ensure the provision of telecommunication support to the center, state and district administration.
- ☆ Co-ordination of the requirement of temporary telecommunication services in the affected areas.
- ☆ Co-ordination for restoration of telecommunication services.

SOP[10] also proposes institutional framework in form of National State and District level committees and their areas of work/responsibilities and their reporting structures to higher authorities at DOT.

Indian telecom market is considered to be amongst the fastest growing telecom markets in the world. NTP 99 fuelled immense growth of the sector in India, while sector was opened to private operators. At present numerous public and private telecom operators are providing telecom services in India. Subsequent part of this chapter gives description of various telecommunication networks owned by central government ministries/departments/public sector undertakings, in the country. This information is being furnished with the intention to explain existing status of public network and exploring possibilities of utilization of same for disaster management and emergency communication purpose.

Telecommunication Networks owned by Government Institutions/ Public Sector Undertakings (PSUs)

There are numerous telecommunication networks presently functional in public domain by various government agencies/institutions/service providers. However many of them are designed for specific purposes and have their own merits and demerits based on their data carrying capacities, point of presence (POP), technology adopted for transmission/reception,coding, bandwidth offered and access to general administration at various levels. Some of them, which are prominent in size and can play a significant role in disaster management and emergency communication, are discussed as under:

Police Telecommunication Network (POLNET)

POLNET is a satellite based specialized wide area network (operating on C-band of INSAT-3E) owned by directorate of coordination - Police Wireless under the Ministry of Home Affairs. This is a specialized network meant for the Police organization. Bandwidth constraints (64Kbps/512 kbps) limit the use of this network for high resolution images or video.

POLNET is an amalgamation of different types of, Very Small Aperture Terminals (VSAT)[11] technologies for example TDM/TDMA[12], SCPC[13]/DMA[14] (Demand Assigned Multiple Access) and DVB-S (Digital Video Broadcast-Satellite)[15]. The HUB of POLNET is installed at New Delhi with 11 meter antenna with a capacity to support total VSAT Network of about 1500 locations for voice, data and fax facilities. TDM/TDMA VSATs work on double hop link through the central Hub at New Delhi for communication with other sites. State Headquarters with Hybrid VSAT terminal (Star and DAMA modes) can communicate directly with better voice quality utilizing single hop SCPC/DAMA Channel. The connectivity from District Headquarters (connected in Star mode) is being extended up to Police Stations/Police Posts through local Radio Network using MART (Multi Access Radio Telephone) system. POLNET also provides connectivity for interlinking NCRB[16] computers to SCRB [17]and District Crime Record Bureau DCRB computers at State/ District HQs for online transaction processing.

At present more than 960 VSATs are operational and network can support up to 1500 VSAT terminals. VSAT Terminals with 3.8 meter antenna are installed at locations at the capital of States/UTs and some of the CPOs locations supporting both TDM/TDMA and SCPC/DAMA scheme for Star and Mesh connectivity while about 820 TDM/TDMA V-SAT Terminals with 1.8/2.4 M Antenna Size are installed at District Headquarters of all states and UTs and other important locations of CPMFs, MHA., NCRB, and CPOs *etc.* Broadband VSAT terminals have been installed at various locations of BSF and ITBP. About 40+ nodes have a video transmission facility.

National Informatics Centre (NICNET) Infrastructure

National Informatics Centre (NIC) is the nodal information technology organization for informatics development and networking in government and related organizations. NIC provides ICT support to the government in all central Ministries/Departments, 35 states/UTs and 630+ districts. NICNET, the nationwide network has more than 60,000 end users. More than 2 million users access NIC portals every day for information and services. The data centers of NIC host thousands of websites of the government. NIC has the largest email service with more than 80 million emails per day. It has largest video conferencing network, which handles more than 25,000 conference hours per month. A high speed NICNET National info-Highway, an incremental overlay over the existing network had been setup with SCPC, TDMA VSATs, DVB Broadband VSAT technology and Wireless WAN to take advantage of Internet technology in India.

NICNET (Guideline, 2012, p. 19)[18] was primarily using terrestrial connectivity with a minimum of 2 Mbps links to connect all districts in the country with their respective state capitals. Most of the Central Government Ministries/Departments in Delhi are connected through Gbps bandwidth capacity. The terrestrial connectivity is used for running all the three services such as data, voice and video conferencing. However, the VSAT connectivity is currently being used at locations where terrestrial connectivity is not available (< 5 to 10 per cent locations) or as a backup during a failure in terrestrial connectivity. The backup for data service has been provided over the DVB technology VSATs and the backup for video conferencing service has been provided using the SCPC DAMA technology VSATs. The leased line network has a central Network Management System (NMS), while each type of VSAT Hub has different NMS to manage the network.

NICNET currently operates at Ku Band VSAT network connecting all districts (barring a few newly created ones).This satellite has shared bandwidth and individual limit of the VSAT is 512 Kbps. Some of the districts in the country are also connected through video-conferencing facility on NICNET through 2Mbps DAMA VSATs. In some States, such as North-east and Sikkim, J&K (under implementation), NICNET connectivity extends up to the block level through the CICs (Community Information Centers) via a 64 Kbps satellite link. Orissa is implementing a VSAT network up to the block level through NICNET. NICNET has about 2500 VSATs operating in districts, blocks of Jammu and Kashmir, Chhattisgarh, Orissa and few North Eastern States.

NIC state centers are functioning 24X7 and NIC district centers are functioning from 8 am to 8 pm for providing ICT support to government. NICNET is connected to central government ministries/departments/institutions, state/UT departments and districts through terrestrial link, offering sufficient bandwidth. It is also having VSAT network as standby in difficult and far flung areas; it is therefore having potential to be suitably utilized for disaster management related activities for pre, during and post- disaster phases.

Disaster Management Support (DMS) Network of ISRO

Department of Space (DOS) had embarked upon the Disaster Management Support (DMS) Programme as a prime application activity, to harness the benefits of the space based technology for the applications in disaster management in the country.

ISRO has set up a satellite based Virtual Private Network (VPN) facilitating secure data access through a dedicated electronic network connecting all the key responsibility centres for disaster management in India. The National Emergency Operation Centre (NEOC) of MHA, Cabinet Secretariat, NDMA, PMO, other key data providing agencies (IMD, CWC, GSI, SOI, INCOIS, NIDM), and the 20 multi-hazard prone State Emergency Operation Centres (SEOCs) are connected to the Decision Support Centre (DSC) of NRSC / ISRO. For a fail-safe communication during emergency, ISRO/DOS has developed INSAT Type-D terminals (portable satellite phones with solar chargeable battery packs). These terminals are positioned in different ISRO/ DOS Centres, for deployment by various States during disasters. INSAT based Distress Alert Transmitter (DAT) is a low cost satellite transmitter for emergency communication of alert messages from fishing boats. DATs are distributed to fishermen though the Indian Coast Guard. The DTH based Disaster Warning Dissemination System (DWDS) is a customized Set-top box developed through BEL, Bangalore for dissemination of Disaster warnings directly to the potential victims. In the DTH based warning system, individual or group of DTH receivers can be selected for sending siren warning followed by voice message.[19]

State Wide Area Network (SWAN)

As the major initiative for ICT support in the country, SWAN is the approved scheme of government of India for interconnecting state headquarters with district headquarters and district headquarters with block headquarters with minimum 2 Mbps leased lines, in order to create secure government network for the purpose of delivering Government-to-Government (G2G) and Government to Center (G2C) services for supporting various e-Governance initiatives being undertaken by the respective states and central ministries.

Under this Scheme, every state/UT headquarter is connected in a tree/star connection up to the block level, with minimum 2 Mbps capacity per link. Link from State Headquarter (SHQ) to District Headquarter (DHQ) upgraded up to 4Mbps. All the government offices are connected with the respective Point of Presence (PoP) at the State/District/Block level. In the states, there are two options for establishment of SWAN namely NIC model and PPP (Public Private Partnership) model. In both cases, private operators are involved either for operation of the network or for Facility Management Services (FMS).

Common Services Center (CSC) Scheme

The CSC is a strategic cornerstone of the National e-Governance Plan (NeGP), approved by the government in May 2006, as part of its commitment in the National Common Minimum Programme to introduce e-governance on a massive scale. Accordingly CSC e-Governance Services India Limited[20] is a special purpose vehicle (SPV) formed on July 16, 2009. The SPV monitors and enables delivery of public and private services across India. CSC ecosystem is expected to eventually comprise of more than one lakh village level entrepreneurs (VLEs), 35 State Designated

Agencies (SDAs), over 20 Service Center Agencies (SCAs) and a host of public and private entities as service providers.

The CSCs provide high quality and cost effective video, voice and data content and services, in the areas of e- governance, education, health, telemedicine entertainment as well as other private services. A highlight of the CSCs is that it will offer web-enabled e-governance services in rural areas including application forms, certificates and utility payments such as electricity, telephone and water bills. Following services can be offered through CSCs:

- Agriculture Services (Agriculture, Horticulture, Sericulture, Animal Husbandry, Fisheries, Veterinary)
- Education and Training Services (School, College, Vocational Education, Employment, *etc.*)
- Health Services (Telemedicine, Health Check-ups, Medicines).
- Rural Banking and Insurance Services (Micro-credit, Loans, Insurance).
- Entertainment Services (Movies, Television).
- Utility Services (Bill Payments, Online bookings).
- Commercial Services (DTP, Printing, Internet Browsing, Village level BPO).

The scheme creates conducive environment for the private sector and NGOs to play active role in implementation of the CSC Scheme there by becoming a partner of the government in development of rural India. The PPP (Public Private Partnership) model of the CSC scheme envisages a 3-tier structure consisting of the CSC operator (called Village level Entrepreneur or VLE) the Service Center Agency (SCA) that is responsible for a division of 500-1000 CSCs and State Designated Agency (SDA) identified by the State Governments responsible for managing the implementation over the entire state.

The CSC network consists of infrastructure that is developed in the rural areas for delivery of citizen centric services and to reach out to the citizens. Various G2C services such as land records, electoral services, birth and death certificates, panchayat services, capturing of information and relevant data and reports for various federal government and local government run schemes, are being offered through the CSCs.

The CSC network can be efficiently utilized, by using CSCs as disaster management and mitigation points for awareness building and media campaigning. The Campaign may include film shows, radio jingles, and awareness campaigns in pre-defined intervals. The CSCs may target schools for audio-visual spots and short documentary films. Important data, information brochures, training modules, related to capacity building programmes for village panchayats and members of other such local governance units/civil defense personals/volunteers/Anganwadi/ Asha workers *etc.*, who can play a definitive role in disaster preparedness, prevention and mitigation measures, may be kept.

Emergency Planning and Response System (for Chemical Industrial Incidents)

Ministry of Environment and Forests of India has established a communication system called 'GIS Based Emergency Planning and Response System' for major cluster of industries developed to assist the various response agencies for planning emergency situation arising out of the spillage of hazardous chemicals. The system was developed in technical collaboration with the Remote Sensing and Geographic Information System (RS and GIS) division of NIC. Geographical Information System (GIS) based emergency planning[21] and response system has been developed for major cluster of industries in selected industrialized states in the country. The system integrates critical information in a spatial form under a single banner through GIS (Geographical Information Systems) and incorporates computer modeling for major chemical accident scenarios. The project is funded by the Ministry of Environment and Forests, Government of India. It's front-end software is supported by data on major accident hazard units, first responders (fire, police, hospitals, *etc.*), sensitive areas, chemical accident scenarios, spatial data (detailed digitized maps with data in 'layers'), Response Information Data Sheets or RIDS packed in a database all under a single banner. Locations of industries, responders and sensitive areas as recorded by GPS are shown on these maps. For any chemical accident scenario, one can view the template on the base map and logical queries can be handled. It can greatly assist response agencies at the local level during both pre-emergency state for planning and rehearsing and also during actual emergency situation.[22] Further, this has been integrated with the Chemical Accident Emergency Planning and Reporting System (CAIRS), maintained by MoEF and NIC jointly involving Districts having Major Accident Hazard (MAH) Units. CAIRS is a web-based system. It provides visual data of chemical accients in India.

The National Knowledge Network

The idea of setting up the NKN[23] was approved in March 2010 by the Cabinet with an outlay of Rs 59900 Million, to be implemented by NIC over a period of 10 years.

NKN with its multi-gigabit capability aims to connect all universities, research institutions, libraries, laboratories, healthcare and agricultural institutions across the country to facilitate paradigm shift in the field of educational research and innovation through efficient knowledge management. Establishing NKN is a significant step towards ushering in a knowledge revolution in the country with connectivity to 1600+ institutions. The leading mission oriented agencies in the fields of nuclear, space and defense research are also part of NKN. NKN is aimed to connect all the knowledge and research institutions in the country using high bandwidth/low latency network. Establishing a high-speed backbone connectivity which will enable: knowledge and information sharing; collaborative research; development and innovation; and advanced distance education in specialized fields such as engineering, science, medicine *etc.* NKN is intended to facilitate an ultra-high speed backbone for e-governance facilitating integration of different sectoral networks in the field of research, education, health, commerce and governance.[24]

The architecture of NKN has been designed for reliability, availability and scalability. The network consists of an ultra-high speed core, starting with multiple 2.5/10 G and progressively moving towards 40/100 Gigabits per Second (Gbps). The core is complimented with a distribution layer covering all districts at appropriate speeds. The backbone of the network starts from 2.5 Gbps and progressively moves onto 10 Gbps connectivity between 7 Super core (fully meshed) locations pan India.

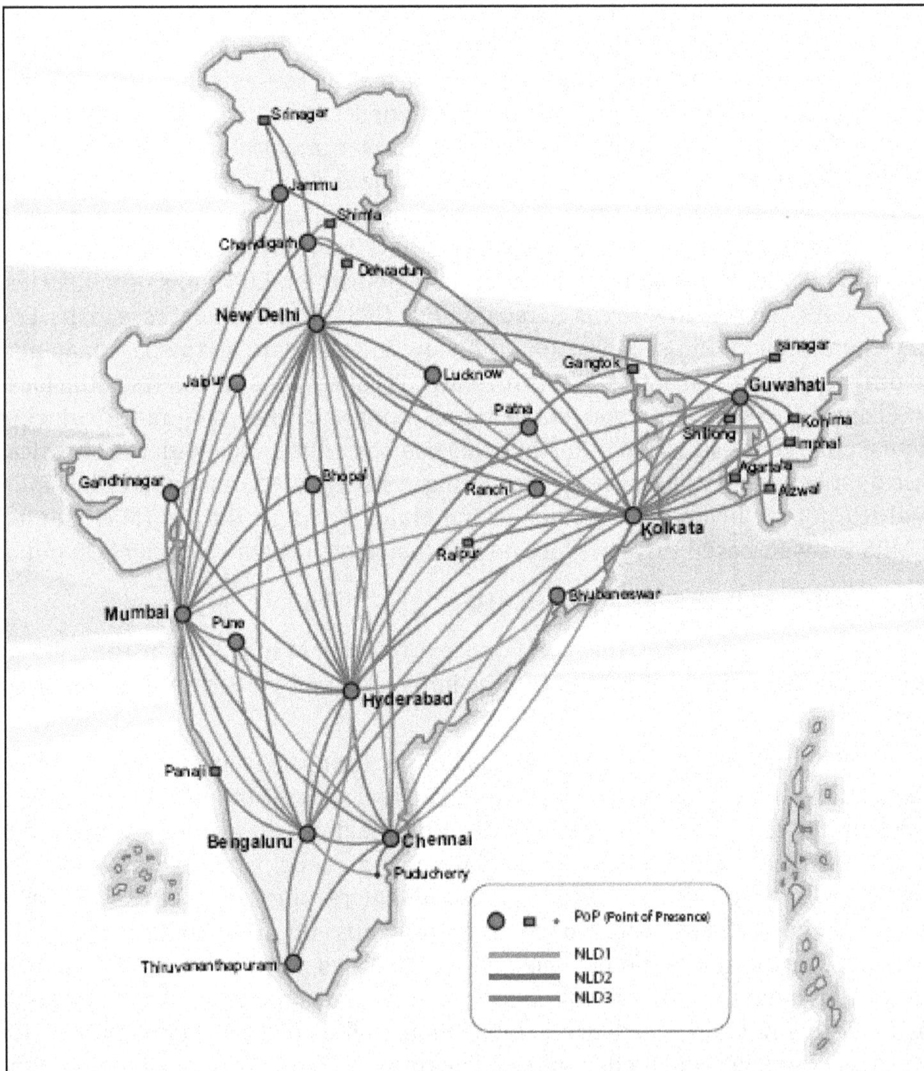

Multiple 2.5/10 G Connecting all the State Capitals & Gigabit Connectivity to all the Districts

Figure 6.1: National Knowledge Network.

The network is further spread out through 26 Core locations with multiple of 2.5/10 Gbps partially meshed connectivity with Super core locations. The distribution layer connects entire country to the core of the network using multiple links at speeds of 2.5/10 Gbps. The end users are being connected up to a speed of 1 Gbps. The network architecture and governance structure allows users with options to connect to the distribution layer as well. NKN enables creation of Virtual Private Networks (VPN) as well for special interest groups. NKN also provides international connectivity to its users for global collaborative research.

Following technical features are envisaged for NKN:

- NKN is a protocol independent network and is designed to carry multiprotocol traffic.
- NKN is capable of offering hierarchical Quality of Service (QoS) for real time traffic (voice and video) and guaranteed bandwidth for business critical applications. NKN governance backs it up by Service Level Agreements (SLAs) for the users.
- NKN design supports IPv6 transport, IPv6 networking and IPv6 MPLS VPN services in addition to the similar facilities based on traditional IPv4.
- NKN design, implementation, management, and control is such that service provisioning is internal to the NKN network and does not depend on telecom service providers from whom "raw" bandwidth or fiber is likely to be leased.
- NKN design supports multicast enabled VPN for running multicast applications, both in IPv4 and IPv6.
- NKN management is capable of handling provisioning for the central services such as multimedia conferencing, e-access, digital library, and central data centre to all users.
- NKN provides access to secure data centre with information assurance.
- NKN offers cloud based services, network services and application services.

NKN is evolving as National Educational Research Network. In view of the high spread and high bandwidth of this network, the National Knowledge Network, may be envisaged to become a driver for paradigm shift in the field of disaster management in India.

Indian Meteorological Department (IMD) Network

The Directorate of Tele-communication was set up in IMD at New Delhi in 1969 to cater the needs of National Meteorological Service and strengthen the Meteorological Telecommunication in India.[25] Since its inception, IMD maintains an extensive telecommunication network for speedy collection of meteorological information, both basic data and processed products, over the globe. The main telecommunication hub of the IMD's telecommunication network is located in headquarter at New Delhi. This is internationally known as Regional Telecommunication Hub (RTH) under the aegis of World Meteorological Organization (WMO). IMD communication network has evolved from manual/ semiautomatic systems during time of its inception to TCP/IP based services and Internet including security aspects, with a view to ensure efficient and safe operation of information systems. It works on automatic decoding and analysis and forecast by using high performance computing system running on highly sophisticated and complex numerical weather prediction models. The output of these models are objectively visualized by the forecasters in highly sophisticated graphic workstations to overlay with satellite and radar products.

The meteorological telecommunication in IMD consists of an integrated network of point-to-point and point to multipoint (MPLS VPN) links and meteorological centers within the country and the world for receiving data and relaying it selectively. It is mainly organized on a two level basis, namely:-

i) The Meteorological Telecommunication Network (MTN) within the Global Telecommunication System (GTS) of World Weather Watch (WWW) program of World Meteorological Organization (WMO);
ii) The National Meteorological Telecommunication Network (NMTN).

The Global Telecommunication System (GTS), the main part of WMO, National Meteorological Telecommunication Centre (NMTC), in Mausam Bhavan, New Delhi, acts at RTH (Regional Telecommunication Hub) of the GTS of WMO.

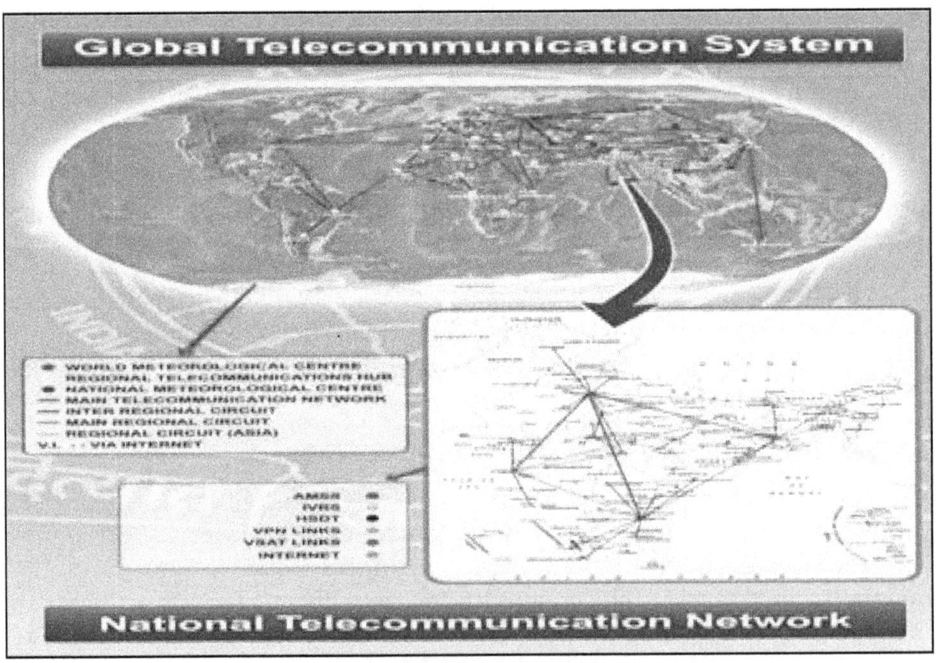

Figure 6.2: GTS Network[26].

New Delhi telecommunication center is a designated as Regional Telecommunication Hub (RTH) located on the Main Telecommunication Network (MTN). The MTN is the core network of GTS which connects three World Meteorological Centers (WMCs) and 14 other RTHs on the MTN.

National Meteorological Telecommunication Network (NMTN)

The National Meteorological Telecommunication Network (NMTN) is designed to collect/receive observational data and to receive and distribute meteorological information on national level. Basic data from the observatories are collected at 19+ collection centers (which are either Meteorological Centre at the State Capitals or Regional Meteorological Centre) by various communication means. The stations included in the list of Global Observing System (GOS), the observational data of which are transmitted on GTS, are mostly connected by point-to-point/point-to-multipoint (MPLS VPN) links.

Satellite based Data Communication in IMD

Satellite based data collection/distribution systems are integrated in the MTN as an essential element at the global, regional and national level. Medium and high speed links connecting RTH New Delhi with Meteorological Telecommunication Centers at the major international airports in India are operating via satellite. A network of 26+ V-SATs have been installed at selected Seismological Observatories, Cyclone Detection Radar stations, Cyclone Warning Centres and Meteorological Centers for reception of observational data utilizing communication transponder of INSAT. Processed INSAT cloud imageries, GTS data and analyzed weather charts are transmitted through satellite every half hour through Digital Meteorological Data Dissemination (DMDD). At present there are about 37 DMDD ground receiving stations in India (IMD) and one each in Maldives, Nepal and Sri Lanka.

Various modes of communication such as dedicated high speed leased line circuits, VPN links, Telefax and internet facility are extensively being used for exchange of observational data within the country. Services of Maritime Telex Stations (NAVTEX) are used for sending warnings/coastal weather bulletins by Area Cyclone Warning Centre (ACWC), MUMBAI for the Arabian Sea, Area Cyclone Warning Centre (ACWC), Kolkata for the Bay of Bengal, Indian Ocean and Southern Hemispheric Analysis Centre (INOSHAC), Pune for the Indian Ocean north of the equator to ships in Met area VIII (N) (north of equator).

IMD warning disemination network is combination of multiple mode of connectivity namely dedicated leased lines, MPLS VPN, IP(FTP) CCTs, Normal Telephone lines, VSAT *etc.* This network is based on links hired from various public and private telecom service providers therefore it may not act as an independent dedicated disaster management network.

Indian Environmental Radiation Monitoring Network (IERMON)[27]

BARC has established a countrywide environmental radiation-monitoring network IERMON. The network has 25 stations across the country. Network is aimed for:

1. Providing on-line information about radiation levels at various stations to the Emergency Control Rooms of Department of Atomic Energy (DAE).
2. Facilitating environmental impact assessment from nuclear emergencies.
3. Establishing background environmental radiation levels and monitoring long term shift in them.
4. Measurement of external radiation dose rates at ground level all the time.
5. Airborne radioactivity assessment on weekly basis.
6. Providing redundancy for critical parameters.
7. On-line information about radiation levels at various locations in the country to emergency control rooms of DAE facilities.
8. Data on background environmental radiation levels and long term shift in the background levels.
9. Data for environmental impact assessment following nuclear emergencies.

The Central Station of the Network comprises one PC for downloading data from remote stations through 'PSTN' connection, and one for maintaining the database. The IERMON is designed for radiation monitoring thus cannot be used for generic disaster management purpose.

Railway Communication Network

Indian Railways had been using various forms of communication technologies namely Microwave, Overhead lines, VHF, OFC systems, Mobile Train Radio Communication (MTRC) for extensive connectivity network to facilitate train operations. With increased demand for a large number of such connectivity, the original analog microwave systems were replaced with digital microwave systems for the major train routes. Subsequently, when IT systems were planned, it was decided to lay Optical Fiber Cables (OFC) as a backbone to meet the demand for bandwidth through a railway owned PSU, RAILTEL Corporation. (Established in September, 2000 to look after the backbone facilities) offering connectivity to Railway establishment through fiber based broadband and multimedia network (using latest SDH/DWDM technology based transmission system and high-end MPLSIP routers) laid along railway tracks.

RailTel has more than 42,000 Route Km of Optical Fiber Cable running along Indian Railway Track in many part of the country. RailTel is having fibers at every station en-route, spaced at 8-10 Kms to meet Railway operations. RailTel is laying fibers in uncovered sections and shall complete 54,000Km of Rail route and covering most of the stations and commercial requirement (5000+) on its backbone. This network is designed in a layered for its efficient utilization and management. The three layers namely Access Layer, Edge Layer and Backbone layer enables required capacity with desired granularity. Access layer provides of STM-1 (155Mbps) capacity at stations, spaced at every 8-10 Kms. The traffic from access layer is

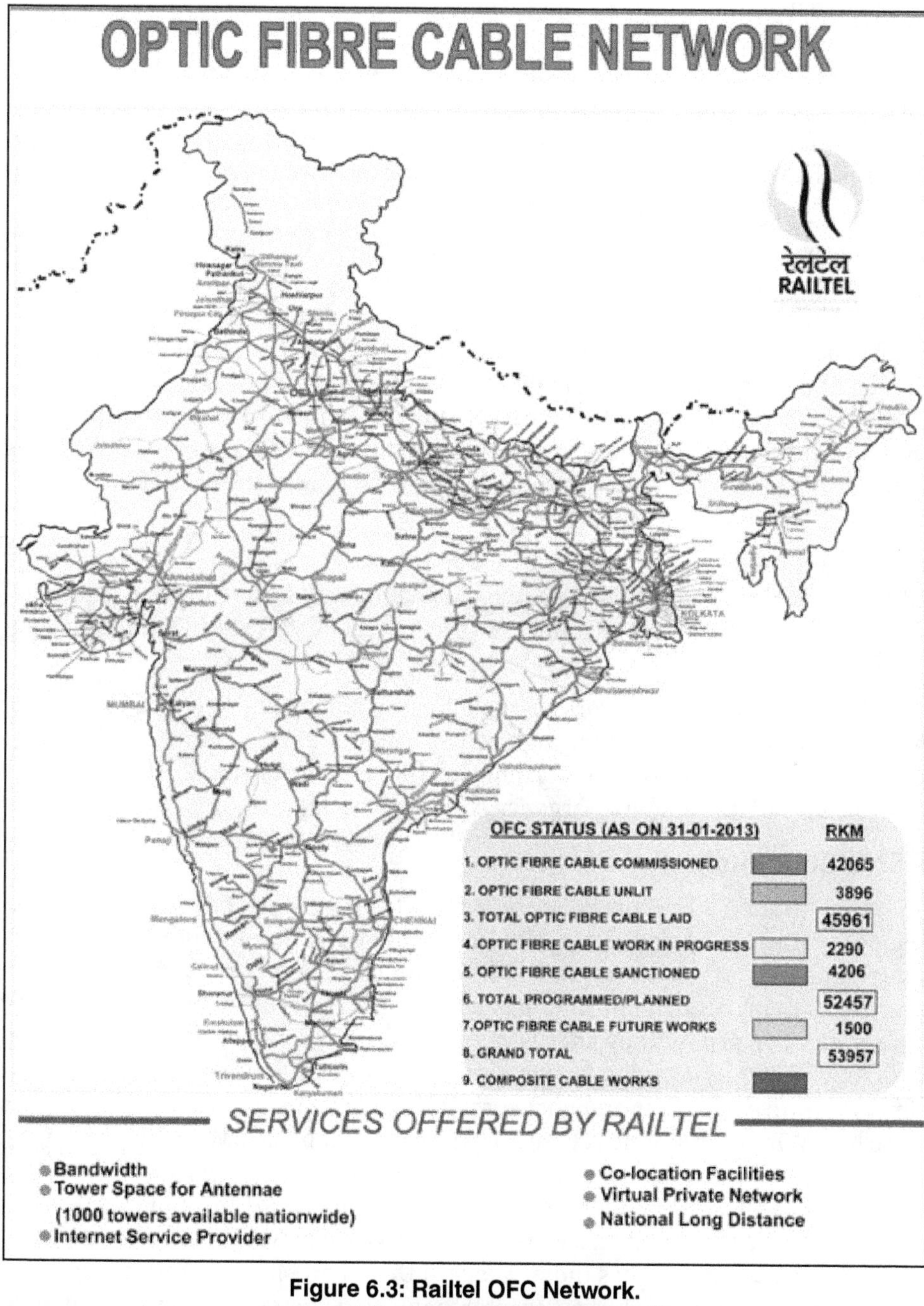

Figure 6.3: Railtel OFC Network.
(*Source*: http://www.railtelindia.com/).

aggregated on to the edge layer having STM-16 connectivity, which is available in important locations at every 50-60 Kms. Collecting the traffic from edge layer is finally aggregated on to the backbone layer with DWDM/STM-16/64 available in big cities/towns situated at every 60-70 Kms. Backbone network have been configured in multiple 'Self-Healing' ring architecture which provides redundancy by automatically redirecting traffic away from failed/de-graded route. At very remote locations, where commercial activities take place, connectivity is provided through V-sat links for which a railway owned HUB station has been established under the Freight Operations Information System (FOIS) with hiring of transponder services. These are being planned to be augmented by having some mobile V-sat units linked to Accident Relief Trains.

To provide whole range of VPN (like Layer 3 and Layer 2 VPN) and broadband internet access, multicast services, RailTel has implemented countrywide MPLS-IP[28] backbone network using high end routers. The complete MPLS network is managed by centralized Network Management System (NMS). RailTel has got the unique advantage to meet the quality bandwidth and service requirements from single network. The state of art network enables point and click provisioning of the bandwidth from anywhere to anywhere in the country. It enables provisioning of traffic of any granularity with the extensive reach from any part of the country to any other part.

In addition to above, RailTel is also rolling out a broadband wireless access (BWA) network across India. BWA will be a standards technology for "last mile" solutions that can connect customers to the RailTel ISP/MPLSPOPs through high-speed wireless access.

RailTel is having a network very big in size, having reached to thousands of railway stations connected through reliable media of OFC. Other than specific conditions of disasters such as earthquake, OFC is disaster safe media. Therefore RailTel may be utilized for emergency communication.

BSNL Communication Network

As per National Telecom Policy NTP 1999, the government decided to separate the policy and licensing functions of Department of Telecommunications (DOT) under ministry of communications from the service provision functions as a precursor to corporatization. Bharat Sanchar Nigam Ltd. (BSNL) was carved out by corporatization of operation and service provisioning wings of DOT. It was formed in October, 2000, as largest telecommunications service Provider Company of that time. BSNL provides comprehensive range of telecom services in India such as Wire line, Wire Less Mobile (2G and 3G) (CDMA[29] mobile, GSM Mobile, WiMax[30]), Internet, Broadband, Carrier service, MPLS-VPN, VSAT, VoIP services, IN Services *etc.* Being a public sector undertaking, BSNL makes efforts to bridge the rural-urban digital divide with its wide network giving services in every nook and corner of country and operates across India except in Delhi and Mumbai. Services in Delhi and Mumbai are provided by having interconnection and roaming agreement with;

MTNL another telecom Service provider public sector undertakings of government of India, under Ministry of Communications.

BSNL has set up a world class multi-gigabit, multi-protocol convergent IP infrastructure that provides convergent services like voice, data and video through the same backbone and broadband access network. It has built a very effective IP backbone network in name of NIB (National Internet Backbone Network of BSNL), which was planned to cover complete country for provisioning of highly effective data services.

BSNL has a long distance network consisting of more than 658000 Root kilometers of optical fiber. This covers a significantly large segment of optical fiber communication in the rural areas. With Dense Wavelength Division Multiplexing (DWDM) technology[31] extensively deployed, the optical fiber communication network provides huge bandwidth capacity in gigabits rate BSNL is instrumental in OFC deployment for Bharat Broadband Network Limited (BBNL) for setting up National Optical Fiber Network (NOFN).

BSNL has wide area coverage of cellular mobile service and extensive connectivity provided through the wire-line network penetrating into rural areas. BSNL offers 2G and 3G voice and data services to its mobile customers. Its mobile network is very effective and wide reaching across the length and breadth of the country.

BSNL also has a satellite based network and it offers services based on satellite communication to its subscribers. BSNL has a very large network spreading across length and breadth of India. It also has satellite based connectivity to far flung areas. BSNL network has full potential to be used for disaster communication purpose.

Power Grid Corporation of India (PGCIL) Network

PGCIL, is a *Navratna* Company and Government of India Enterprise incorporated in1989 for transmitting bulk power across the country. It has been contributing significantly towards development of Indian power sector. It is the central transmission utility and one of the largest electric power transmission utilities in the world with a vast network of more than 74000 circuit- kms of transmission lines, wheeling 45 per cent of India's total power generation.

Using its power transmission infrastructure to spread benefit of revolution in communication reaching across the length and breadth of the country, PGCIL has diversified into telecom by establishing about 41,899 km of high capacity Optical Fiber Cable, out of which more than 20,000 km is over its transmission infrastructure (overhead) across the country with 290+ PoPs POWERTEL, the telecom arm of PGCIL has the unique distinction of being the only telecom player with a countrywide presence of optical fiber network, using overhead OPGW (Optical Ground Wire) technology. It offers connectivity to remotest, most difficult of terrains and far flung areas.

OPGW network is on its transmission lines (overhead) which are sturdy, secured, vandalism proof and all weather networks. OPGW does not require any routine maintenance, which is an essential criterion for networks in such far flung areas. In case of disasters like earthquake, floods *etc.* power transmission is restored using Emergency Restoration System (ERS). Same benefit can also be availed in restoring communication network on OPGW.

In (September: 2014) Jammu and Kashmir Flood situations telecommunication networks were not functioning; as inundation of water caused failure of terrestrial links. In this catastrophic situation, PGCIL network which has overhead fiber links, were used for restoration of communication in the valley.

Therefore with multiple benefits of OPGW technology, network reach in far flung areas it has immense potential to be used for emergency communication purpose.

Conclusion

- ✰ There exist several ICT networks operational both in public and private domain, but they all are operating in a 'Stand Alone' mode and need to be integrated appropriately. For example, existing police network operates as Intranet and is inaccessible to an outsider user; even through such accessibility is essential during disaster situations.
- ✰ Keeping in mind the time sensitivity; and volume of data transfer needed for effective and reliable communication to the right stakeholder at the right time, various technological options need to be explored.Multiple technological options are to be adopted for various phase of DM. To ensure that early warning (EW) and forecasting reach the community in the likely affected area, all the technological options available with public and private ICT operators, including the services of individual members/ NGOs have to be utilized to achieve disaster management objectives.
- ✰ The present availability of last mile connectivity (LMC), especially for the affected community and disaster managers, is inadequate. In addition, very limited provisions exist for providing effective emergency communications to the affected community at the time of disaster.
- ✰ Existing Service Provider networks are mostly based on terrestrial communications which, are invariably inadequate in their design to withstand earthquake, wind forces, water inundations, or not located at safe heights to avoid damage due to floods. As a result, such networks are vulnerable for break down at the time of disaster. Hence, satellite connectivity is an effective alternative to provide communication links during disasters.
- ✰ POLNET if suitably upgraded for enhancement in data transmission rate may be utilized as satellite communication based disaster management network of India. It has connectivity to all police control rooms which can be converted in to EOC or any other important office as per requirement

of IRS (Incident Response System) during emergency. However some of terrestrial networks are having capacity to serve as DM network in most of the disaster situation. BSNL network reaches to most of difficult areas, and with some system overhauling for enhancement of capacity and power availability, may be used as nationwide DM network. NICNET is connected to central Government Ministries/Departments/Institutions, State/UT departments and District headquarters through terrestrial links, offering sufficient bandwidth. NICNET also has VSAT network as standby in difficult and far flung areas; it is therefore having potential to be suitably utilized for disaster management related activities in pre-disaster, during and post-disaster phases. Similarly other network such as POWERTEL, National Knowledge Network (NKN), RailTel have capacity and potential to be used for emergency communication. Specifically POWERTEL and RailTel networks are very resilient networks.

- Apart from major networks mentioned above there are multiple private telecom service provider networks having wider coverage and are equipped with latest technologies such as 3G/4G/LTE which offer greater wireless bandwidth. These networks can also be used for emergency communication.
- Mechanisms such as use of telecom licensing condition, prior service level agreements with state governments, invoking of Disaster management Act 2005 by state and local administration *etc.* may be used to wield power for utilization of private and public telecom networks during crisis situations.
- All existing telecom networks, public or private, do not offer facility of priority call routing. During the time of emergency, communication traffic exceeds traffic handling limits of these network; this results in breakdown of telecommunication networks. Therefore to handle such situation it is necessary that traffic need to be controlled and prioritized. Traffic from pre-identified telephone numbers of all important institutions, administrative officers need to be assigned priority during emergency.
- The CSC network can be efficiently utilized, by using CSCs for awareness building and media campaigning on disaster preparedness and mitigation non structural measures. The Campaign may include film shows, radio jingles, and awareness campaigns in pre-defined intervals. The CSCs may target schools for audio-visual spots and short documentary films.
- Though a number of disaster related Websites have been established at National and State levels, these are not interactive in nature. Additionally, all State level Web-sites need to be multi-lingual (Guideline, 2012, p. 37). [32]There is immense scope of information sharing and building a mechanism for knowledge Management on the subject. Even existing websites of apex institutes such as NDMA, NIDM, and MHA lack this capability.

Notes

1. TechTarget 2004; UNESCAP 2008; IBRD 2010; Nonita, 2011.
2. Number of telephone connections per 100 people in a specified geographic area. Teledensity is often used to compare the level of access to voice and data communications services between metropolitan and rural areas, or between one country and another. The governments of many emerging economies are focused on increasing teledensity as an economic enabler.
3. Source: TRAI report on The Indian Telecom subscription data (13 July 2017).
4. www.trai.gov.in› Acts and Policies
5. Ibid
6. http://www.dot.gov.in/
7. http://usof.gov.in/usof-cms/
8. http://deity.gov.in/content/national-e-governance-plan
9. Last mile refers to the part of the telecom network through which users are connected to the main network to have access to telecom services.
10. http://www.dot.gov.in/dataservices/standard-operating-procedure-sop-2017-telecommunication-services-responding-disasters.
11. VSAT (Very Small Aperture Terminal) is a satellite communications system that serves home and business users. A VSAT end user needs a box that interfaces between the user's computer and an outside antenna with a transceiver. The transceiver receives or sends a signal to a satellite transponder in the sky. The satellite sends and receives signals from an earth station computer that acts as a hub for the system.
12. Time-division multiplexing (TDM) is a method of putting multiple data streams in a single signal by separating the signal into many segments, each having a very short duration. Each individual data stream is reassembled at the receiving end based on the timing.
13. SCPC Single channel per carrier refers to using a single signal at a given frequency and bandwidth. Most often, this is used on broadcast satellites.
14 Demand Assigned Multiple Access (DAMA) is a protocol used in satellite communications, particularly Very Small Aperture Terminal (VSAT) systems. DAMA enables efficient and instantaneous assignment of transponder channels on a first come, first served basis according to data priority.
15. Digital Video Broadcasting (DVB) is a set of standards that define digital broadcasting of data (voice, video, television signals *etc.*) using existing satellites.
16. The National Crime Records Bureau, abbreviated to NCRB, is an Indian government agency responsible for collecting and analyzing crime data as

defined by the Indian Penal Code (IPC). As an attached office of Ministry of Home Affairs (MHA), Government of India, NCRB is headquartered in New Delhi.

17. State Crime Record Bureau.
18. NDMA Guideline: National Disaster management Information and communication System
19. https://www.isro.gov.in/applications/infrastructure-disaster-management
20. http://csc.gov.in//images/CSC-Manual.PDF
21. Chapter 32 of Compendium of e-Governance Initiatives in India "GIS Based Emergency Planning and Response Systems" by Indrani Chendrasekharan Project Leader and Diector NIC and MoEF, NCT of Delhi
22. Ibid
23. http://www.nkn.in/
24. http://www.nkn.in/
25. http://www.imd.gov.in/section/rth/static/country_net.htm
26. Picture Source: http://www.imd.gov.in/section/rth/
27. Retrieved from dae.nic.in
28. Multiprotocol Label Switching (MPLS) is a standards-approved technology for speeding up network traffic flow and making it easier to manage. MPLS involves setting up a specific path for a given sequence of packets, identified by a label put in each packet, thus saving the time needed for a router to look up the address to the next node to forward the packet to.
29. CDMA code Division Multiple Axis a wireless technique for voice data transmission
30. WiMAX (Worldwide Interoperability for Microwave Access is a wireless communications standard
31. Dense wavelength division multiplexing (DWDM) is a technology that puts data from different sources together on an optical fiber, with each signal carried at the same time through separate light wavelength.
32. NDMA Guideline on National Disaster management Information and communication System 2012

Chapter 7

CONCLUSIONS AND RECOMMENDATIONS

Introduction

The very purpose of this book relates to prognostic analysis of existing early warning systems and disaster communication facilities in India; finding gaps in prevailing disaster management practices, suggesting pragmatic long-term and short-term solutions. The contents and contexts are important not only for India but for other developing, and under developed countries in the world.

It has been practically seen and been proved over time that losses in disaster aftermath can be substantially minimized if proper measures are taken with right spirits in pre-disaster phase. Pre-disaster preparedness such as formulations of proper policy, plans and SOPs; awareness generation; training and capacity building of all stakeholders; resource creation and inventory management; creation of ICT infrastructure for information and data creation and sharing; are imperatives for effective disaster management. Proper handling of pre-disaster and during disaster situations translate into effective post-disaster recovery, rehabilitation and reconstruction. Holistic management of disasters, therefore, calls for assigning priority to the prevention, mitigation and preparedness activities during the pre-disaster scenario within the disaster management continuum. Strengthening of efforts for faster and more efficient response through properly trained personnel equipped with advanced equipment/instruments and building better appropriate during rehabilitation and reconstruction period can be culmination of pre-disaster efforts.

During disasters, information is as much a basic necessity such as water, food, shelter, medicine, and clothes.[1.] Responding effectively to disasters demands rapid access to reliable and accurate data; mechanism to convert data in to useful ground information; and the capacity to assimilate and analyze data and timely decisions making. It requires speedy communication between administration, stakeholders and key actors: to organize resources (for example food, water, medical supplies, and other key emergency relief supply materials); to mobilize, deploy and coordinate emergency personnel; and to coordinate response activities among all agencies involved. Proper communication and information sharing are therefore very critical for effective management of disasters.

A Case: Proposed Application of ICT for Drought Risk and Vulnerability Research in "Bundelkhand" in India.

Incidences of drought are recurring phenomena in India. It should be noted that drought is different from aridity and seasonal aridity since it is an irregularity of the climatic condition of the afflicted region. Bundelkhand region has a long standing history of droughts and famines. Historically, the region was thickly forested but is now characterized by bare hilly terrain with sparse vegetation. Regular failures of the monsoon have severely affected available water reservoir in river systems of the area. The resulting diminishing water levels in surface water sources as well as depletion of groundwater tables have not only decreased the availability of drinking water for people and domestic animals, but have also impacted natural vegetation and growing grasses (crucial as fodder). Most tribal populations inhabiting forests areas adjacent to rivers have no choice but to continue to exploit forests for their survival which cause further over exploitation of already depleting resources.

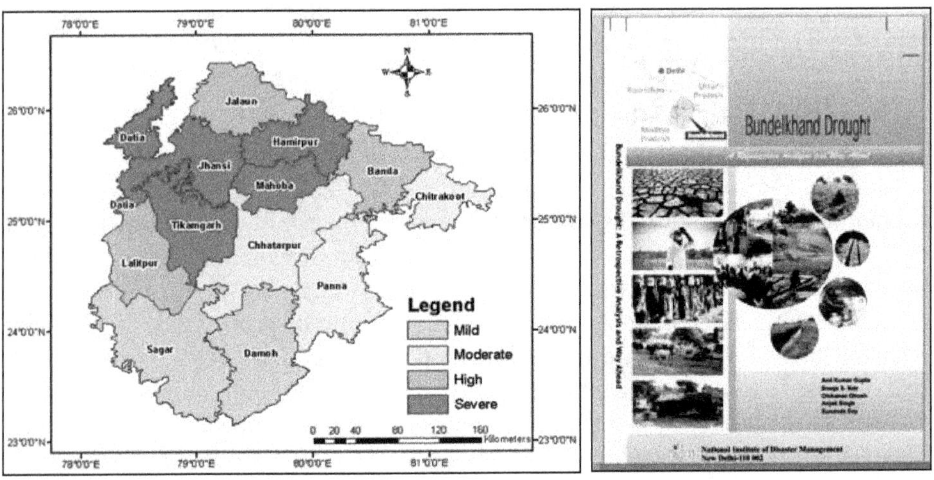

Figure 7.1: ICT Application in Drought Risk and Vulnerability Research in India.

A scientific study of the Bundelkhand drought has been carried out using space based data along with meteorological and hydrological data, for understanding of meteorological and hydrological patterns; and resultant agricultural drought in 13 districts from Madhya Pradesh and Uttar Pradesh (Gupta *et al.,* 2013). The indices included deciles of Precipitation, Standard Precipitation Index, Percent By Normal, Standardized Water Level Index, Normalized Difference Vegetation Index, Vegetation Condition Index, and using geographical information system to collate all the hazard vulnerability maps to arrive at composite risk maps at district level which also shown varying patterns of different types of drought. Based on results and inferences of ibid study, a new tool called as Mitigation Analysis has been suggested and evolved to capture the impact of government and community interventions in form of specific developmental programmes/actions related to natural resource management. Social and livelihood promotion for their drought mitigation impacts which often has a spatial dimension but rarely captured in assessments, are also taken care of.

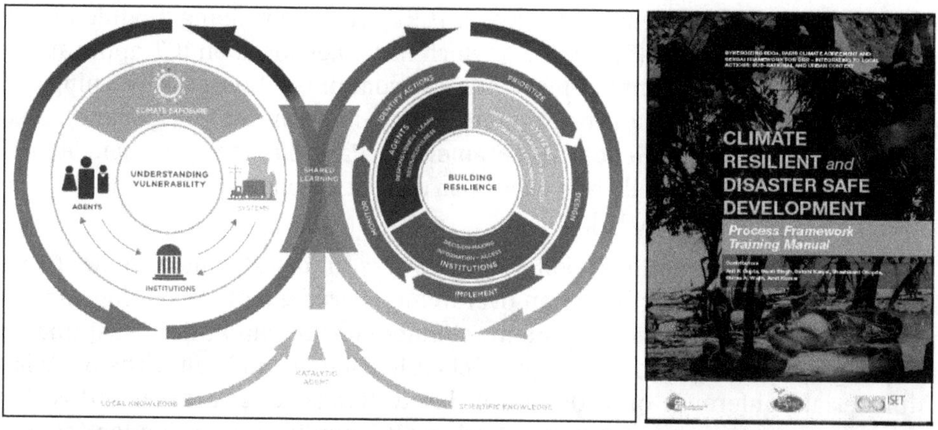

Figure 7.2: Climate Resilience Framework and Extrapolation Integrating DRR at Sub-national Level (Gupta *et al.*, 2016).

A Case: Risk and Vulnerability Assesment in Gorakhpur

A principle of climate and disaster risk factoring into developmental planning process at level of Governments and people's action at sub-national and local levels, more particularly districts, urban local bodies and panchayats/villages has been recognized as fundamental to disaster risk reduction. An approach called shared learning dialogue (SLDs) was used in participatory and action research process in understanding the flood risk and vulnerability of key departments and resources in eastern Uttar Pradesh with case study of Gorakhpur to draw lessons for state level and national level DRR mainstreaming process, with support of Climate Development Knowledge Network (CDKN), State Government, Gorakhpur Environmental Action Group, Institute of Social and Environment Transition (USA) and National Institute of Disaster Management New Delhi. Information and Communication Technology related departments have come to the fore of DRR for two specific aspects:

(a) ICT as critical emergency system that plays key role in early warning and risk communication and for effective coordination of emergency response, relief, rehabilitation *etc.*

(b) ICT as critical infrastructure which itself is vulnerable to the situations of natural hazards such as flood, earthquake (and its infrastructures/ resources and functions get affected).

(c) ICT infrastructure should take care of existing vulnerabilities and should help in further decreasing disaster risks.

The approach of risk and vulnerability assessment developed in Gorakhpur pilot was further extended at State level with district pilots to Odisha (Puri district – coastal, multihazard) and Uttarakhand (Almora district – hilly, multihazard), using the Climate Resilience Framework (CRF) and the training module as key outcomes of the project (Figure 7.2), which has a key focus on ICT applications and database management for improving risk communication, early warning and disaster preparedness. The process offers lessons-for mainstreaming of integrated efforts for implementing Sendai DRR Framework, Paris Climate Agreement at local levels in urban context and SDGs have been considered in developing the training strategies (Gupta *et al.,* 2016).

The call of international policy frameworks on disaster management that emerged in 2015 further brings an overarching paradigm of integrating various approaches of disaster management, climate change adaptation, sustainable infrastructure, ecosystem based approaches, livelihood and health. The 6th Asian Ministerial Conference on DRR, which led to Bangkok declaration and HFA-2 strategic input documents, gave emphasis on ICT applications for understanding changing ecosystems and land use patterns along integration of technology mediated projections and predictions for understanding risk and vulnerability scenarios. India's current Prime Minister Mr. Narendra Modi who himself has been in forefront of disaster management in his Chief Ministerial tenures managing mega disasters like Bhuj earthquake, coincided his experience and expertise by giving a 10 point agenda on Disaster Management (described in following sections).This gives emphasis on ICT, space based and mobile technology applications.

Disaster Management: The Paradigm Shift

Disaster management is being talked about during past two decades, however the art of managing disaster risk is yet to be mastered. Vulnerability to disasters has always been a challenge to humankind. For reducing such vulnerabilities, landmark initiatives at the international level, such as International Decade for Natural Disaster Reduction (IDNDR), Yokohama Strategy, International Strategy for Disaster Risk Reduction, Tampere Convention, Kyoto Protocol, Hyogo Framework, Sendai Framework for Disaster Risk Reduction, *etc.,* were adopted in recent decades to effectively integrate disaster risk reductions into sustainable development planning. There is pressure on developed and developing economies, to put emphasis on climate change adaptation and green growth agenda by forming appropriate policies and having rules and regulatory frame work in place.

At the international scenario, India is a country constantly rated as a high-disaster-risk nation. In the past four to five decades, vulnerability to disasters, both natural as well as human-made, has been increasing due to rising population, haphazard urbanization, structural development in high risk zones, environmental degradation, climate change, *etc.* In many areas of the country, disaster losses tend to outweigh the development gains. Economic and social costs on account of losses caused by natural disasters continue to mount year after year as disasters occur with unfailing regularity encompassing every segment of national life.

Almost every state of India is prone to one or other type of disasters (such as, earthquake, flood, cyclone, drought, tsunami, *etc.*), it is only in recent years that disaster management (DM) has acquired priority in India. After enactment of Disaster Management Act in 2005, National Policy (2009) was framed and institutional mechanisms at national, sub-national and local levels were created. At national level, National Disaster Management Authority (NDMA), National Institute of Disaster Management (NIDM), National Disaster Response Force (NDRF) Disaster Management Divisions in Ministry of Home Affairs and Ministry of Agriculture and Farmers Welfare have been created as per the provision of the DM Act. These are supported by similar institutions (SDMA, SDRF and Disaster Management faculties in State Administrative Training Institutes) at state (Local) levels. These premier institutions are formed to facilitate paradigm shift in disaster management in India. Other institutional set ups that exist at the all India level includes National Crisis Management Committee, Crisis Management Group, and National Executive Committee. In consonance with the arrangements at the National and state levels, corresponding arrangements are also constituted at district level (A district is an administrative division of an Indian state or territory.) which include Disaster Management Authority (DDMA). The battery of national and state level institutions, are trying to bring about necessary focus on disaster management at ground level.

Environmental Policies and Disaster Management

'Economic efficiency' and 'disaster risk management' are two complementary objectives of an environmental policy or law on its strategy for 'sustainable development', which is defined as *"Development that meets the needs of the present without compromising the ability of the future generations to meet their own needs"* (Our Common Future, World Commission on Environment, 1972) with three main principles *viz.* (a) Inter-Generational Equity, (b) The Precautionary Principle, and (c) Polluter Pays Principle. This provides a strong vision for disaster risk reduction and green-relief-recovery approach within the framework of environmental sustainability. Although, most environmental laws may not use the terms 'hazard, vulnerability, disaster, emergency or response' but a blended approach based on science and human ecology to the interpretation of their provisions shall help significant opportunities for DRR tools and actions. Nation's initiatives on integrating environment and disaster management are therefore, right steps towards sustainability of lives, livelihoods and development.

(a) International Law

International law is considered the supreme body of law by international tribunals and in international relations. International law may be considered persuasive in interpreting constitutional or statutory provisions. The jurisprudence of international tribunals also can be considered in this context. In *Andhra Pradesh Pollution Control Board-II v. Prof. M.V. Nayudu and Others* [2001] 4 LRI 657, Sup. Ct. India, the Court referred to the Declaration of the United Nations Water Conference, the International Covenants on Civil and Political and Economic, Social and Cultural Rights, and the Rio Declaration on Environment and Development as persuasive authority in implying a right of access to drinking water as part of the right to life in the Indian Constitution. The main principles of international environmental law are found in treaty law (The Brundtland Commission). 'Legal regimes are rapidly outdistanced by the accelerating pace and scale of impacts on the environmental base of development.' Law must be reformulated to keep human activities in harmony with the unchanging and universal laws of nature (Brundtland, 1987). On occasion, courts have looked to treaties for the meaning of undefined terms in national law. In Ramiah and Autard v. Minister of the Environment and Quality of Life (Mar. 7, 1997), the Mauritius Environment Appeal Tribunal looked to the Ramsar Convention for a definition of wetlands, even though the convention had not yet been ratified by Mauritius. The Ministry of Environment agreed that the Convention provided guidance on the issue.

The sources of international law that may become domestic law through incorporation generally include those sources as listed under Article 38 of the Statute of the International Court of Justice. The Statute refers to (a) international conventions, (b) international custom (c) general principles of law, and, (d) judicial decisions and doctrine, as subsidiary persuasive sources. Some of the most important global treaties are: the International Whaling Convention, 1946, Ramsar Convention, 1971; Stockholm declaration on Human Environment, 1972, World Heritage Convention, 1972, United Nations Convention on the Law of the Sea, 1982, Vienna Convention for the Protection of the Ozone Layer, 1985, and its 1987 Protocol, Basel Convention, 1989, Climate Change Convention 1992; Biodiversity Convention 1992, Rio declaration on Environment and development 1992, Rotterdam Convention (1998), The Johannesburg declaration on Sustainable Environment 2002, Rio+20 Earth Summit (2012).

(b) The United Nations Framework Convention on Climate Change

UNFCCC is concerned with greenhouse warming. The Protocol to the UN Framework Convention adopted in Kyoto on December 11, 1997 specified different goals and commitments concerning emission of greenhouse gases. The potential adverse effects of climate-change have been characterized as changes in the physical environment or biota which have significant deleterious effects on the composition, resilience or productivity of natural and managed ecosystems or on the operation of socio-economic systems or on human health and welfare

(*U.N. Framework Convention on Climate Change* (New York 9th May 1992), Article 1. IPCC formation). Article 3(3) of the Convention refers to minimizing the cause and mitigating adverse effects, with obligations (articles 4 and 12) for inventory of emissions, sinks and reservoirs, technology transfer, coastal zone management and research cooperation.

(c) The Rio Declaration

Rio Declaration states that the only way to achieve long-term economic progress is to link it to environmental protection. Therefore, nations must establish a new and equitable global partnership involving governments, populations and key sectors of societies and build international agreements that protect the integrity of the global environment and the development system. The Rio declaration thus reaffirms and builds upon the declaration of the United Nations Conference on the Human Environment, adopted in Stockholm 1972 where there was a presence of many world leaders from 179 countries. The UNCED (United Nations Conference on Environment and Development), in Rio underlined that thinking of environmental, economic and social development as isolated fields is no longer possible. At the Earth Summit major international treaties and agreements were made on issues of global climate change, biological diversity, deforestation, and desertification. In addition the Rio Declaration contains fundamental principles on which nations can base their future decisions and policies, considering the environmental implications of socio-economic development. Principle 6 states that the special situation and needs of developing countries, particularly the least developed and those most environmentally vulnerable, shall be given special priority. International actions in the field of environment and development should also address the interests and needs of all countries. Principle 7 states that "States shall cooperate in a spirit of global partnership to conserve, protect and restore the health and integrity of the Earth's ecosystem". According to Principle 13, "States shall develop national law regarding liability and compensation for the victims of pollution and other environmental damage. States shall also cooperate in an expeditious and more determined manner to develop further international law regarding liability and compensation for adverse effects of environment." Principle 18 states that "States shall immediately notify other States of any natural disasters or other emergencies that are likely to produce sudden harmful effects on the environment of those States. Every effort shall be made by the international community to help States so afflicted."

Agenda 21 was adopted at the Earth Summit 1992 in Brazil by nations representing over 98 per cent of the Earth's population, it is the principal global plan to confront and overcome the economic and ecological problems of the late 20th Century. It provides a comprehensive blueprint for humanity to use to forge its way into the next century by proceeding more gently upon the Earth. As its sweeping programs are implemented world-wide, it will eventually have an impact on every human activity on our planet. Deep and dramatic changes in human society are proposed by this monumental historic agreement. Understanding those changes is essential to guide us all into the future on our fragile planet. Seven Central Themes of Agenda 21 include: 1. The Quality of Life on Earth, 2. Efficient use of the Earth's Natural Resources, 3. Protecting our Global Commons, 4. Management of Human Settlements, 5. Chemicals and the Management of Waste, 6. Sustainable Economic Growth and 7. Implementing Agenda 21. It also deals with efficient use of the Earth's Natural Resources deals with different types of resources, detail reasons for protecting them, and benefits associated with resource protection. Areas of concern include: sustainable agriculture, water, energy, bio-diversity, and bio-technology.Forests, Deserts and drought and Mountain ecosystems are given special importance.

(d) Rio +20 Earth Summit

The world once again came together at Rio De Janerio in June 2012, popularly known as Rio+20 Earth Summit, to discuss issues around Sustainable Development, but the summit did not result in any concrete outcome and failed to recreate the history of 1992. However concern for environment is always there, which is evident from various treaties, protocols and conventions drafted from time to time. The salient principles of "sustainable development " as culled out from Brundtland Report, Rio- declaration and Agenda 21 are (i) Intergenerational Equity (ii) Use and Conservation of Natural Resources (iii) Environmental Protection (iv) The Precautionary Principle (v) Polluter Pays Principle (vi) Obligation to Assist and Cooperate (vii) Eradication of Poverty and (viii) Financial Assistance to developing countries. These principles are important in achieving disaster risk reduction as well.

(e) National Policies on Environment and Disaster Risk Reduction

(i) The National Environmental Policy, 2006 adopts an integrated approach including coastal zone management, management of wetlands and river systems; conservation and development of mountain ecosystems; land use planning; watershed management and reducing the impacts natural hazards like, flood, landslides, storm surges and climate-change. EIA notification envisages for Hazard Mapping, Vulnerability and Risk Assessment Report as a part of environment management plan of the projects. It is a party to the key multilateral agreements, and recognizes the interdependencies among, and trans-boundary character of, several environmental problems. The National Environment Policy (NEP) is intended to be a statement of India's commitment to making a positive contribution to international efforts.

(ii) National Forest Policy has several provisions related to DRR. The policy envisages the maintenance of environmental stability through preservation and restoration of the ecological balance that has been adversely disturbed by serious depletion of the forests of the country. The policy recommends checking soil erosion and denudation in the catchment areas of rivers, lakes, and reservoirs in the "interest of soil and water conservation, for mitigating floods and droughts and for the retardation of siltation of reservoirs and controlling the extension of sand-dunes in the desert areas of Rajasthan and along the coastal tracts. This policy, by conserving forests, helps in reducing risks of floods, landslides and enhanced the livelihood options of the people depending on forest. Disaster Risk Reduction measures are basic in the objectives of the policy.

(iii) Objectives of the National Water Policy of India are planning, development and management of water resources. The policy, by addressing the need of conserving and effectively managing water resources, is paving way to mitigate impact of hydro-meteorological hazards and climate change. Water policy recognised water as a key environmental resource for sustaining all life forms. 'Water is part of a larger ecological system' (1.3). Disaster risks like floods and drought are of concerns in the water policy of India. Socio-economic aspects and issues such as environmental sustainability, appropriate resettlement and rehabilitation of project-affected people and livestock, public health concerns of water impoundment, dam safety *etc.* are envisaged in the policy while planning and implementation of water resources projects. In the policy it is recognised that problems of water logging and soil salinity have emerged in irrigation command areas, leading to the degradation of agricultural land. Besides the physical issues the social problems like equity and social justice in regard to water distribution are required to be addressed. Considering the decline in quality and quality of ground water due over exploitation in certain parts of the country the concern and need for judicious and scientific resource management and conservation is envisaged in the policy. The policy recommended water zoning of the country and the development activities including agricultural, industrial and urban development should be guided and regulated in accordance with availability and constraints of water resources. Both in urban and in rural areas adequate safe drinking water facilities should be provided to the entire population.

(iv) Water environment management with watershed, forestry, soil conservation, agriculture and a river-basin approach is the central philosophy of water policy (3.3 to 3.5). It also prescribes for drainage and irrigation management (6.6), groundwater (7.4), water quality (14), zoning (15), with devoted sections like – floods (17), sea or river erosion (18), drought (19) and use of science-technology and environmental impacts (25).Water Policy is giant leap towards reducing the impact of hazards and vulnerability reduction by assuring drinking and irrigation water supply through structural and non-structural measures. The policy also recommends integration of concerns of water resource management in developmental activities.

(v) India's agriculture sector accounts for 18 per cent of GDP, and employs around 60 per cent of the workforce. National Agriculture Policy (NAP) aims to attain an annual growth rate of 4 per cent in the agricultural sector over two decades (2000-2020). NAP is another important policy to address the issues pertaining to protection and management of natural resources and ecosystems in rural areas in particular for hydro-meteorological disasters and climate-change issues. The new Agriculture Policy, 2000, of India emphasized on Sustainable agriculture, Food and nutritional security, Technology generation and transfer, risk management and organizational framework. Agro-ecosystems include mainly the man-made ecological production systems – farms, plantations, ponds, *etc.* also the natural systems used for bio-productivity purposes, covering the purposes of food, dairy, fisheries, and other livestock, *etc.*

(vi) India's Urban Sanitation Policy was notified in 2008. Sanitation is a key aspect of disaster management due to its significance in pre-disaster phase in relation to water and garbage related health disasters (epidemics), as well as in the post-disaster risks in context of hydro-meteorological disasters like flood, cyclone and drought, *etc.* during response and relief phase. While this policy pertains to management of human excreta and associated public health and environmental impacts, it is recognized that integral solutions need to take account of other elements of environmental sanitation, *i.e.* solid waste management; generation of industrial and other specialized/hazardous wastes; drainage; as also the management of drinking water supply. Key Sanitation Policy Issues like addressing poor and served people, integrated approach, technology choice, occupation and organizational aspects of sanitation. The policy prescribes for a city-sanitation plan and its implementation. Policy also emphasized the need of emergency preparedness and response aspects of sanitation management in cities.

(vii) National Disaster Management Policy, 2009, is a major policy guideline in India. Definition of Disaster as per the DM Act of India recognises damage to environment as a disaster. Introduction to disaster risks in India (Refer 1.2.1), recognize environmental degradation and climate change as factors increasing people's vulnerability. Objective of the policy (2.4.1) encourage mitigation measures based on technology, traditional wisdom and environmental sustainability. Section on 'Environmentally Sustainable Development' (5.1.6) policy recognises that environmental considerations and developmental efforts need to go hand in hand for ensuring sustainability. Restoration of ecological balance in Himalayan regions and raising coastal shelter belt plantations need to be incorporated wherever necessary in disaster management plans. Ecosystems of forests, islands, coastal areas, rivers; agricultural, urban and industrial environment are also to be considered for restoration of ecological balance and sustainable development. Zonal regulations should ensure preservation of natural habitats. Climate Change Adaptation (5.1.7) with focus on glacial reserves, water balance, agriculture, forestry, coastal ecology, bio-diversity and health in order to reduce disaster risks and vulnerability (6.3.1). Environmental and hazard data for formulation of alternative land-use plans for different geographical and administrative areas with a holistic approach Institutional arrangements (12.2.1) emphasised the need of close interaction with Central Ministries and Departments of Agriculture, Atomic Energy, Earth Sciences, Environment and Forests, Health, Industry, Science and Technology and Space; and with academic institutions.

Disaster Management: An Integrated Process

It has been mentioned at various stages of this book that effective disaster management comprises of efficient prevention, preparedness, mitigation substantiated with response rehabilitation, reconstruction and recovery. Various activities involved in the disaster management process, roles and responsibilities assigned to and performed by and a number of institutions, actions taken by affected community and other stakeholders during various disasters in India have been thoroughly analysed and compared to arrive at this statement. Analysis of existing setup based on selected indicators states that the government, the community, the NGOs, the traditional panchayats[2] *etc.*, did not lay required focus on preparedness. It can be found that though various players have been involved in managing recent disasters(Sikkim: 2011, Uttarakhand :2013, Jammu and Kashmir: 2014) in India, their activities were not completely coordinated as there is no existing mechanism to get everyone connected through a disaster proof communication network. It has been repeatedly established that disaster management activities, if performed in a piecemeal manner resilience from disasters would not at all be possible. Set of activities, as defined in chapter 2 of this book, as components of DM continuum that is prevention, mitigation, preparedness, relief, response, recovery are integrated activities. These activities cannot be segregated, and therefore when disaster management is attempted in a piecemeal approach, cascading effect of failure of one activity catch on other activities resulting in over all failure.

It has been observed in almost all disaster's situations that while the role of the community is very strong as the first responders to disasters, the government has to play its role effectively in the phase of prevention, mitigation, preparedness, recovery and rehabilitation. The local government needs to act as system integrator for resources mobilization and on-ground activities by various stakeholders. While the government facilitates the role of various NGOs and other organizations, the NGOs and other donor organizations are to be involved in the actual provision of services such as construction of houses, organizing awareness campaign and training programmes, restoring education of children, providing necessary safety and security to women, facilitating social bonding, *etc.* The traditional Panchayats and the elected representatives have to be instrumental in maintaining social order in the villages at the time of distribution of relief measures. The study reveals that various stakeholders that got into action after a disaster are not involved in initiating preparedness measures before the event. There is no mechanism of regular communication amongst them.

It has been reported that in cloud burst (June-2013) of Uttarakhand, the major bottleneck during response was non-availability of suitable communication mechanism amongst responders, community leaders, government functionaries and victims. This disaster has extracted a very high price in social, psychological and economic terms. Its impact was felt in trails of destruction, families torn apart, children orphaned, livelihoods destroyed, and communities traumatized. The Comptroller and Auditor General (CAG) of India in its performance audit report on the disaster management mechanism in the country, submitted to Parliament, had highlighted that the Uttarakhand Disaster Management Authority (SDMA), constituted in 2007, had not formulated any rules, regulations, policies or guidelines for disaster management in the state.[3] Effect of this disaster could have been substantially reduced had: people been well informed; hazard and vulnerability analysis been properly carried out and updated; DM plans been in place; SOPs and well structured Incident Response System been in place.

A very good example which shows what preparedness can do is case of cyclone Phailin in Odisha and Andhra Pradesh [4]

Cyclone Phailin hit the Bay of Bengal on the eastern coast provinces (states) of Andhra Pradesh and Odisha on Saturday the October 12, 2013, with winds over 200 kmph and heavy rainfall. The Red Message, the highest alert from the Indian Meteorology Department in Delhi, was not only concise and to-the-point, but was also usable to a large extent by the local district officials. The alert also enlisted where and what type of damage was expected to shelter and infrastructure.

The Orissa State Disaster Management Authority (OSDMA) team and the ministry for disaster Management managed the largest-ever evacuation exercise in the state and planned large-scale relief preparations. Various departments took initiatives in preparedness. The panchayati raj ministry had issued detailed guidelines for preparedness for the impending cyclonic storm as early as on October 9, 2013. Control rooms were set up in Ganjam and other nine districts, mobile phone numbers were updated and verified, leaves were cancelled to have almost all the staff on stand-by, and food and relief stocks were kept in readiness. Block Development Officers visited cyclone shelters to keep them ready for the likely influx of those fleeing from the cyclone. The Odisha government ensured that there was required coordination among the various ministries, from panchayati raj to social welfare. A plan of action on which department would do what and when once the cyclone strikes was drawn up in detail. The state government held meetings with international NGOs to draw up a roadmap to tackle the aftermath of the cyclone.

The National Disaster Management Authority facilitated local efforts in Odisha by mobilising rescue teams and by sending equipment to possible hot spots. The NDMA deployed nearly 2,000 personnel of the National Disaster Response Force in Andhra Pradesh, Odisha and West Bengal. The teams were equipped with satellite phones and wireless sets to maintain smooth communication.

PRIME MINISTER'S AGENDA 10

India's Prime Minister, based on his long administrative experience which also includes dealing with major disasters, opened with outlining of an Agenda 10 on Disaster Risk Management, during the 7th Asian Ministerial Conference (fist after Sendai Framework) in November 2016. Disaster risk, much talked but less understood, particularly how hazards proceed into disaster, is still a valid question to explore, as evident from Priority 1 of Sendai Framework of Disaster Risk Reduction, *i.e.*, 'Understanding Disaster Risk'. The paradigm shift for disaster risk management, encompassing climate resilience factors and sustainable development focused, therefore, calls for an honest introspection of the journey and efforts at country level and to delineate a prudent roadmap for attaining the goals of three major agreements India is signatory to, *viz.* SFDRR, Paris Climate Agreement and SDGs. Asian Ministerial Conference Inaugural speech of the Prime Minister delineated the framework for simple and practical Roadmap for achieving sustainable development through DRM. The 10 point agenda delineated by the India's Prime Minister were as following:

(i) All development sectors must imbibe the principles of disaster risk management.

(ii) Work towards risk coverage for all-starting from poor households to SMEs to multi-national corporations to nation states.

(iii) Encourage greater involvement and leadership of women in disaster risk management.

(iv) Invest in risk mapping globally. For mapping risks related to hazards like earthquakes we have accepted standards and parameters.

(v) Leverage technology to enhance the efficiency of our disaster risk management efforts.

(vi) Develop a network of universities to work on disaster issues.

(vii) Utilize the opportunities provided by social media and mobile technologies.

(viii) Build on local capacity and initiative.

(ix) Opportunity to learn from a disaster must not be wasted. After every disaster there are papers on lessons that are rarely applied.

(x) Bring about greater cohesion in international response to disasters.

People knew about the coming storm and were able to take some measures to reduce their exposure. Early warning messages sent out through cell phones and TV channels helped saving lives even in the most remote rural areas.

The death toll of the cyclone was astonishingly low considering the 12 million people who lived in the storm's path. By virtue of an efficient early warning system and rapid evacuation measures deployed by national and local governments, nearly 500,000 people were evacuated in time and moved to higher grounds and safer cyclone shelters.

Summary of Disaster Communication and Early Warning Scenario in India

There are many public communication networks operational in India, however a completely sustainable fault proof arrangement for emergency communications does not exist. In almost all recent catastrophes in India (Leh Cloudburst : 2010, Sikkim: 2011, Uttarakhand :2013, Jammu and Kashmir: 2014), lack of fool proof dedicated emergency communication facility was a common factor in all situations. Whenever a disaster strikes, depending upon nature of incidence, communication facilities go haywire, as a result communication facilities are not available during the time of utmost need. There is an immediate need of concrete efforts for bridging the existing capacity deficit by all concerned departments/authorities.

There are many early warning systems in India meant for various kinds of disasters. Agencies creating and disseminating early warning in India such as Indian Meteorological Department (IMD), Central Water Commission (CWC), Geological Survey of India (GSI), Indian National Centre for Ocean Information Services (INCOIS), Snow and Avalanche Study Establishment (SASE) are not connected through a common network. Every organization uses their own set up for dissemination of warnings. Early warnings being issued by various organizations are discretely communicated on hotlines and MPLS VPN circuits, Internet, fax, emails *etc.* to concerned federal and local administration. All EW organizations in India are having different setups for communication of warnings. Except for Indian Meteorological Department, organizations involved in managing early warning systems are not equipped with latest state of art communication technologies. In addition to this, sensitization and training of local staff for creation of effective warnings is the need of the hour Most of these organizations do not possess their own communication setup and are therefore dependent on public and private telecom service providers, as a result there warning setups are as vulnerable to disasters as these communication networks are. An integrated disaster early warning network systems connected to all early warning systems does not exist in India. Besides, the last mile connectivity of the community at risk, with the core early warning system is a major problem.

Mobile BTS towers and Telephone Exchange buildings, when designed and constructed by government telecom service provider, comply building norms for earthquake and other hazard resistant structures, however when buildings are hired, such norms are not complied. Besides, for private players no such norms are specified. Noncompliance to specified building norms, for telephone exchange buildings, result in failure of services in the time of need.

Telecommunication officers (Sub divisional, District, state Telecom operational Heads), workers, who play very important role for disaster management are not sensitized and trained on emergency communication preparedness. There is no coordination mechanism amongst various telecom service providers for restoration of communication facilities during the time of disaster.

Moreover local administration is not involving telecom personnel as member of District Disaster Management Committees/Teams. Technical abilities of telecom officers are not leveraged during pre-disaster phases of prevention and preparedness.

Recommendations and Suggestions

NDMA has already prepared National Disaster Management Policy and National Guidelines on varied DM related subjects, yet these have not seen practical on-ground implementations. Disaster Management measures undertaken in India, are yet to be holistic in nature; when a disaster strikes only piece meal strategies are activated. Tsunami (2004) led to enactment of DM Act 2005 and creation of several national and state level institutes. However a multi-stakeholder partnership arrangement for disaster management does not exist in India. India repeatedly faces multiple disasters yet lessons learnt from those incidents are not being implemented at fast pace. Active roles have been played by various actors during a catastrophic event but the momentum could not be maintained for a longer period, as coordination among such players does not sustain due to multiple reasons such as lack of transparency, leadership, *etc.*

For a much talked about paradigm shift for DM in India and other least developed and developing countries, disaster management mechanism has to completely leverage science and technology for DRR, disaster prevention and mitigation and fail safe emergency communication. All important documents and frameworks (HFA 2005-2015, UNISDR strategy documents, Sendai Framework on DRR) lay emphasis on using science and technology for early warnings, preparedness and mitigation interventions, risk and vulnerability reduction, community empowerment, effective n coordinated response through fail safe emergency communication. Following parts of this chapter provides for important and pragmatic measures to be taken for effective disaster management.

A Dedicated Disaster/Emergency/PPDR Communication Network

All disasters have great impact on telecommunication services and as they create additional temporary traffic loads during disasters. When a disaster strikes, due to anxiety people start calling each other, this is the time when telecom infrastructure also get affected and thus at a time of reduced availability networks get overloaded. In such situations, the operating agencies have not only to make best use of available telecommunication networks but may have to scale up prevailing capacity of the communication networks. The impact of disaster on telecommunication infrastructure is manifold, for example the telecom buildings collapse, power supply gets disrupted, roads are blocked, fire breaks out, telephone exchanges get overloaded, towers get twisted, cables get washed away *etc.* In addition to this always there exist huge gaps between required network capacities during disasters and capacities actually available on the ground. This gap arise due to the fact that normal communication networks are designed to cater to peak loads regularly faced by the network, however during disaster communication requirements exponentially expand, which is much beyond normal traffic handling capacity of the network. Thus during disaster immediate need is to fill this gap to the maximum extent possible.

India, due to its geography is a Sub- Continent by itself. Infrastructural development in India, in the field of ICT varies from region to region especially in the context of mountains, high altitude and coastal and other difficult terrains where the ICT infrastructure is yet to develop fully. In addition, different technologies at different points of time are placed for infrastructural support. Therefore, at national level, existence of such disparities in communication infrastructure calls for planned and conscious integration of existing functional ICT infrastructure. (Guideline, NDMA,National Disaster management Information and communication System 2012) Therefore there is an immediate need of a dedicated Communication and IT structure to support Disaster Management in India.

Satellite communication is considered to be a disaster fail proof technology; a dedicated, satellite based, nationwide communication network need to be established for emergency communication. This dedicated network may be designed in such a way that it specifically caters to risky and vulnerable parts of the country. Considering the fact that satellite communication is very costly, proposed dedicated disaster communication network may be an optimum combination of satellite communication and terrestrial modern wireless technologies, based on geographies risks and vulnerabilities of the terrain to be covered. It may be designed in a way to ensure utilization of already existing ICT infrastructure (governmental/private) and requisite integration with new proposed ICT structure, should be taken care of. The aim should clearly be to plug the gaps while keeping in mind the futuristic requirements. The design and implementation of such dedicated emergency communication network should be centralised and under the control of federal government to avoid any interoperability, information sharing and integration issues. It, therefore, becomes evident that dedicated network architecture should be centrally engineered, managed and controlled.

All early warning systems in the country, meant for creation and dissemination of early warnings related to various kinds of disasters, should have a common communication system. An integrated system ensures that all warnings can be delivered through common platform, with common warning protocols to maximum number of people. An integrated system may also promote and enhance last mile connectivity for early warnings. Besides, these early warning systems should be integrated with dedicated emergency communication system. An integrated emergency communication and early warning system offers end to end solution of all problems related to disaster warnings and communications required during various phases of DM continuum.

Indian Police communication network POLNET is a satellite based network but it is not interoperable with other networks thus cannot be used as interagency communication network during disasters. Public safety agencies normally do not have adequate mechanism to communicate and share information with other agencies/organizations. Though they have the communication network and technology in place to do so within their own organization, however their networks mostly are not inter-operable/compatible with the networks of other agencies, thus making inter-department coordination difficult. Lack of a comprehensive communication structure can result in delays in action during emergency. Therefore all public safety agencies should also be connected to ibid proposed network and it should act as **Public Protection and Disaster Relief (PPDR) communication networks.** PPDR applications which involves transmission of high resolution images and real time video, requires larger bandwidth. Therefore to meet out such bandwidth intensive requirements broadband access networks with technologies such as LTE with improved data and multimedia capabilities may be deployed. An adequate amount of spectrum may need to be made available on a national basis to meet these growing needs. Accordingly dedicated frequency for wireless ground access may be allocated for such services.

A National Dedicated Emergency Communication Network (hereafter will be referred as NDECN) may inter-alia have following requisite design and implementation considerations:

- ✰ The emergency communication infrastructures can build-up in a graded manner. This requires uninterrupted voice, data and video transmission with built-in broadcast capabilities. In effect, therefore, there is need to ensure building and deploying emergency ICT systems with desired scaling from routine day to day operational requirements to time sensitive critical communication needs posed by disasters, till the time of restoration of all ICT supports in the disaster affected area. Redundancy and diversity are key factors to achieve complete network resilience for an impeccable emergency communication system.
- ✰ The dedicated network should be designed with open ended and non-blocking architecture so as to cater for futuristic addition, expansion and alteration. At no stage, it should be engineered on proprietary based technologies.
- ✰ The infrastructure of dedicated NDECN should be such that built up and software generation, implementation and upgradation should be carried out in parallel with technological advancements. It is this facet, with

related application sensitive requirements for DM, which needs very clear understanding and focused implementation particularly at central and local/field of action levels.

- Integration of all existing early warning/forecasting mechanisms/ networks with the NDECN architecture should be given utmost priority.
- NDECN design should be such as to take care of needs of users functional requirements; whether during routine functioning or during management of live disasters, on occurrence of which these very systems need not to be re-engineered they should be able to cater to enhanced communication requirements of relief and response phase.
- Infrastructural establishment and subsequent 24 X 7 basis availability with desired responsiveness requires proper planning and implementation with on-ground available support measures. It is not only the initial NDECN set up (which is fundamental to Disaster Management,) it is its subsequent round the clock maintenance, security, availability and dynamic upgradation in an evolutionary manner and in tune with the emerging technologies are of paramount importance. Establishment and subsequent 24x7 availability of NDECN on continued basis with the desired responsiveness require planned and on ground available support measures. Some of these in brief are enumerated below":- (Guideline, NDMA,National Disaster management Information and communication System 2012)

  (i) Specified control and administration mechanism under central government organization authorized for operation and maintenance of NDECN.

  (ii) Assured availability of suitable operational organization for necessary network management control and dynamic reconfiguration wherever required.

  (iii) Network logistic support separately for technical maintenance and regular operation and testing of standby communication and IT equipment.

  (iii) Training and continued education of users and management/control staff/data administrators. This is going to be a major human resource management issue for DM work.

  (iv) Stringent power backup arrangements with prolonged durations would dictate provision of diesel generator sets of adequate capacity, desired levels of fuel stocking/movement and also, charging back-up batteries for communications and allied equipment. Green energy sources such as solar cells, wind mills can also be planned.

  (v) Application based Software with particular reference to Early Warnings and Disaster Management and creation of associated forms and formats for appropriate data base management.

(vi) Ensuring network system security, harmonising and media support.

(vii) Dynamic and regular database updating.

(viii) Protocol based data sharing. Appropriate data mining and storage tools along with simulation platforms assume importance.

(ix) Memorandum of Understanding (MOUs) and Service Level Agreements (SLAs) with various Service Providers; or issue of "Communication Support Directives" and Standard Operating Procedures (SOP) through the telecom licensing policy.

(x) State and Local level capacity building for maintenance of a national network will be required and to be taken care of by state/local level training institutes for training of maintenance staff. Trainings of course will be initiated at national level institutes which will develop master trainers for further trainings at state and local levels

Considering the requisite size, cost, technical design, infrastructure and complexities involved in building such network, which certainly is a LONG TERM solution, it is strongly suggested that it should be implemented in a phased manner. Till the time NDECN is put in place, some SHORT TERM and efficient emergency communication mechanism can be employed by using existing telecom infrastructure. Following are some suggestive measures:

- In India and also in most of the developing countries there are several ICT networks operational both in public and private domain, but generally they operate with different capacities and mostly in a 'Stand Alone' mode. To cater to increased communication demands, which arise during a disaster, existing networks need to be integrated appropriately. This requires established links between disaster coordinators, telecommunications licensing and spectrum allocation authorities, telecommunication regulators, and service providers at each level. Government needs to take important steps to undertake a review of contemporary telecommunications legislation to incorporate suitable clauses for introducing such mechanisms.
- In United States of America (USA), public safety agencies have united together to design, develop and deploy ICT infrastructure to support policing, criminal justice, public safety and homeland security[5]. Such interagency collaboration has lead to creation of Public protection and Disaster relief (PPDR) communication networks. PPDR communication networks are scalable and rapidly deployable networks for specific situations where network capacity is to be enhanced. PPDR services (law enforcement, emergency medical service, firefighting, search and rescue, border security *etc.*) are provided by PPDR agencies. PPDR agencies are first responders employed for SAR functions. These agencies are also responsible for day-to-day public protection such as police, fire brigades, emergency medical response, Armed forces *etc.* For disasters response, the

scope of responders may increase to local volunteers, local government personnel, community leaders or civilians. Such network can be developed by multi agency/telecom operator resource pooling already existing in the telecom market of the country. UK is using commercial network to provide mobile broadband to public safety agencies. EE (mobile network operator) has won the contract to provide mobile (access) services to public safety agencies using its existing commercial 4G network. South Korea, US, Australia, Qatar, Thailand, France functions as hybrid model, wherein dedicated spectrum is used for typically dedicated coverage in urban areas while the rural coverage cost can be reduced by sharing radio access with mobile operators. Even biggest public/government owned telecom operator such as BSNL in India may be assigned such responsibility to design and maintain such network out of their existing resources and having memorandum of understanding signed with other private operators through proper service level agreements.

- A centralized database at telecom circle/state level: containing information about various types of communication resources available in different parts of the telecom circle/licensing area; GIS information supplemented with GPS tools and digital maps; availability of standby facilities; network details of other operators; history of past disasters in that area and corrective actions taken during and post- disasters; staff duties and training schedule, list of supporting agencies/industries/ individuals *etc.* should be maintained, by telecom operators with access to Central/Corporate office. This information database should also be accessible to district administration. Further disaster management planning requires integration of services by various agencies, other relief and rescue aspects such as information about geographical maps, food supplies; needed material supply should also be included in the database for helping other agencies and human resources to run the emergency services. This requires distributed databases for fast recovery and restoration. The database should be structured in such a way that once place of disaster is fed, the complete details of resources available around the region for quick response are provided. The mechanism for accessing such databases should be very simple, user friendly with the feature of multi-user working. Considering the multi-vendor scenario in India, suitable mandatory agreements should be in place between all the service providers for providing access to such databases during emergency situations.
- There should be an integrated Disaster Management Information System (DMIS) build at national and state levels. Besides, for efficacy of any such system, data fusion at national level along with GIS based applications is of paramount importance. A national level DMIS with arms of data centers at State and district levels at respective Emergency Operation Centers, is very important for effective disaster management.

- In India PGCIL has a very big nationwide network POWERTEL. Besides it has connectivity established to Bhutan over Optical Ground Wire Technology (OPGW) on full redundancy. PGCIL has hands on experience in dealing with emergency conditions, for example during Gujarat Earth quake or during Indian Ocean Tsunami PGCIL had to restore back power transmission in very limited time. Entire PGCIL network is monitored 24X7X365 and managed through Network Management Systems (NMS) established at its National and Regional Control Centers. In case of disasters like earthquake, floods *etc.* power transmission is restored using Emergency Restoration System (ERS). Same benefit can also be availed in restoring communication network on OPGW overhead. Besides, availability of telecom network over transmission system is higher than 99.97 per cent. Considering multiple benefits of OPGW based POWERTEL network, wider network reach in distant and difficult areas; and network resilience, this network has immense potential to be used for emergency communication.
- The disaster management requires cooperation among a number of relief agencies thus MOU among such agencies should be made mandatory as a part of telecom licensing conditions by licensing authorities (Department of Telecommunication in India). These special conditions should be mentioned that they automatically come into force immediately after occurrence of a disaster. Some suggestions for inter-agency MOUs to be signed for emergency communication purpose are as follows:
    1. DoT and Telecom Operators should have MOU for emergency communication with amateur radio clubs.
    2. Telecom Operators should have MOU with other service providers such as cable TV, ISP, GMPCS, Cellular Mobile Telephone, HAM, VSAT *etc.* They should also have such MOUs among themselves to share the facilities in case of emergencies. The licensing conditions of all service providers should be suitably amended to mandatorily share the infrastructure in area of disaster.
    3. Through agreements between Operators and Airlines, immediate airlifting of telecom equipment during disaster is possible. Further, as per requirement, Helicopter service for carrying telecom equipment to the affected area may be availed of. These measures are critical since telecom connectivity is a prerequisite for effective relief and rescue work.

  The licensing conditions should have provision for telecom operators to enter into such MOU within a specified time period (say 2-3 years) of launching of the services in a circle, failing which may result in stringent penalties as pre defined conditions to be prescribed in licenses by licensing authorities.

- In order to ensure reliable and effective operation of the communication systems during and after disaster incidence, and to avoid network congestion, frequencies and channels may be kept reserved and dedicated to disaster relief operations.
- Establishment of call centers and websites (multi-lingual) will be major assets for effective DM. Creation of multi-lingual websites at the national and state levels would also go a long way in ensuring the right disaster management support and environment. In today's context, establishment of a call centre based 'Helpline', particularly at District and State levels and also, at national level will be of key help and support during management of live disasters. Such help-lines (during routine functioning) can be used for e-governance purposes and during management of disasters these can be duly upgraded for disaster management and these would have to be multi-lingual.
- Keeping in mind the time sensitivity and volume of data transfer needed for effective and reliable real time communication with stakeholders, various technological options need to be explored.Multiple technological options are to be adopted, for various phase of DM so as to ensure that early warning and forecasting reach the community in the likely affected area. Every technological option available with public and private ICT operators, including the services of individual members/NGOs is to be explored and pushed to the service.
- Coordination is the cornerstone for any organized effort. The institutions involved in a given activity perform better with synergy and coordination in place. The collaborative approach and network theory demonstrate certain key points that are essential for effective coordination. But such coordination presently does not exist for prevention, mitigation and preparedness. There is an urgent need to build an integrated collaborative platform for interagency information sharing.
- Disaster communications networks should be exercised regularly under simulated emergency conditions, such as loss of power mains or base stations; reports of strengths, weaknesses and suggested improvements resulting from such exercises be distributed to interested parties.
- During disaster situations, adequate provisions for interaction with the media are not available. There are no institutionalized arrangements for setting up 'Help Lines' at various levels, at the time of disaster. This is extremely important in today's context of active media involvement during such contingencies
- International agencies, specifically International Telecommunication Union(ITU) and Asia Pacific Telecommunity (APT) have done significant work in the field of disaster communications. ITU Recommendation "ITU-R F.1105" on "Transportable Fixed Radio Communication Equipment

for Relief Operation" deals with various types of transportable wireless equipment and their broad characteristics. ITU-D has also issued a Handbook on Disaster Communication against Question No. 16/2. This document deals with this subject in a comprehensive manner with specific needs of developing countries. Technical solutions suggested in these documents may be suitably adopted by India and other developing countries.

Climate Change and Disaster Management

Climate change related disasters have emerged as the most impending challenges of this century. The Fourth Assessment Report (AR4) of the Intergovernmental Panel on Climate Change (IPCC); IPCC's AR4 and AR5 have provided compelling evidences that climate change is advancing rapidly as a global risk with impacts far beyond just the environment. Recent projections and analytical studies indicate that the increasing global temperatures, arctic and glacial melt down, sea level rise and other climate change induced environmental degradation will give rise to extreme weather events and overstretch adaptive capacities of cities and rural areas, industries, governments and communities in general, posing greater burden on disaster management systems. Concomitantly a global review of the statistics of disasters by the Centre for Research on the Epidemiology of Diseases (CRED) indicates that natural disasters more particularly the hydro-meteorological ones are increasing in terms of frequency, complexity, scope and destructive capacity. Hitherto disasters were linked to climate change through short term natural variability manifesting in extreme weather events such as cyclones, storms, floods, droughts, heat waves, windstorms and other natural hazards with potential for catastrophic loss of human lives, damages to infrastructure and environment. A few examples of most likely impacts of climate change due to changes in extreme weather and climate events to the mid- to late 21st century are (Parry *et al.*, 2007) are following.

- ☆ Warmer and fewer cold days and nights, warmer and more frequent hot days and nights over most land areas - virtually certain (greater than 99 per cent chance)
- ☆ Heat waves frequency increases over most areas - very likely (90–99 per cent chance).
- ☆ Heavy precipitation events frequency increases over most areas- very likely (90– 99 per cent chance).
- ☆ Area affected by drought increases- likely (66–90 per cent chance).
- ☆ Intense tropical cyclone activity increases- likely (66–90 per cent chance).
- ☆ Increased incidence of extreme high sea level (excludes tsunamis)-likely (66–90 per cent chance).

Climate Change Early Warning

Climatic projections and predictions for effective and prudent early warning of climatic changes and more particularly of the extreme events are important pre-requisites for disaster preparedness and response coordination. Lessons learned from big disasters like Uttarakhand flash flood of 2013 and on the other hand the successful tracking of the cyclone Phailin are the background for emphasizing the needs of early warning of climate related disasters. It required practical and integrated framework of risk management with reliable impact assessment of future scenarios of risk and vulnerability. Impact assessment of climate change-disaster – development nexus is complex as it includes geophysical, environmental, socio economic and governance issues. However tools used for these assessments need to be continuously evolved. Climate change research, monitoring and modeling is an emerging area and the assessments are in their nascent stage with efforts of downscaling climatic projections.

Sendai Framework of DRR calls for achieving quantitative targets, and hence, quantitative indices form an important tool in climatic and disaster risk management. International initiatives and research on developing risk indices for disaster and climate sensitivity are under process. Such indices need to be developed using the components, parameters and assessment endpoints of vulnerability attributes across environment and ecosystems, land, physiography, socio-economic and human development settings, besides the aspects of physical processes of weather phenomenon and related incidences or dynamics that leads to hazards. Outputs of General Circulation Models (GCMs), and regional model, *viz.,* HadRM2 with a resolution of 50 X 50 Km, have to be strengthened to ascertain simulation accuracy along the use of appropriate RCM for downscaling the scenarios. This is important for developing risk management strategies and warnings for sub-national programmes focused on local resilience. Climatic risk vulnerability mapping is also a critical requirement that is expected to assimilate the dynamic aspects of vulnerability and climatic hazards context with probabilities of disaster scenarios and possible loss and damages.

Climate Finance Green Bonds

According to World Bank [6] "*The year 2017 has seen many catastrophic climate events. Floods brought by rains across South Asia affected more than 41 million people and killed at least 1,200. Hurricanes claimed lives and destroyed infrastructure across several island states in the Caribbean and in towns and major cities in the southern United States. Drought gripping Africa put 20 million people in Somalia, South Sudan, Nigeria, and Yemen at risk of starvation, for the third year in a row, countries with coral reefs - from East Asia and the Pacific to northern Africa have seen massive sections of their reef systems die due to heat stress from unsustainably high ocean temperatures. All these extreme events across the globe demand for immediate climate actions as per goals decided in the Paris Agreement*".

Since the signing of the Paris Agreement in 2015, many positive steps are being taken to cope up with climate change on a global scale. "*According to the International Energy Agency, global energy-related carbon dioxide emissions were flat for the third straight year in 2016 even as the global economy grew. This is a very good sign that emissions are no more directly linked to developmental activities. Global markets are undergoing a significant transformation led by renewable energy due to rapidly falling costs of solar and wind power*".[7]

Attainment of Paris Agreement's climate goals will require huge money. That is why mobilizing finance for climate action will be one of the key focus areas for the World Bank Group at COP23. World Bank and UN has recently launched " ***Invest4Climate"*** an initiative designed to bring together national governments, financial institutions, investors, philanthropists, and multilateral banks to find ways to support transformational climate investments in developing countries. Green Bonds are also part of the finance solution by raising funds from the capital markets for climate investments in developing countries. The World Bank Group, through the World Bank and IFC, are working with countries to put in place frameworks for sovereign and private sector green bond issuances.

A key element of the Paris Agreement was the country-level commitments to action, known as Nationally Determined Contributions or NDCs. Each NDC commits countries to a range of actions aimed at reducing their emissions and build resilience to climate change impacts. For its part, the World Bank Group is supporting around 300 initiatives across 77 countries related to NDC implementation through investments in areas including energy, agriculture, and transport. And through the Climate Action Peer Exchange or CAPE; finance ministries from developing countries are sharing knowledge with each other on meeting the fiscal challenges of implementing NDCs.[8]

Machine to Machine Communications (M2M)

With rapid technological growth in the field of Machine to Machine popularly known as M2M communication, ICT is envisioned to play major role in all facets of disaster management continuum. Broadly, Machine to machine (M2M), [9] refers to technologies that allow both wireless and wired systems to communicate with other devices (machines) of similar abilities. M2M uses a device (such as a sensor/meter/actuator[10]) to capture an event, which is relayed through a network (wireless, wired or hybrid) to an application, that translates the captured event into meaningful information. There are various definitions given by Global Standardization organizations for M2M. European Telecommunications Standards Institute (ETSI)[11] has defined M2M as 'Physical telecommunication based interconnection for data exchange between two ETSI M2M compliant entities, like: device, gateways and network infrastructure.' Accordingly M2M may be described as machines are connected to the Internet, using a variety of fixed and wireless networks and communicate with each other and with the wider world.

M2M is also being inter-changeably used with various other terms - Internet of Things (IoT), Internet of Everything (IoE), Smart systems (Homes, Cities, Meters, Grids *etc.*). However M2M is only a subset of IoT. IoT is a more encompassing term which refers to addition of communications and sensing capabilities to a wide range of physical objects. It includes Human to Machine (H2M), Radio Frequency Identification (RFID), Location Based services (LBS), Robotics, Telematics *etc.* In the coming decade it is expected that millions of IoT devices will be deployed in electric meters, parking meters, thermostats, car components, roads *etc.* IoT devices will send data directly using protocols such as Wi-Fi and Bluetooth and via mobile networks, specialized networks and over the global network.[12]

International Telecommunication Union (ITU) had established various Groups (Foucus Groups, study Group) with the objective of developing recommendations relevant to M2M from telecom/ICT perspectives. Study Group 20 (SG 20) is currently working to address the standardization requirements of Internet of Things (IoT) technologies, with an initial focus on IoT applications in smart cities and communities (SC and C). Mandate of SG20 is to develop international standards to enable the coordinated development of IoT technologies, including machine-to-machine communications and ubiquitous sensor networks. SG20 core objectives inter alia include standardization of end-to-end architectures for IoT, and mechanisms for the interoperability of IoT applications and datasets employed by various vertically oriented industry sectors.[13]

Advancement in M2M technology will add entirely a new dimension to disaster early warnings. Technologically empowered sensors, gauges, meters will take measurements, *viz.* hydrological measurements will talk to their respective devices at back end which in turn will inputs to weather prediction models and more accurate forecast/nowcast/early warnings will be generated. This machine trails would be extended to provide necessary triggers to mitigation and risk aversion mechanical models as well. PPDR operations can obtain considerable benefits by having access to wide variety of information, including informational databases, access to instant messaging, high-quality images and video, mapping and location services, remote control of robots, and other applications. In future, large deployments and proliferations of robotics, Machine-2-Machine (M2M) communication, Internet of Things (IoT) will have a significant impact on routine PPDR operations and emergency rescue operations.

Implementation of Disaster Knowledge Management System (DKMS)

It is very much essential to have reliable data, of uniform quality in a usable format for effective Disaster Knowledge Management System (DKMS). DKMS is imperative for all phases of DM. Lack of reliable historical records of past disaster events make hazard mapping difficult to undertake prevention, preparedness and mitigation planning with any degree of confidence. The domain experts have very limited involvement in developing the solution and this may be one reason for slow progress of the developmental/implementation work. Lack of interactive relationship between the end users first hand responders and community leaders and the plan developers/DM policy makers create implementation gaps in such plans. Interactive relationship is required throughout plan development and implementation lifecycle.

Whenever disaster strikes immediately all stakeholders come in action for response and relief functions and once this operation is over other teams join for recovery and rehabilitation. With passage of time, immediate actions taken by first responders and learning involved starts fading out from systems memory. Therefore it is essential to have a system in place to properly document and record such knowledge for future reference.

Similarly it is very important that officials who collect vital post disaster information create a comprehensive and structured data base so that this information can play an important role for incorporating improvements in overall post disaster short term and long term recovery and rehabilitation planning system.

Adoption of Common Alerting Protocol (CAP)

Presently warning systems are a disorganized combination of technologies and procedures. None of the existing early warning systems are entirely suited to all types of needs. As a result, many different technical and operational warning systems are working simultaneously. There is an immediate need of adopting standardization for early warning messages. The Common Alerting Protocol (CAP) is designed to facilitate interoperability across multiple technologies, and is a content standard for early warning messages. CAP is an open source, XML-based standard that provides a semantic structure for composing warning messages, which can then be quickly and consistently relayed across multiple technology platforms. The Common Alerting Protocol enhances government's "situational awareness" at the state, regional and national levels by decoupling the diverse elements of the national warning infrastructure. The Common Alerting Protocol allows technology developers and sponsors to expand, upgrade or even replace existing facilities without disrupting entire systems. [14]

The Common Alerting Protocol (CAP) is a simple but general format for exchanging all-hazard emergency alerts and public warnings over all kinds of networks. CAP increases warning effectiveness and simplifies the task of activating a warning for responsible officials, by normalizing alert data across threats, jurisdictions, and warning systems. From a procedural perspective, CAP reinforces a research-based template for effective warning message content and structure. The ITU-D Sector has been addressing the issue of disaster telecommunications for several years. During the study period 2006-2010, "ITU-D Study Group 2" published a report on guidelines for using a content standard for alerts and notifications in disasters and emergency situations (2008) regarding implementation of ITU-T recommendation X.1303 on the Common Alerting Protocol (CAP).

There are many interacting parties involved in the provision of telecommunication services to modern societies, at all levels from local to national, international and global. These parties are affected by a broad range of law, policies, rules and business practices. Adoption of the content standard for public warning will proceed most quickly and effortlessly if common policies are built to promote all-hazards and all-media approach to public warnings. Potential sources of public alert messages in CAP format include government agencies and other Early warning institutes and alerting authorities responsible for detection of natural and man-made hazard events. Other Potential intermediaries of public alert messages in CAP format include organisations operating on the local, national, international and global scales, including: news organisations; commercial television and radio networks; amateur radio operators; satellite broadcasters; cable networks; and a range of telecommunication service providers, including internet service providers as well as wired or wireless telephone and related services. Potential providers of infrastructure for public alert messages in CAP format include: authorization and authentication service providers; ICT hardware and software vendors; ICT service providers; and vendors of mapping software, location-aware technology, and geographic information systems, among others. Authorization and authentication service providers may develop products and services that provide mechanisms for end-to-end assurance of authenticity and evidence of authority, validated for handling of alert messages in CAP format during emergency situations. ICT hardware, software and services vendors may add features that handle CAP alert messages, with mechanisms for authenticated delivery and automated processing where appropriate ICT devices and software could feature geographic display of emergency information by targeted area using streams of alert messages in CAP format.

Certain communities may have an especially strong interest in the implementation of CAP. For instance, alert messages in CAP format are being used to create specialized warning messages for persons with disabilities. Other actors with a role in disaster management who may be involved to some degree with public alert messages in CAP format include: politicians, building code regulators, city planners and academic programmes in research areas such as emergency management, public policy, social science, information technology and communications technology. The immediate challenge is to educate these other actors about the emergence of a standards-based, all-media, all-hazards public warning infrastructure on the national, regional, and global scales. There are many opportunities to leverage the warning infrastructure for commercial gain as well as for the public good.

Therefore new policies to be designed to promote the all-hazards, all-media approach to public warning systems, and specifically the dissemination of public alert messages in CAP format. In most cases, the immediate challenge will be to educate key organisations and agencies as potential users and beneficiaries of this approach. Implementation of dissemination of public alert messages in CAP format should also include one of the common mechanisms to assure authenticated delivery and evidence of authority. It is important that sources are quite clear as to the extent of authority for issuing alerts to particular localities. [15]

Once a dedicated communication network is established and integrated with early warning systems, progression to CAP is a natural evolution path for proposed disaster communication system in India.

Integrated Emergency Communication and Response System (IECRS)

In an unforeseen crisis situation that requires immediate response and help by the concerned agencies like police, fire brigade, ambulance *etc.* there should be a communication network in place which facilitates the people in distress to access these agencies from anywhere, at any time and get a proper and immediate response. In order to provide quick and timely intervention by these agencies, an Emergency Response System (ERS) should be always active and ready for immediate action. Presently in India, different emergency communication and response systems are in place for police, ambulance, fire brigade, civil defense, disaster management *etc.* These systems are accessed through different numbers such as 100(Police), 101(Fire), 102(Ambulance) and 108 (Emergency Disaster Management). A number of states have also notified various helpline numbers for assistance to special category of citizens like "Woman in distress-181(Delhi)", "Missing children and women-1094 (Delhi)", "Crime against women -1096 (Delhi), Anti-extortion -1097(Delhi)" Police Headquarter helpline-1090(Uttar Pradesh) *etc.* (Consultation paper On Universal Single Number Based Integrated Emergency Communication and Response System by TRAI, New Delhi vide Consultation Paper No.3/2013: 15th March 2013.))

These multiple number emergency system prevailing in India has multiple disadvantages such as:

- ☆ The caller needs to dial relevant number depending upon the type of emergency situation. In case the caller does not know the correct emergency number to dial or is confused between the host of emergency numbers like whether to dial '101' or '102' for fire brigade, he will be either deprived of any help or will get the help after avoidable delay. Time is the essence in emergency response.
- ☆ In the existing system, the location information is provided by the caller and the emergency helping unit rely on this information. In this scenario, if the caller is either not able to convey his location or the location conveyed by him is vague or erroneous, there is a possibility that his request for help is handed over to a responding unit that may not be in the immediate vicinity of the caller resulting in loss of crucial time in the response. Moreover location specific jurisdiction issues may also arise in current system if calls made to emergency numbers land on Police

control rooms/Ambulance services which may not have jurisdiction at the location of caller.

- There may be situations where multiple agencies need to be contacted. For example, an accident may result into fire and some persons may be seriously burnt. Such a situation would require calling police, fire as well as ambulance. Getting the right contact numbers and explaining the same situation to each of them result in delay in a situation where one would want instant response within minutes.[16]

Therefore there is an immediate need for integration of all existing emergency number mechanism to a single emergency number system. Some of the countries (Australia, New Zealand, UK, and USA *etc.*) have put in place an Integrated Emergency Communication and Response system (IECRS). Such integrated system is accessed through universal single number. For example in USA, number '911' is used; in most of the European countries number '112' has been provided for emergency response. On dialing the emergency number, the call gets routed to a kind of call center, a Public Safety Answering Point (PSAP), located at several places within the country or State. The PSAPs are manned by personnel, who are well trained to seek specific information from callers and based on the requirement of the caller, and the address information that gets transferred from an integrated public number database in case of landline calls; or on the approximate location coordinates that gets transferred from mobile networks in case of mobile calls, these PSAPs route the emergency calls further to nearest local emergency responding agencies like medical, fire, law enforcement *etc.* In case of unavailability of location information, the answering personnel ask the caller for location information.

Implementation of IECRS may give rise to the need of maintenance of a proper and up-to-date subscriber database by the service providers. In a centralized emergency response model, one of the key features is transmission of identification and location information by telecom service providers to the PSAP. Identification information may be the information such as subscriber or company name. Location information in case of wired line connections can be street name, house number, area or town, postal code at which the connection is installed. In case of mobile originated calls, the location information transmitted by the mobile networks can be either the address of the cell site closest to the caller or location derived by triangulation method or with the help of GPS. In some countries like USA, there are rules that require the information regarding the latitude, longitude or position of the subscriber's equipment to be transferred from the service provider's network to the PSAP.

In India apart from natural calamities, people die due to several anthropogenic reasons, such as fire incidences, chemical hazards, biological hazards, accidents, industrial mishaps, train accidents, road accidents and riots. Road-traffic accidents are increasing at an approximate annual rate of 3 per cent. A trauma-related death occurs in India every 2 minutes. The majority of fatal road traffic accident victims are pedestrians, two wheeler riders and bicyclists. Till date no credible data is available on outcome of trauma victims. Even by rough estimates, mortality in serious injuries is six times worse in a developing country such as India compared to a developed country. 90 per cent of all Road Traffic Deaths occur in Middle and Low Income Countries like India.[17]

In addition to excess mortality there is a tremendous burden of disability in developing nations. It is estimated that India Loses approximately 3-3.2 per cent of its GDP to only road traffic injuries. The World Health Organization (WHO) has projected that by 2020 road accidents will be a major killer in India accounting for 546,000 deaths and 15,314,000 disability-adjusted life years lost.

A CASE STUDY

The very idea of implementation of such system is essential for safety of people and with collaborative efforts of government and corporate sector, this can be implemented. GVK EMRI (Emergency Management and Research Institute), a private organization is very successfully managing "108 Ambulance Service" across more than ten states of India. This service has become very popular as a medical emergency response system. EMRI's "108 Ambulance Service" is a Public Private Partnership (PPP) model between state governments and EMRI. Call to the number '108' is a toll free service (in local area) accessible from landline or mobile.[18] GVK EMRI envisions responding to 30 million emergencies and saves 1 million lives annually. A little over nine years since inception, GVKEMRI has come a long way as largest emergency service provider in India, with over 7500 fleet strength. Transporting the patient from the site of emergency to the nearest suitable hospital requires the operator to maintain the rolling stock free from break downs, besides, objectives of reliability, comfort, clean, hygiene, timely, and financially viable are to be met with utmost precision. The ambulances run by GVK EMRI are designed to afford safety, comfort, and avoid aggravation of the patient's injury or illness. GVK EMRI has gained sufficient expertise in fleet management and sustained the position as largest single ambulance operator in the world. The best maintenance practices adopted by GVK EMRI has given it national recognition and applaud from the public[19]. Apart from the acceptability, the endeavor to reach the people in distress, providing pre hospital care and admitting within the "Golden Hour" is well focused with a healthy transport system.

There are multiple, telecom regulatory, and operational issues for implementation of a single number IECRS system in India. Nevertheless based on huge success of public private partnership framework of GVK EMRI ambulance, Government may come out with policy on implementation of such system at national level.

A significant proportion of mortality and morbidity could be prevented by the implementation of early and appropriate response at the emergency site. Many deaths can be prevented by early implementation of medical and surgical interventions such as proper airway control, limitation of blood loss, treatment of crush injury, and treatment of shock *etc.* Even patients with minor injuries need to be clinically observed. Medical personnel should be aware of the unique problems that accident victims pose, and need to be trained in trauma care. An immediate and appropriate intervention can only be assured through an efficient emergency response mechanism.

Priority Call Routing Mechanism in Public Communication Networks

In order to facilitate a mechanism wherein important functionaries engaged in 'response and recovery' work during emergencies get the calls on priority, TRAI had issued a pre-consultation paper on 8th November, 2011 on 'Priority call routing' in Telecom networks for persons engaged in 'response and recovery' work during emergencies', in which stakeholder's inputs/comments/views/papers were solicited[18].

All disasters/emergency situations are immediately followed by an intense burst in telecommunications traffic which can congest even the best-managed networks resulting in call-blockages and lost-messages. A crisis situation expands human need for communication, it is required to coordinate response activities, to convey news and information about affected groups and individuals, and as a panic reaction to crisis situation by affected people and their well wishers. For cost optimization, most of the communications networks are engineered for peak load at levels well beneath the demands placed on them during disasters. When heavy traffic start blocking network, people in panic try harder to get access to network, repeated failed attempts further increase traffic and finally it results in choking of the network. During disaster when communication becomes the lifeline for response operation, its unavailability may further deteriorate a situation. Thus it is very important that during crisis situation, all agencies/personnel involved in search, rescue, response and recovery functions are able to communicate. It demands for priority treatment to be given to their calls. The possibility of giving priority to certain users like police, hospitals, departments involved in disaster response, blood banks *etc.* can be explored so as to ascertain that during such an incidence calls to/from certain key agencies and persons get priority on the network. The likelihood of getting network resources for such users to complete the call will then be much higher.

International Telecommunication Union (ITU) also came out with recommendation E.106[19] that was approved on 31st October 2003 by ITU-T Study-Group2 (2001-2004) under the "World Telecommunication Standardization Assembly (WTSA) Resolution1. This Recommendation describes an international preference scheme for the use of public telecommunications by national authorities for emergency and disaster relief operations. The International Emergency Preference Scheme (IEPS) for disaster relief operations is needed when there is a crisis situation causing an increased demand for telecommunications when use of the international telephone service may be restricted due to damage, reduced capacity, congestion or faults. In United Kingdom, the Priority call routing system named as Mobile Telecommunication Privileged Access Scheme (MTPAS) is based on access category/priority that is given for certain users for intra-operator and inter-operator network calls. Similarly in USA Wireless Priority Service (WPS) and Government Emergency Telecommunications Service (GETS) are the priority call routing systems for certain users that are based on 'access-code' dialing. (Telecom Regulatory Authority of India Consultation paper On Telecom network failures during Emergencies/Disasters Priority routing of calls of persons engaged in 'response and recovery' released on 10th May, 2012). TRAI had already taken up this important issue and came out with consultation paper on the subject, which Department of telecommunication has accepted for further implementation

Important Cross-Cutting Issues

Promotion of Research and Innovation

The empirical research in disaster management, both at the micro and macro level facilitates disaster effectiveness in the area. There are multiple ways of doing empirical research in the area of disaster management. The research may be based on scientific data base, wherein deep rooted causes behind natural and anthropogenic disasters are identified, tested and documented. These researches may pave ways for better prevention, preparedness and mitigation techniques. Other methods may involve formation of hypotheses, collection of evidences based on past incidences and current observations and comparing them on preset indicators for validating hypotheses. Such kinds of researches are very useful for nonstructural measures of disaster mitigation, capacity building and preparedness. The preparedness phase of emergency management is a cyclical process that focuses on proactive behaviors and actions aimed at lessening the impact of a disaster. It has been repeatedly seen in case of disasters that preparedness encourages the development of emergency response capabilities on several fronts. This development should take place at the individual, organizational, community, and state/federal levels. In fact, the Federal Emergency Management Agency (FEMA) IS-1 tool kit suggests that preparedness allows us to offset the complexity of most disasters, including those that are known for their rapid onset, by determining desired actions beforehand. These actions include establishing authorities and responsibilities for emergency actions, accumulating and having control of the resources to support them, and developing plans to link these areas so that emergency preparedness plans can be implemented.

Following is an indicative list where research in the field of disaster management may help:

- Preparation of DM Plans (for example mitigation, prevention, capacity building plans);

- Hazard Risk Vulnerability Analysis (HRVA) of an area;
- Risk minimizing behavior for resilient communities;
- Structural measures for earthquake resistant building design;
- Production of building material suitable for building construction in disaster prone areas;
- Design of GIS and early warning systems;
- Design of emergency communication systems;
- Integration of various emergency management systems;
- Regional and urban risk sectors assessment and mitigation planning;
- Post disaster urban/rural loss assessments;
- Macro-form and urban pattern risk analyses;
- Design of several microfinance product and services for vulnerable group living in disaster prone areas;
- Non-compatible and hazardous uses, and open space deficiencies;
- Location of emergency facilities, risks in infrastructure and accessibility networks;
- Comprehensive urban regeneration methods;
- Land use planning;
- Environmental design guidelines and standards to facilitate emergency management and cost/benefit analyses in risk mitigation;
- Reconstruction planning;
- Organizations and legal structures in disaster management;
- Participatory methods in disaster mitigation;
- Cost-effectiveness in mitigation;
- Economics of disaster relief;
- Effective insurance design for disaster victims;
- Post-disaster social collaboration;
- Role of Communications and media during different phases of disasters;
- Public awareness and capacity building;
- Psychology and behavior of survivors and provision of psychological support;
- Role of NGOs and other donor organizations;
- Design of short term and long term recovery plans;
- Rehabilitation of disaster survivors

Capacity Building and Training

National level training institute National Institute of Disaster Management (NIDM) is having a rich knowledge base on DM and it has been sharing its wisdom through trainings of Government Officers at national and state levels. NIDM has been doing training of trainers as well. However capacity building of first responders demand promulgation of skill-based training to local bodies' representatives and some indentified community members from each village/ward. The said task can be undertaken by a separately created Disaster Management Institute or through already envisaged SDMA/DDMA, Block and Village level Disaster Management Committees. The traditional wisdom and coping mechanism available with the local people need to be properly harnessed and supplemented with latest scientific knowledge for better results. The traditional knowledge can be documented by local people. A pool of resource persons can be created by local institutes for creation of knowledge base. All measures should be taken at local level for enhancing community awareness on disaster management.

Besides classroom trainings and field level trainings it is very important to hold mock trills. NDMA has been holding mock drills for many years these drills should be multiplied at every vulnerable district.

Disaster awareness, knowledge of risk and vulnerability and its management should be part of school and college education. This knowledge should reach to grass root level to empower communities.

Disaster Management should be mandatory part of all science and technology, social sciences streams to cover scientific, technical, social, economic, psychological and behavioral aspects of disasters and its coping mechanisms.

Many community projects are implemented by: world bank in collaboration with national and local government agencies; by national and international NGOs; various arms of UN at different disaster affected parts of the world; knowledge material created under these projects are very useful a handbook on Do's and Don'ts on various disasters particularly carries a lot of useful information for everyone. Efforts to be made by all national, state level institutes for multiplication of such projects/trainings and utilisation of training resources created out of such project extensively.

Promotion of Amateur Radio

In times of disaster many a times it happens that local terrestrial communication get disrupted and amateur radio is often used as a means of emergency communication as wireline, cell phones and other conventional means of communications fail. The ITU has recognized national/international use of radio communication in disasters. ITU has recommended that governments and telecom administrations should include the amateur radio services in their national disaster plans and telecommunication assistance information inventories. It mentions that "*Administrations are encouraged to reduce and, where possible, remove, barriers to the effective utilization of the amateur services for disaster communications. Amateur and disaster relief organizations are encouraged to develop memoranda of understanding (MOU) between themselves and with administration as well as to cooperate with each other*". Therefore amateur services should be recognized as disaster communications resources, particularly useful during the first few days of a disaster prior to the arrival of relief organizations and particularly before regular communication servcies are restored. They are capable of being reconfigured as required for short-range and international radio communication, independent of other telecommunication services; and are capable of providing qualified operators during and in the aftermath of disasters.

Advocacy and Public Awareness

In all disasters most affected and at the same time first responders are people who actually suffer that incident. Building community resilience is the key for effective disaster management. Capacity building of community is cardinal to disaster risk reduction. Therefore public awareness programmes; hazard risk vulnerability analysis and advocacy thereof; training of community leaders; building a robust volunteer mechanism; education and of women child old disables and other vulnerable group members and their participation in preparedness plans; are some of many important steps to be taken for effective DM. Federal and local government has a key role in promulgating such measures.

It is proposed that in India a network which connects Common Service Centers (CSC) can be efficiently utilized for the purpose of disaster preparedness and mitigation initiatives, for disaster awareness building and media campaigning. The Campaign may include film shows, radio jingles, and awareness campaigns in pre-defined intervals. The CSCs may target schools for disaster education through audio-visual spots and short documentary films. Important data, information brochures, training modules related to capacity building may be kept with CSC. Such initiatives for mass education of village panchayats and members of other such local governance units/civil defense personals/volunteers/Anganwadi/Asha workers *etc.*, can play a definitive role in disaster preparedness, prevention and mitigation measures.

Risk Communication and Crisis Communication

Risk Communication

Risk communication is meant to effectively inform people at risk about hazards to their environment, or their health, or their other belongings due to potential problems. Risk communication not only inform but to educate and make them aware in a manner that promotes goodwill, disseminates information, and communicates potential crisis and emergency situations well, encouraging prudent action and reducing panic.

For example:

- ✰ *How good is to use crackers?*
- ✰ *What is the pollution level?*
- ✰ *How polluted is the air we breathe?*

- *How safe are our rivers?*
- *Is water level of rivers rising above safety marks?*

Risk communication thus prepares the ground for trust between the community and the organization dealing with the risks involved. Risk communication should be based on strategies and plans with specific objectives, such as providing information to the public, motivating individuals to act, or contributing to the response to a conflict. It is important to check frequency, message, mode of communication *etc.* for risk communication and regular feedback of clientele population be taken for its efficiency enhancement. All precautions should be taken to make such communication objective and to avoid subjectivity.

Crisis Communication

Crisis communication deals with things that have gone wrong or is going wrong or impending to go wrong and a crisis situation is imminent. An emergency will be best served if a crisis communication strategy is made beforehand and is very much in place to communicate to decision makers and the public. Crisis communication is meant to be followed by a long series of actions, therefore it has to be specific, factual, clear, should indicate immediate course of action, to be targeted to all stakeholders to be involved and should not trigger panic. Therefore the best way to handle a crisis communication situation is to have a plan in place for managing a crisis situation. A well planned media management should be an integrated part of crisis communication.

Mainstreaming of DM Initiatives

All developing countries are developing new infrastructures, progressing towards better education, moving towards better financial inclusion and so on. It is therefore right time for them to mainstream disaster management initiatives such as preparedness, prevention, mitigation and capacity enhancement in all new developmental activities to actually ***build better.***

Emergency Operation Centers and Inter-organizational Coordination

When a disaster strikes, immediately an operation centre needs to be established which is to act as epicenter of all response and relief activities. This centre should have all communication amenities to coordinate with all stakeholders. This operation center should have connectivity to a permanent establishment meant for handling emergencies *i.e.* Emergency Operation Centers (EOC). EOC should have facilities for environment scanning, intelligence gathering, operations and logistics management under one roof for holistic and efficient management of disasters at the National state and district levels. EOCs should be always operational 24 X 7 X 365 basis. There should be schedule of operating procedure, detailed chain of commands to be followed during emergency and requisite training schedule of staff supported with simulation and mock drills.

In disasters, communication difficulties are often hard to separate from coordination difficulties, and the greatest coordination difficulties are inter-organizational. Therefore, many of the communications problems are those related to interagency information sharing. Frequently, the means for communication exists, but for a number of reasons, persons are hesitant to communicate with members of other organizations. "Inter-organizational communication is fostered by those factors which promote trust in other organizations and familiarity with how they function. These include: informal contacts, joint planning and training sessions, preplanned agreements for the division of disaster responsibilities, and the use of similar terminology, procedures, and performance criteria. Inter-organizational communication/computer networks contribute to effective communications". (Edward H. Young, January-April 2007)[20]. Therefore a very strong mechanism for interagency communication needs to be established. This may be in form of shared ICT infrastructure, supported by a collaborative resource sharing platform for all stakeholders. This kind of system should be in place for pre disaster DM continuum, so that processes flow seamlessly during disaster and in post-disaster situations.

Intra-organizational and Inter-organizational communication requires adoption of standard communication pattern to avoid ambiguities. Organization, therefore, should be trained to prepare their crisis communication plan. This aspect is very important for medical teams involved in emergency management, as lack of proper communication manual, may lead to increase in number of fatalities.

Standard Operating Procedures (SOP)

Disaster/Emergency management plans duly supported by SOPs are essential for efficacies of disaster management. There should be properly laid down guidelines and practices for preparing DM plans and SOPs for all working areas of disaster management which may for example be for school safety, community preparedness, search and rescue operation handling, volunteer enrolment and management, Post disaster need assessments, mass casualty management *etc.*

Implementation of Tampere Convention

Tampere 1998, At the Intergovernmental Conference on Emergency Telecommunications (ICET- 98) at, Finland, held on 18 June, 1998, a legally binding international treaty on deployment of Telecommunication Resources was unanimously adopted by 75 countries and it was recognized that the effective, timely deployment of telecommunication resources for rapid, efficient, accurate and truthful information flows are essential for reducing loss of life, human suffering and damage to property and the environment caused by natural disasters. India became signatory to this Tampere Convention on 29th November, 1999.The Tampere Convention calls on States to facilitate the provision of prompt telecommunication assistance to mitigate the impact of a disaster, and covers both the installation and operation of reliable, flexible telecommunication services. Regulatory barriers that impede the use of telecommunication resources for disasters are waived. These barriers include the licensing requirements to use allocated frequencies, restrictions on the import of telecommunication equipment, as well as limitations on the movement of humanitarian teams. The Convention describes the procedures for request and provision of telecommunication assistance, recognizing the right of a State to direct control and coordinate assistance provided under the Convention within its territory. It calls on States to provide prompt telecommunication assistance to mitigate the impact of a disaster, and to ensure the installation and operation of reliable telecommunication resources. It requires States to make an inventory of the resources both human and materials available for disaster mitigation and relief, and to develop a telecommunication action plan that identifies the steps necessary to deploy those resources.[21] Signatory states may come together while sharing above details and making necessary regulatory telecom policy changes to implement recommendations of Tampere Conventions for mutual and multilateral benefits.

Disaster Communication and Disaster Management Best Practices in India

India in year 2005 has enacted Disaster management ACT (DM Act 2005). India has also put proper institutional mechanism in Place. There are National Disaster Management Authority, State and District Disaster Management Authorities. There are National (NDRF) and State Disaster Response Forces (SDRF) which are specifically raised for combating disasters. NDRF is capable of handling difficult search and rescue functions and has proved its mettle in many disasters in India as well as in the Region such as Japan earthquake and tsunami and Nepal earthquake. There is a National Institute of Disaster Management for training and capacity building. There is a National Disaster Management Policy which entails all administrative, techno-legal aspects of disaster management in India.

National Disaster Management Authority (NDMA) has so far prepared about 27 guidelines and some reports on management of various types of disasters which covers almost all types of disasters. NDMA has also issued templates on preparation of state DM plans, school safety plans, hospital safety plans and holding mock drills *etc.* Similarly National Institute of Disaster Management (NIDM) is very efficiently handling capacity building and trainings on DM at national level and doing handholding of all state level administrative institutes for trainings on DM. NIDM has many publications, journals, periodicals on its accord. NIDM also holds National Platform on Disaster Risk Reduction. These form a very informative knowledge repository on disaster management in India. Detailed information can be accessed at **http://www.ndma.gov.in** and **http://nidm.gov.in**.

Radio Division of Telecommunication Engineering Centre (TEC), under Department of Telecommunication, Ministry of Communication New Delhi, had issued guidelines on Disaster Management (With reference to telecommunications) in year 2006. This guideline refers to redundancy in telecom infrastructure so that telecom services remain uninterrupted during disaster. The guideline touches upon aspects such as transmission and access planning, telecom building infrastructure, outdoor plant, power plants and alternate power supply such as engine alternator sets *etc.* Redundancy of communication systems is essential for disaster management, while emergency power supplies and back-up systems are critical in order to avoid the collapse of communication systems after disasters occur. Subsequently, Department of Telecommunications (DOT) has prepared an exhaustive guideline on Standard Operating Procedure (SOP-2017) for Telecommunication Services for responding to disasters[22]. Going further, DOT has created a functional pool of telecom professionals placed at all Telecom Licensing Area headquarters to take all necessary steps to enforce these measures.

India has a very robust early warning system in place. Primary Nodal agencies for monitoring and early warning of disasters in India are:

- ✰ IMD (Cyclones, Floods, Drought, Earthquakes)
- ✰ Central Water Commission under Ministry of Water Resources (Floods)
- ✰ Geological Survey of India (Landslides)
- ✰ Ministry of Earth Sciences through Indian Tsunami Early Warning Centre at the Indian National Centre for Ocean Information Services (INCOIS), Hyderabad. (Tsunami)

Indian early warning agencies are also disseminating important information to neighbor countries in the region and to several similar agencies in Indian Ocean and Asia pacific region. Indian EW system is also a part of Global Telecommunication System (GTS) of World Weather Watch (WWW) program of World Meteorological Organization (WMO). http://www.imd.gov.in, http://www.incois.gov.in, http://www.nrsc.gov.in, http://www.india-water.gov.in, http://www.india-wris.nrsc.gov.in, http://drdo.gov.in/drdo/labs/SASE are some important websites which provide information on EW system and services in India.

National Remote Sensing Centre under Indian Space Research Organization (ISRO) and other organizations such as GSI, BIS, CWC have done zonation of India on the basis of hazard vulnerability. These maps are very useful for pre-disaster planning, prevention and mitigation activities. "BHUVAN" is Indian Geo-Platform of ISRO which provides extensive range of services based on GIS maps.

Telecom Regulatory Authority of India (TRAI) has given its recommendations on implementation of single emergency number in India. This has been accepted and number "**112**" has been allocated to this service. DOT has issued necessary instructions to telecom service providers for implementation. TRAI has also given recommendations on Priority Call routing which has been accepted. Now TRAI is doing consultation on "Next Generation Public Protection and Disaster Relief (PPDR) communication networks" which will pave ways for efficient PPDR network in India. Information available at **trai.gov.in.**

Reserve Bank of India the central banking institution of India has notified to all banks that " *It is in the interest of lenders to ensure that physical assets created through their financing remain safe and disaster resilient, the guidelines prepared by NDMA can be adopted by banks and made applicable to new constructions as well as additions, modifications, extensions or alteration of houses financed by them. Further, depending on the nature of the asset and the vulnerability of the location to any of the disasters, banks could insist on ensuring that the disaster resistant features of NDMA guidelines are incorporated in the actual construction before the loan is sanctioned or disbursed so that the disaster management features are built in at the design stage itself."* This instruction has far reaching impacts, as it ensures that all big construction projects abide by building bylaws and incorporate disaster resistant features.

Corporate Disaster Resource Network (CDRN) a web based supply chain management system was designed under guidance of NDMA to help relief agencies, response agencies and local government access and feed in real time information on products and services required for relief operations, so that it can be used as a disaster resource inventory data base to be used by agencies during response and relief operations.

Indian Disaster Resource Network (IDRN), a web based on-line information system, (That is.www.idrn.gov.in) was designed with a purpose of managing nationwide data base for disaster management related resource inventory it has been implemented at state and district levels. The data base needs to be updated on a regular basis by local administration so that during the time of a disaster updated information on resource inventory is available.

India's Climate Action on Paris Agreement

The historic Paris Agreement provides an opportunity for countries to strengthen the global response to the threat of climate change by keeping a global temperature rise in this century well below 2 degrees Celsius and to pursue efforts to limit the temperature increase even further to 1.5 degrees Celsius. It entered into force on 4 November 2016. 171 Parties have ratified of 197 Parties to the Convention. India has a comprehensive framework of legal and institutional mechanisms to respond to environmental challenges. It has initiated several climate-friendly measures, particularly in the area of renewable energy.

India is committed to invest in tackling climate change while addressing poverty, food security and access to healthcare and education. India's Intended Nationally Determined Contributions (INDC) targets on its goal of installing 175 gigawatts (GW) of renewable energy capacity by 2022, out of this, 100 GW has been allocated to solar and 60 GW to wind. It has set a new target to increase its share of non-fossil fuel-based energy from 30 per cent today to about 40 percent by 2030 and committed to reduce its emissions intensity per unit GDP by 33 to 35 per cent below of year 2005 by 2030 and create an additional carbon sink of 2.5 to 3 billion tonnes of CO_2 through additional tree cover. India's action plan prioritizes building resilience to climate change impacts, with indication of financing necessary-projects and initiatives to reach its goals.

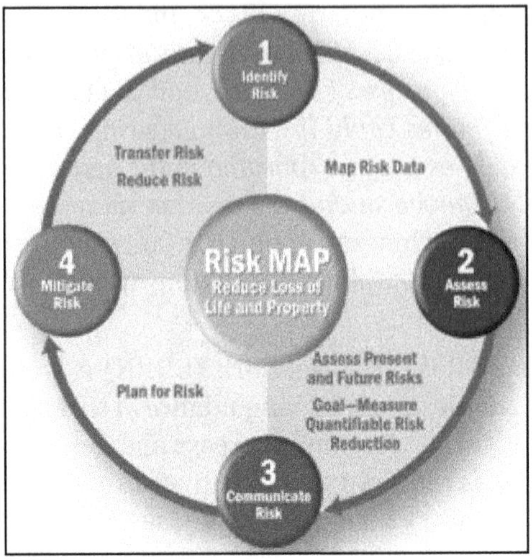

Figure 7.3: Framework for Risk Mapping and Application.

Under India's Constitution, climate action is the mandate of states and union territories. More than 30 states developed action plans on climate change (SAPCCs) under the National Action Plan on Climate Change. With new national goals on climate change as per INDC, the SAPCCs would need to be redefined to address the new commitments. India's plan to reach the INDC needs to be strongly gender-equitable and community based. Community-based adaptation approach and DRR must be the first line of action for integration of adaptation in policies and planning. India recognizes importance of aggressively restoring forest cover, in a manner consistent with supporting livelihoods. Creating an additional carbon sink of 2.5 to 3 billion tonnes of CO_2 would require average annual carbon sequestration to increase by at least 14 per cent over the next 15 years relative to the 2008-2013 period. Apart from Green India Mission expected to deliver 50-60 per cent. It remains to be seen how India will achieve the rest. However, clarity on emissions intensity in the base year (2005) and target year (2030), as well as the scope and coverage of the intensity target and the methodologies for measuring it is crucial for monitoring progress towards targets and for determining contribution to global goal of limiting temperature rise to 2 degrees C. As a country exceptionally vulnerable to climate change, there is heavy focus on adaptation and resilience.

Addressing the effects of climate change in form of increasing and complex climatic hazards and extreme events is a core issue in adaptation and resilience that calls for concerted and proactive efforts for saving people's lives and gains of decades of developmental process which are increasingly at risk. Utilization of ICT applications in risk assessments, planning, programming and mainstreaming besides implementation and monitoring of actions and strategies is critical (Figure 7.3).

International Solar Alliance (ISA)

The International Solar Alliance (ISA) is an alliance of more than 121 countries, most of them being sunshine countries, which lie either completely or partly between the Tropic of Cancer and the Tropic of Capricorn. The primary objective of the alliance is to work for efficient exploitation of solar energy to reduce dependence on fossil fuels. This initiative was first proposed by Indian Prime Minister in November 2015. This alliance is based on treaty between member countries and was declared in Paris on 30th November, 2015. Countries that do not fall within the Tropics can join the alliance and enjoy all benefits as other members, with the exception of voting rights. Alliance aims to undertake joint efforts required to reduce the cost of finance and the cost of technology required for massive deployment of solar energy, and pave the way for future technologies to achieve targets of Paris Climate Change Agreement.[23,24]

It was envisioned that if solar resource rich countries act in a coordinated manner, with strong political impulse and resolve, and that better harmonizing and aggregating the demand for inter alia solar finance, technologies, innovation or capacity building, across countries, will provide a strong lever to lower costs, increase quality, and bring reliable and affordable solar energy within the reach of all.[25]

It is to be headquartered in India. Interim Secretariat at the National Institute of Solar Energy (NISE) in Gurugram, India was inaugurated in January 2016. The Indian government has promised for requisite land and fund for construction of headquarters in the NISE campus.

Conclusion

Disaster Management; Disaster Risk Reduction; Climate Change Mitigation and Adaptation; Inter linkage of Climate Change and Disaster Risk Mitigation measures; Role of ICT in overarching Scheme of Preparedness and Mitigation, Disaster Communication and Early Warnings as an imperative for DM are some of many important aspects This book has touched upon. The book also does prognostic analysis of existing disaster communication and early warning systems in India as a case and outcome of this analysis has been presented in form of recommendations and way forward. The contents and contexts are very important not only for India but for most of the developing and under developed countries in the world. This book may be considered as a good reference document on the subject matter.

Notes

1. Disaster management developing country communities and climate change The Role of ICT: NONITA T. Y. 2011.
2. Panchayat is a decentralized form of government in which each village would be responsible for its own affairs.

3. http://www.oneindia.com/2013/06/24/u-khand-tragedy-cag-submits-report-1244367.html
4. Southasiadisasters.net special issue October 2013, http://reliefweb.int/sites/reliefweb.int/files/resources/99per cent20Cycloneper cent20Phailin.pdf
5. TRAI consultation paper on 'Next Generation Public Protection and Disaster Relief (PPDR) communication networks' (2017).
6. http://www.worldbank.org/en/news/feature/2017/11/06/a-small-island-country-steps-forward-to-unite-the-world-behind-climate-action?cid=ECR_E_NewsletterWeekly_EN_EXT
7. Ibid.
8. http://www.worldbank.org/
9. http://www.dot.gov.in/sites/default/files/Nationalper cent20Telecomper cent20M2Mper cent20Roadmap.pdf
10. An actuator requires a control signal and a source of energy and is responsible for moving or controlling a mechanism or system. The control signal may be low energy electric voltage or current, pneumatic or hydraulic pressure, or even human power. (https://en.wikipedia.org/wiki/Actuator)
11. TR 102 725 V1.1.1 (2013-06)
12. http://trai.gov.in/sites/default/files/Consultation_Paper_M2Mper cent20_18_October_2016.pdf
13. Retrieved from http://www.itu.int/en/ITU-T/about/groups/Pages/sg20.aspx
14. Retrieved from Asian Disaster Management News January-April 2007
15. ITU study group 2: Guidelines for using a content standard for alerts and notifications in disasters and emergency situations (2008) regarding implementation of ITU-T Recommendation X.1303 on the Common Alerting Protocol (CAP), Report on use of remote sensing for disaster prediction, detection and mitigation (2008), and Guidelines for Implementation of Satellite Telecommunications for Disaster Management in Developing Countries (2009)
16. Consultation paper On Universal Single Number Based Integrated Emergency Communication and Response System by TRAI, New Delhi vide Consultation Paper No.3/2013: 15th March 2013
17. Project report of Capacity Building for Advanced Trauma life Support in India, implemented by JPNATC, Delhi, in association with NDMA.
18. Telecom Regulatory Authority of India Consultation paper On Telecom network failures during Emergencies/Disasters Priority routing of calls of persons engaged in. 'response and recovery' released on 10th May, 2012).
19. http://www.itu.int/rec/T-REC-E.106-200310-I/en.
20. Asian Disaster Management News January-April 2007
21. http://www.itu.int/ITU-D/emergencytelecoms/tampere.html

22. http://www.dot.gov.in/sites/default/files/SOPper cent202017per cent20depttper cent20ofper cent20Telecom.pdf
23. http://isolaralliance.org/
24. https://en.wikipedia.org/wiki/International_Solar_Alliance
25. http://isolaralliance.org/

BIBLIOGRAPHY

Abbasi, S. (2007) '*Role of Telecenters in Gender Empowerment*', paper presented at the Karlstad Symposium, the Royal Holloway, University of London. http://www.humanit.org/PID/karlstad_symposium/presentations/Salma_Abbasi.pdf

Abraham, A. (2007). '*Tsunami Swamps Aid Agency Accountability: Government Waives Requirement*', Australian Accounting Review , vol 17, no.1 , pp. 1-9. [Online] Available: http://onlinelibrary.wiley.com/doi/10.1111/j.1835-2561.2007.tb00449.x/pdf

Act ahead - Partnership in Community Based Disaster Management in Asia, (2004). Report based on proceedings of the International Conference, 24-26 August 2004, New Delhi, India.

ADB, UN and WB (2007). *Tsunami, India – Two years after*. Retrieved from http://www.tn.gov.in/tsunami/digitallibrary/ebooks-web/66%20TSunami%20_India_-Two_%20Years%20_After.pdf

ADB, WB and UN (2005). *India, Post tsunami recovery program, Preliminary damages and needs assessment.* Retrieved from http: //www.tnrd.gov.in / externallyaidedprojects /Tsunami_rehabilation /india-assessment-full-report[1].pdf

ADPC. (2000). *Post-disaster damage assessment & needs analysis reporting*. Retrieved from http:// unpan1.un.org /intradoc /groups/ public/ documents /APCITY /UNPAN019544.pdf

ADRC. (2005). *Total disaster risk management: Good practices.* Retrieved from http : // www.adrc.or.jp /publications/TDRM

Agranoff, R., & McGuire, M. (2001). Big questions in public network management research. *Journal of Public Administration Research and Theory, 11 (3)*, 295-326.

Agranoff, R., & McGuire, M. (2003). Inside the matrix: Integrating the paradigms of intergovernmental and network management. *International Journal of Public Administration, 26 (12),* 1401-22.

Ahmad, A. (2004). *Disaster management: Through the new millennium.* New Delhi. Anmol Publications:

Ahmed, K. I. (2007). *Emergency Telecommunications and Early Warning Systems for Disaster Preparedness in Chittagong,* Bangladesh. ITU Dhaka.

Akhtar, A.S. (2006). *Making of a disaster. Economic and Political Weekly, 41(06),* 479-480.

Albala-Bertrand, J.M, (1993), *Political Economy of Large Natural Disasters; With Special Reference to Developing Countries,*, New York, Oxford University Press.

Alexander, D. (2002). *Principles of emergency management and planning.* New York: Oxford University Press.

Allen, K. (2006) *'Community-based Disaster Preparedness and Climate Change Adaptation: Local Capacity Building in the Philippines',* Disasters. vol 30, no 1, pp. 81-101.

American Red Cross. (1998). *Are you ready for an earthquake and Are you ready for a* flood. Washington. Retrieved from http://www.redcross.org

Anderson, M.B., & Woodrow, P.J. (1989). *Rising from the ashes: Development strategies in times of disaster.* Boulder, Colorado: West view Press.

Andharia, Janki, (2003), *Disaster Risk Reduction through Development Orientation in Disaster*

Anthony M. Townsend and Mitchell L. Moss (2005). *Telecommunications infrastructure in disasters : Preparing Cities for Crisis Communications.*

Apikul, C. (2010) 'Overview', *ICT for Disaster Risk Reduction.* APCICT, Incheon City, Republic of Korea, pp. 9-50.

Arnstein, S. (1969). *A ladder of community participation.* American Institute of Planners Journal , 35, 216-224.

Aronoff, M., & Valerie. (1992*).* Defining disaster: Local constructions for recovery in the aftermath of chemical contamination. *Social Problems, 39(4),* 345-365

Asplund, M., Nadjm-Tehrani, S. & Sigholm, J. (2008) *Emerging Information Infrastructures: Cooperation in Disasters* [Online]

Available: 'http: // www.springerlink.com/content/l2031v5282075497/fulltext.pdf.

Available: http://technologysalon.org/2010/04/technology-in-disaster-response.html

Awasthy, A. (2009). *Disaster management: Warning response and community relocation.* New Delhi: Global India Publication.

Bahadur, J. (1998). *Water resources management in the himalayan region.* New Delhi: Vigyan Prasar.

Bahuguna V.K. and Singh S. (2002). Fire Situation in India. IFFN, 26: 23-27

Barnum, A. (1994) . *"Bay Area firms took heed after Loma Prieta".* San Francisco Chronicle

Barton, A. H., & Merton, R. K. (1963*). Social organization under stress: A sociological review of disaster studies.* Washington D.C.: National Academy of Sciences.

Basu, R.L. (2011). *Kautilya's Arthasastra (300 B.C.): Economic Ideas.* Retrieved from http://www.smashwords.com/books/download/65088/1/latest/0/0/

Baumwoll, J. (2008). The value of indigenous knowledge for disaster risk reduction: A unique assessment tool for reducing community vulnerability to natural disasters. Retrieved from http: //www.islandvulnerability.org/m /baumwollm.pdf

Bell, S. (2002). Institutionalism: Old and New. In John Summers (Ed.), *Government, politics, power and policy in Australia (*7th ed.). (pp. 363-380). New South Wales, Australia: Pearson Education Australia.

Benini, A.E. (1991) 'Computer Simulation as a Means of Dialogue between Local Committees and Relief Agencies: A Case from Southern Sudan', *Disasters,* vol 15 no. 4, pp. 331-339.

Bennett, J., Bertrand, W., Harkin, C., Samarasinghe, S., & Wickramatillake. (2006). *Coordination of international humanitarian assistance in tsunami affected countries.* London: Tsunami Evaluation Coalition.

Bennis, W.G., & Slater, P. E. (1968). *The temporary society.* New York, NY: Harper Colophon.

Benson, C., & Twigg, J. (2007). *Tools for mainstreaming disaster risk reduction: guidance notes for development organizations.* Geneva: International Federation of Red Cross and Red Crescent Societies.

Berke, P. R., Kartez, J.D., & Wenger, D.E. (1993). Recovery after disaster: Achieving sustainable development, mitigation and equity. *Disasters,* 17(2), 93-109.

Berke, P.R., & Beatley, T. (1997). *After the hurricane: Linking recovery to sustainable development in the Caribbean.* Baltimore, Maryland: Johns Hopkins University Press.

Bertalanffy, L.V. (1950). An outline of general system theory. *British Journal of Philosophy of Science, 1,* 139-164.

Best, M. (2003) 'The Wireless Revolution and Universal Access', *Trends in*

Bhattacharya, T. (2012). *Disaster science and management.* New Delhi: Tata-Mcgraw Hill.

Bhatti ,A. & Ariyabandhu, M.M. (2002) .*Disaster Communication a media Kit.*. New Delhi: A Duryog Nivaran Publication.

Bhatti, A. and Ariyabandhu, M. M. (2002). Disaster Communication : A Resource Kit for Media.

Blackburn J., & Holland, J. (1998). Who changes: Institutionalizing participation in development. London: Intermediate Technology Publications.

Blantz, E. (2010) *Technology in Disaster Response: ICT in Haiti and Beyond* [Online]

BMPTC. (1994). Disaster mitigation & vulnerability atlas of India: A paradigm shift from post-disaster reconstruction & relief to pre-disaster pro-active approach. New Delhi. Retrieved from http://www.bmtpc.org

Bolin, R. (1982). *Long-Term Family Recovery from Disaster*. Boulder, Colorado: University of Colorado.

Boo, H-W. (2008). *A meta-synthesis of emergency network management strategies and analysis of hurricane katrina – Doctoral Thesis.* Retrieved from http://scholar.lib.vt.edu/theses/available/etd-07042008-095344/unrestricted/HWBoo.pdf

Bose, B.C. (2006). *Modern encyclopaedia of disaster and hazard management.* New Delhi: Rajat Publications.

Bosher, L., Dainty,A., Carrillo, P., & Glass, J. (2007). Integrating disaster risk management into construction: A UK perspective. *Building Research and Information, 35(2),* 163-177.

Brooks, N. (2003). Vulnerability, risk and adaptation: A conceptual framework. *Working Paper no 38.* Norwich: Tyndall Centre.

Bryant, E. (2001). *Tsunami: The under-rated hazard, New York*: Cambridge University Press.

Buckland, J., & Rahman, M. (1999). Community-based disaster management during the 1997 red river flood in Canada. *Disasters* 23(2), 174-91.

Bunker, D., & Smith, S. (2009) *'Disaster Management and Community Warning Systems; Inter-organisational Collaboration and ICT Innovation'*, paper presented at the 2009 Pacific Asia Conference on Information System. [Online] Available: http: //aisel.aisnet.org /cgi /viewcontent.cgi

Buzna, L., Issacharoff, L., Peters, K., Rosalo, V. & Helbing, D. (2007) '*How Can Future ICT Enhance Disaster Relief and Recovery?' ICT-FET Workshop*, n.p., 6 Oct [Online] Available:ftp://ftp.cordis.europa.eu/pub/fp7/ict/docs/fet-proactive/cosiict-ws-oct07-03_en.pdf

Canton, L.G. (2007). *Emergency management: concepts and strategies for effective Programs.* New Jersey, Hoboken: John Wiley & Sons.

Carole, L. J. (2007). Roots of abysmal administrative failure. *Public Administration Review, 67(s1),* 22-23.

Carr, L. (1932). Disasters and the sequence-pattern concept of social change. *American Journal of Sociology, 38,* 207–218.

Carreno, M.L., Cardona, O.D., & Barbat, A.H. (2005). *Evaluation of the risk management performance.* Retrieved from http: //emi.pdc.org /DRMlibrary /Bogota/Evaluation-Risk-Management-index.pdf

Carter, W. N. (1992). *Disaster management: A disaster manager's handbook.* Manila: Asian Development Bank.

CBSE. (2005). *Towards a safer India – Education in disaster management.* New Delhi: Central Board of Secondary Education.

Census of India. (2011). Retrieved from http:// censusindia.gov.in /2011 -prov-results /prov_results_paper1_india.html

Centre for Development Studies. (2005). *Towards post tsunami livelihood security for fishing communities in Tamil Nadu.* Retrieved from http://www.ukabc.org/TamilNadupostTsunami.pdf

Centre for Public Health and Disasters. (2005). *Writing a disaster plan: A guide for health departments.* Retrieved from http: //ftp.cdc.gov /pub /phlpprep /Legal %20Preparedness%20for%20Pandemic%20Flu/8.0%20-%20Non-Governmental% 20Materials/8.6%20Writing%20a%20Disaster%20Plan%20 -%20UCLA.pdf

Centre for Research on the Epidemiology of Disasters (CRED). (2011). *Annual disaster statistical review 2011: The numbers and trends.* Brussels.

Chambers, R. (1997). Responsible well-being-A personal agenda for development. *World Development, 25 (11),* 1743-54.

Chambers, R., & Conway G.R. (1992). Sustainable livelihoods: Practical concepts for the 21st century. *IDS Discussion Paper No. 296,* Susses: *Institute for Development Studies,* University of Sussex.

Chauhan, A., R. Chauhan, S. Semwal (2013). Early Detection of Forest Fire Using Wireless Sensor Network. Int. J. Engineering Research & Applications. www.ijera.com 3(4):163-168.

Chen, L., Liu, Y., & Chan, K. (2006). Integrated community-based disaster management program in Taiwan: A case study of Shang - an village. *Natural Hazards, 37 (1-2),* 209-223.

Cherian, J. (2010, Jan 30-Feb 12). Haiti's horror, *Frontline,* 27(3).

Choguill, M. (1996). *A ladder of community participation for underdeveloped countries.* Habitat International , 20 (3), 431-444.

Cigler, B.A., Stiftel, B., & Burby, R.J. (1987). Rural community responses to a national mandate: An assessment of floodplain land-use management. *Publius,* 17(4), 113-130.

Coburn, A.W., Sspence, R.J.S., & Pomonis, A. (1994). *Vulnerability and risk assessment.* Retrieved from http: //worldbank.mrooms.net /file.php /356 /2234 / Introduction%20Reading%20-%20 Vulnerability And Risk Assessment Guide. pdf

Cohen, J.M., & Uphoff, N.T. (1980). Participation's place in rural development: seeking clarity through specificity. *World Development, 8(30),* 213-235.

Coleman, J. (1990). *Foundations of social theory.* Cambridge: Belknap, Harvard University Press.

Collin, T.D. (2000). *Disaster management and preparedness.* Boca Raton, USA: Crc Press.

Comfort, L. K. (1994). *Risk and resilience: Interoganization learning following the northride earthquake of January 17, 1994.* Berkeley: University of California.

Comfort, L. K. (1999). *Shared risks: Complex systems in seismic response.* Oxford: Pergamon/ Elsevier Science.

Comfort, L.K. (1985). Integrating organizational action in emergency management; strategies for change. *Public Administration Review, 45,* 155-164.

Comfort, L.K., Waugh, W.L., & Cigler, B.A. (2012). Emergency management research and practice in public administration: Emergence, evolution, expansion and future directions. *Public Administration Review, 72(4),* 539-547.

Coppola, D. P. (2007). *Introduction to international disaster management.* Oxford: BH Publishers.

Cordeiro, V. J. and Deshpande, S. S. (2005). Research Report on Disability and the Tsunami in India. Part I. International Disability and Development Consortium, Brussels.

Crisis and Emergency Risk Communication (2002). Centre for Disease Control and Prevention . Available at http://emergency.cdc.gov/cerc/resources/pdf

D'Souza, J.B. (2005). Disastrous management, *Economic and Political Weekly, 36,* 3900-3901.

Das, B. (2002). Some aspects of disaster mitigation and management in India. *Indian Journal of Public Administration, 48(2),* 183-196.

Das, B.K. (2009). Flood disasters and forest villagers in sub-himalayan Bengal. *Economic and Political Weekly, 44(4),* 71-76.

Date, V. (2005). Mumbai floods: The blame game begins, *Economic and Political Weekly, 40(34),* 3714-3716.

Dave, Gupta, S., Parmar, N.K., & Kant, S. (2003). *Emergency medical services and disaster management: A holistic approach.* New Delhi: Alpha Science Publishers.

Davis, G. K. B. (1999). Electronic Communication Systems.

Denning, P. J. (2006). Hastily formed networks. *Communication of the ACM, 49 (4),* 15-20.

Department of Earthquake and Engineering. (2000). *A report on Chamoli earthquake of March 29, 1999.* Roorkee: University of Roorkee.

Department Of Telecommunications (2008), "*Disaster Management Guidelines for Telecommunications*" issued by the Telecom Engineering Centre, Ministry Of Communication & It ,Government Of India. New Delhi.

Department Of Telecommunications (2015) Ministry Of Communication & It Government Of India," *Standard Operating Procedure (Sop-2015) For Telecommunication Services for Responding to Disasters*" New Delhi. Retrieved from http://dot.gov.in/sites/default/files/

Depoorter, B. (2006). Horizontal political externalities: The supply and demand of disaster management. *Duke Law Journal, 56,* 101.

DFID. (2004). *Learning lessons from the cyclone – A study of DFID's support for post-cyclone livelihoods rehabilitation in Orissa.* New Delhi.

Dhar-Chakrabarti P.G. & Bhatt, M. R. (2005) "*Mincro-Finance and Disaster Risk Reduction"* edited .New Delhi

Dilling, J. (2008). *Locus of an emergent field:Framing emergency management knowledge, values, and priorities within current academic curricula* - Doctoral Thesis. Florida: Florida State University.

Din, Z.U. (2005). South Asia earthquake: Civilian and government responses. *Economic and Political Weekly, 40(44/45),* 4666-4667.

Dinham, B., & Sarangi, S. (2002). The Bhopal gas tragedy 1984 to? The evasion of corporate responsibility. *Environment & Urbanization, 14(1),* 89-99.

Dombrowsky, W. R. (1981). Another step toward a social theory of disaster. *Preliminary Paper 70.* Newark DE: University of Delaware Disaster Research Center.

Dombrowsky, W. R. (1998). Again and again: Is a disaster what we call a disaster? In E. L. Quarantelli, (Ed.), *What is a disaster? Perspectives on the question* (pp.13-24). London: Routledge.

Douglas, M. (1986). *How institutions think.* Syracuse, NY: Syracuse University Press.

Drabek T.E. & Key, W.H. (1976) 'The Impact of Disasters on Primary Groups', *Mass Emergencies,* vol 1, pp. 89-105.

Drabek, T.E., & Evans, J. (2007). *Emergency management and homeland security curricula: Contexts, cultures and constraints.* Retrieved from http://www.training.fema.gov/emiweb/edu/docs/Drabek%20-%20EM %20and%20 HS%20Curricula-contexts, %20Cultures,%20and%20Constra.doc

Drennan, L.T. (2007). *Risk and crisis management in the public sector.* London: Routledge.

Drucker, P. (1969). *The age of discontinuity*. New York, NY: Harper & Row.

Dynes, R. R. (1970). *Organized behavior in disasters*. Lexington, Massachusetts: Heath Lexington Books.

Dynes, R. R. (1991). Disaster reduction: The importance of adequate assumptions about social organization. *Preliminary Paper. 172*. Delaware: Disaster Research Center, University of Delaware.

Dynes, R. R., & Quarantelli, E. L. (1968). Group behavior under stress: A required convergence of organizational and collective behavior perspectives. *Sociology and Social Research* , 416-429.

Eadie, C. (1991). *Phases of earthquake response and recovery planning*. Santa Cruz, California: Santa Cruz Redevelopment Agency.

EMA (2009). *What is Emergency Management?* Retrieved from: http://www.ema.gov.au/agd/ema/emaInternet.nsf/Page/RWP685D60F554675173CA256C83

Enarson, E., & Chakrabarti, D. (2009). *Women, gender and disaster: Global issues and initiatives*. Sage Publications.

Epstein, P.R., & Mills, E. (2005). "Climate change futures: Health, ecological and economic dimensions. Boston: Center for Health and the Global Environment.

Erramilli B. P. (2008). Research on "*Disaster Management in India: Analysis of Factors Impacting Capacity Building*" Political Science Dissertations Paper 15. Available at http://scholarworks.gsu.edu/cgi/viewcontent.cg

EWC III (2006) *The Third International Conference on Early Warning (EWC III)* held in Bonn, Germany from 27-29March 2006.

Fagan, B. (1999). *Floods, Famines, and Empires*. New York: Basic Books.

Faiola, A. and Reed, T. (1994). L.A. communications in chaos. Miami Herald N Willing. December 29, 2004. "*Tsunami telecom recovery continues*". [http://www/lightreading.com]

Fakhruddin S.H.M. (2007) article on "Emergency Communications for Disaster Management " in ADPC magazine " Asian Disaster Management News January-April 2007 Vol. 13".Available at www.adpc.net

Fakaruddin, S H M (2007). Emergency Communications for Disaster Management. ADPC. Asian Disaster Management News. Jan-April 2007.

FAO., NACA., SEAFDEC., & BOBP-IGO. (2005). *Impacts of the tsunami on fisheries, aquaculture and coastal livelihoods in India*. Retrieved from *ftp://ftp.fao.org/FI/DOCUMENT/tsunamis_05/india/impact/05-03-16-India.pdf*

Farazmand, A. (2001). Crisis and emergency management: An introduction. In Ali Farazmand (Ed.), *Handbook of crisis and emergency management* (pp.1-9). New York: Marcel Dekker.

Farazmand, A. (2003). Chaos and transformation theories: A theoretical analysis with implications for organization theory and public management. *Public Organization Review: A Global Journal, 3,* 339–372.

Feedback Business Consulting Services. (2005). *Post tsunami: Focus on livelihoods going forward.* New Delhi.

FEMA. (2003). *IS 230 - Principles of Emergency Management.* Retrieved from: http://www.training.fema.gov/EMIWeb/IS/is230lst.asp

Fengler, W., Ihsan, A., & Kaiser, K. (2008). Managing post-disaster reconstruction, finance international experience in public financial management. *Policy Research Working Paper 4475.* Retrieved from http :// elibrary.worldbank.org /docserver /download/4475.pdf?expires=1382377930&id=id & accname=guest &checksum = 34D1E19C80D578BC995F6A1CDAE5023D

Filho, W.L. (2012). *Climate change and disaster risk management.* Springer Publication.

Flyvbjerg, B. (2006). Five misunderstandings about case-study research. *Qualitative Inquiry, 12(2),* 219-245.

Forrest, T. R. (1974). *Structural differentiation in emergent groups.* Retrieved from http://udspace.udel.edu/bitstream/handle/19716/1318/RS15.pdf?sequence=1

Forrest, T. R. (1978). Group emergence in disasters. In E.L.Quarantelli (Ed.), *Disasters: Theory and research* (pp. 105-128). Sage Publications.

Francaviglia, R.V. (1978). Xenia rebuilds: Effects of predisaster conditioning on post-disaster redevelopment. *Journal of the American Institute of Planners, 44(1),* 13-24.

French, S.P., Ewing, C.A., & Isaacson, M.S. (1984). "Restoration and recovery following the coalinga earthquake of may 1983. *Working Paper No. 50.* Colorado: Natural Hazards Research and Applications Information Center, Institute of Behavioral Science, University of Colorado.

Friesema, H. P., Caporaso, J., Goldstein, G., Lineberry, R., & McCleary, R. (1979). *Aftermath: Communities after natural disasters.* California: Sage Publications.

Fritz, C. (1961). *Disasters.* In R. Merton, & R. Nisbet (Eds.), *Social problems.* New York: Harcourt Brace.

Fukuyama, F. (1995). *Trust: The social virtues and the creation of prosperity.* New York: The Free Press.

Gadgil, S., Joseph, P.V., & Joshi, N.V. (1984). Ocean-atmosphere coupling over monsoon regions. *Nature,* 312, 141-143.

Ganapathy, G.P., & Hada, C.L. (2012). Landslide hazard mitigation in the Nilgiris district, India – Environmental and societal Issues. *International Journal of Environmental Science and Development, 3(5),* 497-500.

Gariott, G.L. (1991) *Low Earth Orbiting Satellites and Internet-Based Messaging*

Services.Development Communication Report #75, Issue 4 [Online] Available: [http://www.isoc.org/inet96/proceedings/g1/g1_1.htm]

Geis, D.E. (1996). *Creating sustainable and disaster resistant communities"*, Retrieved from http://www.smartcommunities.ncat.org/articles/CREATING.shtml

Geis, D.E. (2000). By design: The disaster resistant and quality-of-life community. *Natural Hazards Review , 1 (3),* 151-160.

General Dynamics (2013), "*White paper on Maintaining Public Safety Communications in Disaster Situations Key Requirements for Uninterrupted, Interoperable Communications*" .Retrieved from http:// www.gdc4s.com/ Documents/ NotInNavDocs/GD_Disaster_Response_Communications_White_Paper_FINAL_26-8-13.pdf

Ghesquiere, F., & Mahul, O. (2007). Sovereign natural disaster insurance for developing countries: A paradigm shift in catastrophe risk financing. *Policy Research Working Paper 4345.* Retrieved from https: //openknowledge.worldbank.org /bitstream/handle/10986/7331/wps4345.pdf?

Gidron, B., Kramer, R. M., & Salamon, L. M. (1992). *Government and the third sector; emerging relationships in welfare state.* San Francisco: Jossey- Bass Publishers.

Gilbert, C. (1998). Studying Disaster: Changes in the Main Conceptual Tools. In E.L. Quarantelli, (Ed.), *What is a disaster? Perspectives on the question* (pp.3-12). London: Routledge.

Godschalk, D. R. (1985). *Rebuilding after hurricane Frederick: Gulf shores' struggle with mitigation policy.* North Carolina : Center for Urban and Regional Studies, University of North Carolina, Chapel Hill.

Goel, S.L. (2007). *Disaster administration and management: Text and case studies.* New Delhi: Deep &Deep Publications.

Goodchild, B. (1990). Planning and the modern/post-modern debate. *Town Planning Review, 61(2),* 119-137.

Government of NCT of Delhi. (2010). State of environment report for Delhi 2010. New Delhi: Department of Environment and Forest.

Government of India. (2009). *Thirteenth finance commission, 2010-2015.* New Delhi: Finance Commission of India.

Government of India. (2011). *Disaster management in India.* New Delhi: Ministry of Home Affairs.

Graziano, T & Lûtteke, G. (2006). *Convergence and Interoperability'*, paper presented at eBusinessWatch Workshop, Brussels

Grist, R. E. (2007). *The Changing paradigm of emergency management: Improving professional development for the emergency manager - Doctoral Thesis.* Retrieved from http://www.training.fema.gov/emiweb/edu/docs/Grist%20-%20DISSERTATION %20-%20Changing%20Pradigm%20of%20EM.pdf

Gupta, K. (2006). *Disaster management in India: Responding internally and simultaneously in neighbouring countries.* Retrieved from http: //training.fema. gov /EMIWeb/edu/ Comparative%20EM%20Book%20-%20Chapter%20 -%20Disaster % 20Management %20and%20India%20-%20Responding%20 Internally%20and% 20 Simulta.doc

Gupta, M.C, Sharma, V.K., Gupta, L.C., & Tamani, B.K. (Eds.). (2001). *Manual on natural disaster management in India.* New Delhi: National Center for Disaster Management, IIPA.

Gurenko, E., & Lester, R. (2004). Rapid onset natural disasters: The role of financing in effective risk management. *World Bank Policy Research Working Paper 327.* Washington D.C: World Bank.

Gürhan Ertur and Argun Yum , " *Risk Mitigation and the Role of the Media*" available [http://info.worldbank.org/etools/docs/library/114715/istanbul03/docs/ istanbul03/18ertur3-n.pdf]

Guzmann, E.M.de. (2005) "*Towards total disaster risk management approach*", Retireved from http ://unpan1.un.org /intradoc /groups public /documents /APCITY/UNPAN009657.pdf

Haas, J. E., Kates, R.W., & Bowden, M.J. (Eds.). (1977). *Reconstruction following disaster.* Cambridge: MIT Press.

Halder, S.R. & Ahmed, T. (2010) '*The Bangladesh Comprehensive Disaster Management Programme and ICTs'*, ICT for Disaster Risk Reduction. Asian and Pacific Training Centre for Information and Communication Technology for Development, Incheon City, Republic of Korea pp. 53-75.

Handy, C. (1991). *The age of unreason.* Boston, MA: Harvard Business Review Press.

Haque, C. E. (2005). *Mitigation* of natural hazards and disasters. New York: Springler.

Hegenbarth, J. L., & Brower, D.J. (1985). *Redevelopment after the Storm: Hazard mitigation opportunities and obstacles in the post-disaster setting.* North Carolina : Center for Urban and Regional Studies, University of North Carolina, Chapel Hill.

Helen Roeth (2009) report on "The Development Of A Public Partnership Framework And Action Plan For Disaster Risk Reduction in Asia UNISDR .Available at http://www.unisdr.org/

Hoffman, D., & Brukoff, P. (2006). Insuring public finances against natural disasters—A survey of options and recent initiatives. *International Monetary Fund Working Paper/06/199.* Retrieved from http :// www.imf.org /external/ pubs/ft /wp/2006 /wp06199.pdf

Hood, C. (1991). A public management for all seasons?. *Public Administration Review, 69(1)*, 3-19.

Howard,K.,& Roth, R.J. (1998). *Paying the price: The status and role of insurance against natural disasters in the United States.* Washington D.C: Joseph Henry

Press.

Howel, P. (2003). Indigenous early warning indicators of cyclones: Potential application of Bangladesh. *Disaster Studies Working Paper. 6.* London: Benfiled Hazard Research Centre.

HPC (2001*). HPC on Disaster management Report* Department of Agriculture and Cooperation Ministry of Agriculture, Government of India: New Delhi. Available at http://saarc-sdmc.nic.in/pdf/india

HPC (2001*)"HPC on disaster management plans",* available , at http: //www.gisdevelopment.net/application/natural_hazards/overview/nho0015.htm

Hurricanes and Coastal Management, (2005), EMAC After-Action Report, 2004: Hurricane Response, Emergency Assistance Management Compact (EMAC) at http://www.emacweb.org/

Hveinden, B. (1994). *Divided against itself: A Study of integration in welfare* bureaucracy. Oslo: Scandinavian University Press.

Hyrapiet, S. (2003). *Emergent phenomena in India after the indian ocean tsunami - MA Dissertation.* Oklahoma: Oklahoma State University.

ICAR. (2005). *Report on assessment of loss due to tsunami to brackish water aquaculture and fisheries sectors in coastal states of Andhra Pradesh.* New Delhi.

IFRC. (2009) *World Disasters Report : Focus on Early Warning, early action*, IFRC, Geneva. Retrieved from http://www.ifrc.org/

IGNOU. (2006). *Course material of post graduate diploma in disaster management.* New Delhi.

IGNOU-NDMA. (2012a). *Conceptual and institutational framework of disaster management.* New Delhi.

IGNOU-NDMA. (2012b). *Disaster preparedness and mitigation.* New Delhi.

IGNOU-NDMA. (2012c). *Responding to disasters.* New Delhi.

IGNOU-NDMA. (2012d). *Disaster recovery and the road ahead.* New Delhi.

IGNOU-NDMA. (2012e). *Handbook for effective disaster management at micro level.* New Delhi.

IIPA. (1993). *India disaster management training country workshop, disaster assessment.* New Delhi: Indian Institute of Public Administration.

Ingelhart, R. (1997). *Modernization and post modernization: Cultural, economic and political change in 43 societies.* Princeton, New Jersey: Princeton University Press.

Institution of Civil Engineers. (1995). *Mega cities: Reducing vulnerability to natural disasters.* London: Thomas Telford.

International Federation of Red Cross and Red Crescent Societies. (2005). *World disaster report, 2005.* Retrieved from http: //www.ifrc.org /Global /

Publications /disasters /WDR /69001-WDR2005-english-LR.pdf

International Federation of Red Cross and Red Crescent Societies. (2007). *World disaster report, 2007.* Retrieved from http: //www.ifrc.org /Page Files /99876 / WDR2007-English.pdf

International Federation of Red Cross and Red Crescent Societies. (2012). *World disaster report, 2012.* Retrieved from http: //www.ifrcmedia.org /assets /pages / wdr2012 /resources/1216800-WDR-2012-EN-FULL.pdf

ITU (2005) "*Handbook on Emergency Telecommunications - Edition 2005 by ITU*" available at http://www.itu.int/ITU-D/emergencytelecoms/publications.html

Izumi Aizu (2011) , in article "*The role of ICT during the disaster – A story of how Internet and other information and communication services could or could not help relief operations at the Great East Japan Earthquake*", Tama University Japan. Retrieved from https://tulaneict4d.wordpress.com/tag/humanitarian-aid

Jain, S. (n.d.) Standing up for trees: Women's role in the chipko movement. Retrieved from http://www.fao.org/docrep/r0465e/r0465e03.htm

Jametti, M., Sternberg, T.V.U., & Cesifo. (2004). *Disaster insurance or a disastrous insurance – natural disaster insurance in France.* Working Paper No. 1303. Retrieved from http://ssrn.com/abstract=614823

Jeggle, T. (2001). The evolution of disaster reduction as an international strategy: Policy implications for the future. In U.Rosenthal, A. Boin, & L.K.Comfort (Eds.), *Managing crises: Threats, dilemmas, opportunities* (pp.316- 342). Springfield, Illinois: Charles C. Thomas Publisher, Ltd.

Jerillos, S. (1999). Fundamentals of disaster risk management: How are southeast asian countries addressing this?. *Risk, Development and Disasters: Southern Perspectives.* Rondebosch, South Africa: Department of Environment and Geographical Sciences, University of Cape Town.

Joint Commission Report (2005) "*Standing Together: An Emergency Planning Guide for America's Communities*" . Available http:// www.jointcommission.org/ assets /1 /18 /planning_guide.pdf

Junqi ZHANG& KUMARASWAMY M.M. Paper on "*Public-Private-People Partnerships (4P) for Disaster Preparedness, Mitigation and Post-disaster Reconstruction*" Dept. of Civil Eng., The University of Hong Kong(HKU) available at [http:// www.iiirr.ucalgary.ca/files/iiirr/B6-1_.pdf]

Jurkiewicz, C.L. (2007). Roots of abysmal administrative failure. *Public Administration Review, 67(s1),* 22-23.

Kanal, A.S. (2013). Disaster resilience through right based and capability perspectives of NGOs: A case study of 2004 tsunami in Nagapattinam. *Dynamics of Public Administration, 28(2)*,215-237.

Kangle, R.P. (1972). *The kautilya arthasastra.* Delhi: Motilal Banarsidass Publishers.

Kapucu, N. (2005). Interorganizational coordination in dynamic context: Networks in emergency response management. *Connections , 26 (2),* 33-48.

Kapucu, N., & VanWart, M. (2006). The evolving role of the public sector in managing catastrophic disasters: Lessons learned. *Administration and Society , 38 (3),* 279-308.

Kapucu, N., Hawkins, C.V., & Rivera, F.I. (2013). *Disaster resiliency: Interdisciplinary perspectives.* Routledge.

Kapur, A. (2005). Insensitive India: Attitudes towards disaster prevention and management. *Economic and Political Weekly, 40(42),* 4551-4560.

Kapur, A. (2009). *On disasters in India.* New Delhi: Foundation books.

Kapur, A., Neeti, Meeta, Deeptima, Roshani, & Debanjali (Eds.). (2005). *Disaster in India, studies of grim reality.* New Delhi: Rawat Publications.

Kaur, J. (2006). Administrative issues involved in disaster management in India. *International Review of Psychiatry. 8 (6),* 553 – 557.

Kennedy G. & Davis B. (1999) *"Electronic Communication Systems"* (McGraw Hill)

Kettl, D. F. (2000). The Transition of governance: Globalization, devolution and the role of government. *Public Administration Review, 60(6),* 488-497.

Kickert, W. J., Klijin, E.-H., & Koppenjan, J. F. (1997). *Managing complex networks: Strategies for the public sector.* Thousand Oaks: Sage Publications.

Kirschenbaum, A. (2003). *Chaos organizations and disaster management.* New York: Routledge.

Knack, S., & Keefer, P. (1997). Does social capital have an economic payoff? : A cross-country investigation. *The Quarterly Journal of Economics, 112(4),* 1251-1288.

Konoorayar, V. (2007). Disasters: Global responses to the challenges. *AALCO Quarterly Bulletin, 4,* 359-384.

Kramer, W., & Schich, S. (2008). Large-scale disasters and the insurance industry. *CESIFO Working Paper No. 2243.* Retrieved from http: //www.cesifo.de / DocCIDL /cesifo1_wp2243.pdf

Kreps, G.A. (1984). Sociological inquiry and disaster research. *Annual Review of Sociology, 10,* 309-330.

Kreps, G.A. (1998). Disaster as systematic event and social catalyst. In E. L. Quarantelli, (Ed.). *What is a disaster? Perspectives on the question* (pp.25-50). London: Routledge.

Kumaran, V.T., & Negi, E. (2006). Experiences of rural and urban communities in Tamil Nadu in the aftermath of the 2004 tsunami. *Build Environment, 32(4),*

375-386.

La Porte, T. M. (2006). Organizational strategies for complex system resilience, reliability and adaptation. In P. E. Auerswald, Branscomb, L. M, L. Porte, T. M, & E. O. Michel-Kerjan (Eds.), *Seeds of disaster, roots of response: How private action can reduce public vulnerability*. New York: Cambridge University Press.

Laituri, M. (2010), 'Geospatial Responses to Disasters: The Role of Cyberspace', *Arc News,* vol 32, no. 2, pp. 1.

Lane, J.E. (2009). *State management: An enquiry into the models of public administration and management.* New York, NY: Routledge.

LaPlante, J.M. & Kroll-Smith, J.S. (1989). Coordinated emergency management: The challenge of the chronic technological disaster. *International Journal of Mass Emergencies and Disasters,* 7(2), 116-132.

Lassa, J.A. (2010). *Institutional vulnerability and governance of disaster risk reduction: Macro, meso and micro scale assessment- Doctoral Thesis.* Retrieved from http://hss.ulb.uni-bonn.de/2011/2451/2451.pdf

Lawal Billa Shattri Mansor and Ahmad Rodzi Mahmud (2004) A paper on "*Spatial information technology in flood early warning systems: an overview of theory, application and latest developments in Malaysia* "

LeHuuti. (2004). *Community-based disaster risk management: Integration to socio-economic development process.* Retrieved from http://www.adpc.net/pdr-sea /publications/ 13SEBklet.pdf

Litan, R.E. (2005). Sharing and reducing the financial risks of future mega-catastrophes. Retrieved from http://www.brookings.edu/views/papers/ litan/20051111.html

Loren B. Legarda, Senator of the Republic of the Philippines UNISDR Regional Champion for Disaster Risk Reduction and Climate Change Adaptation for Asia Pacific and UNEP Laureate comments as given in *World Bank Report on Disaster, (2009).*

Macamo, E. & D. Neubert, 2008: Erwartungen an Sicherheit. Subjektive Katastrophenwahrnehmungen und Bedingungen der Bewältigung am Beispiel Mosambiks und Deutschlands Frankfurt a.M.: Retrieved from http://geo.aau. at/images/stories/pdf/

Maldonado, E. A., Maitland, C. F., & Tapia, A. H. (2010). Collaborative systems development in disaster relief: The impact of multi-level governance. *Inf Syst Front, 12,* 9-27.

Marks, G. (1993). Structural policy and multi-level governance in the european community. In A. Cafruny, & G. Rosenthal (Eds.), *The state of the european community: The maastricht debate and beyond* (pp. 391-411). Boulder, Colorado: Lynne Rienner.

Matthew L. Collins and Naim Kapucu , " A journal on Early warning systems and disaster preparedness and response in local Government " University of Central Florida, Orlando, Florida, USA

May, P.J. (1982). Formulating disaster relief when needs are unknown. *Journal of Policy Analysis and Management, 2(1),* 39-54.

Maykut, P., & Morehouse, P.S. (1994). *Beginning qualitative research: A philosophic and practical guide.* London: Falmer Press.

McConnell, A., & Drennan, L. (2006). Mission impossible? Planning and preparing for crisis. *Journal of Contingencies and Crisis Management , 14 (2),* 59-70.

McEntire, D. A. (2004). Development, disasters and vulnerability: A discussion fo divergent theories and the need for their integration. *Disaster Prevention and Management , 13 (3),* 193-198.

McEntire, D. A., Fuller, C., Johnston, C. W., & Weber, R. (2002). A Comparison of disaster paradigms: The search for a holistic policy guide. *Public Administration Review , 62 (3),* 267-281.

McLeod, D. (2003). Post modernism in urban planning. *Planning & Environmental Information.* Retrieved from http:// www3.sympatico.ca /David.macleod / PINFO.HTM

McMillan D.W, & Chavis, D.M. (1986). Sense of community: Definition and theory. *Journal of Community Psychology, 14, 6-23.*

Mehta, A. (1999). Local governments in disaster management. *Asian Urban Disaster Management News*, Bangkok.

Menard, C., & Shirley, M.M. (2008). *Handbook of new institutional economics.* New York, NY: Springer Publishing.

Mileti, D. S. (1999). *Disasters by Design: A Reassessment of Natural Hazards in the United States.* Washington D.C.: Joseph Henry Press.

Milles, J. G. (2006). Managing after a disaster: Or there and back again. *Buffalo Legal Studies Research Paper Series Paper No. 2006-020. Legal Reference Services Quarterly*, 25, 59-75.

Milward, H. B., & Provan, K. G. (2006). A manager's guide to choosing and using collaborative networks' in networks and partnership Series. Retrieved from http: //www.aihealthsolutions.ca/rtna/conference-files /IBM %20Milward %20 Report_JR_V4_single.pdf

Ministry of Home Affairs. (2003). *Relief and rehabilitation measures in natural calamities.* Retrieved from http://tsunami.icsf.net/images/stories/tn0102.pdf

Mishra, D.K. (2008). Bihar floods: The inevitable has happened. *Economic and Political Weekly, 43(36),* 8-12.

Mishra, P.K. (2006). Towards excellence in disaster management: Governance and sustainability of post-disaster initiatives. *Indian Journal of Public*

Administration, 52(3).

Miskel, J.F. (2008), *Disaster response and homeland security what works, what doesn't?* Stan Ford, U.K: Stand Ford University Press.

Misra, G.K., & Mathur, G.C (Eds.). (1995) Natural disaster reduction. New Delhi: Reliance Publishing House and IIPA.

Moe, T. L., & Pathranarakul, P. (2006). An integrated approach to natural disaster management: Public project management and its critical success factors. *Disaster Prevention and Management, 15(3),* 396-413.

Moore, H. (1956). Toward a theory of disaster. *American Sociological Review, 21,* 734–737.

Mukhopadhy, A.K. (2005). *Crisis and disaster management: Turbulence and aftermath.* New Delhi: New Age Publishers.

Myers, M.F. (1994). Women and children first: Introducing a gender strategy into disaster preparedness. *Women and Emergencies, 2(1),* 14-16.

Myers, M.F. (1997). *Insights emerging from the assessment of research and applications for natural hazards in the United States.* Disaster Preparedness Resources Center. Vancouver, BC: University of British Columbia,

Nagar, R. (2010). *Disaster management.* New Delhi: A.P.H Publishing Corporation.

National Governars' Association. (1979). *Comprehensive emergency management: A Governor's guide.* Retrieved from http://www.training.fema.gov/emiweb/edu/docs /Comprehensive%20EM%20-%20NGA.doc.

Natural Hazard Working Group. (2005). *The role of science in physical natural hazard assessment: Report to the UK government.* Retrieved from http://www.preventionweb.net/files/29922_theroleofscienceinphysicalnaturalha.pdf

NCDM. (2001). *The report of high powered committee on disaster management.* Government of India. New Delhi.

NCRC. (2007). *Mapping and study of coastal water bodies in Nagapattinam district.* Retrieved from http://www.ncrc.in/Publications/NCRC/ Study Report of Coastal Water BodiesCoastalWaterBodiesinNagapattinamDistrict.pdf

NCRC. (2005). *Civil society input to the permanent shelter rehabilitation and GOVT-NGO partnership policies for the tsunami-affected communities in Tamil Nadu.* Nagapattinam: NGO Co-ordination and Resource Centre.

NCRC. (2005a). *Snapshot of damages due to tsunami: Village-wise in Nagapattinam district.* Nagapattinam: District Collectorate.

NDMA (2012). *NDMA National Disaster Management Guidelines on Disaster Management Information and communication System* New Delhi: Ministry of Home Affairs, Government of India. Released in February 2012.

NDMA. (2006). *Report on revamping of civil defence in the country.* New Delhi:

Ministry of Home Affairs, Government of India.

NDMA. (2007). *Guidelines on chemical (industrial) disasters.* New Delhi: Ministry of Home Affairs, Government of India.

NDMA. (2007a). *Guidelines on management of earthquakes.* New Delhi: Ministry of Home Affairs, Government of India.

NDMA. (2007b). *Guidelines on medical preparedness and mass casualty management.* New Delhi: Ministry of Home Affairs, Government of India.

NDMA. (2007c). *Guidelines on preparationof state disaster management plans.* New Delhi: Ministry of Home Affairs, Government of India.

NDMA. (2008a). *Guidelines on management of biological disasters.* New Delhi: Ministry of Home Affairs, Government of India.

NDMA. (2008b). *Guidelines on management of cyclones.* New Delhi: Ministry of Home Affairs, Government of India.

NDMA. (2008c). *Guidelines on management of floods.* New Delhi: Ministry of Home Affairs, Government of India.

NDMA. (2009 a). *Guidelines on management of landslides and snow avalanches.* New Delhi: Ministry of Home Affairs, Government of India.

NDMA. (2009 b). *Guidelines on management of nuclear and radiological emergencies.* New Delhi: Ministry of Home Affairs, Government of India.

NDMA. (2009 c). *National Policy on Disaster Management.* New Delhi: Ministry of Home Affairs, Government of India.

NDMA. (2010). *Guidelines on incident response system.* New Delhi: Ministry of Home Affairs, Government of India.

NDMA. (2010a). *Guidelines on management of drought.* New Delhi: Ministry of Home Affairs, Government of India.

NDMA. (2010b). *Guidelines on management of urban flooding.* New Delhi: Ministry of Home Affairs, Government of India.

NDMA. (2010c). *Report on role of NGOs in disaster management.* New Delhi: Ministry of Home Affairs, Government of India.

NDMD. (2004). *Disaster management in India: A status report.* New Delhi: Ministry of Home Affairs, Government of India.

NDMD. (2005). *ICT for disaster risk reduction – The Indian experience.* New Delhi: Ministry of Home Affairs, Government of India.

NDMD. (2005). *Knowledge management in disaster risk reduction: The Indian Approach.* New Delhi: Ministry of Home Affairs, Government of India.

NDMD. (2005). *Prevention/protection and mitigation from risk of tsunami disaster-A concept note.* New Delhi: Ministry of Home Affairs, Government of India.

NDMICS (2011). *NDMA National Disaster Management Guidelines on Disaster Management Information and communication System* New Delhi: Ministry of Home Affairs, Government of India.

New Economics Foundation. (2007). *Up in smoke? Asia and the Pacific, The threat from climate change to human development and the environment.* Retrieved from http://pubs.iied.org/pdfs/10020IIED.pdf

NIDM. (2012). *India disaster report, 2011.* New Delhi: Ministry of Home Affairs, Government of India.

Nigg, J.M. (1995). *Disaster recovery as a social process in Wellington after the quake: The challenge of rebuilding cities.* Wellington, New Zealand: Wellington Earthquake Commission and the Centre for Advanced Engineering.

NONITA T. Y. (2011) *"Disaster Management, Developing Country Communities & Climate Change: The Role of ICTs".* University of Guelph, Canada. Available at http://www.manchester.ac.uk/cdi

Norris, F.H., Galea, S., Friedman, M.J., & Watson, P.J. (2006). *Methods for* **disaster** *mental health research.* New York: Guilford Press,.

North, D.C. (1990). *Institutions, institutional change and economic performance.* Cambridge: Cambridge University Press.

North, D.C. (1998). Economic performance through time. In Mary C. Brinton & VictorNee (Eds.), *The new institutionalism in sociology* (pp.247-257). Stanford: Stanford University Press.

North, D.C., & Thomas, R.P. (1973). *The rise of the modern world: A new economic history.* Cambridge: Cambridge University Press.

NTP 2011 & 2012 "National *Telecom Policy*" New Delhi: Department of Telecommunications, Ministry of IT and Communications, Government of India.

Olson, G., Malone, T., & Smith, J. (2001). *Coordination theory and collaboration technology*. Hillsdale, New Jersey: Lawrence Erlbaum Associates.

Ostrom, E. (1990). *Governing the commons: The evolution of institutions for collective action*. Cambridge: Cambridge University Press.

O'Toole, L. J.Jr. (1997). Treating networks seriously: Practical and research-based agendas in public administration. *Public Administration Review, 57(1),* 45-52.

Oxfam. (2007). *Tamil Nadu state emergency contingency plan.* Retrieved from http://www.contingencyplanlive.org/files/tn_cp.pdf

Ozerdem, A., & Jacob, T. (2006). *Disaster management and civil society: Earthquake relief in Japan, Turkey and India.* New York: Tauris.

Palakudiyil, T., & Todd, M. (2003). *Facing up to the storm: How local communities cope with disaster: Lessons from Orissa and Gujarat.* London: Christian Aid Publications.

Pande, R.K., (2006). Participation in practice and disaster management: Experience of Uttaranchal (India). *Disaster Prevention and Management, 15(3)*, 425-428.

Parasuraman, S., & Unnikrishnan (Eds.). (2000). *India disasters report: Towards a policy initiative*. New Delhi: Oxford University Press.

Parasuraman, S., & Unnikrishnan (Eds.). (2013). *India disasters report II: Redefining disasters*. New Delhi: Oxford University Press.

Parry M.L., O.F. Canziani, J.P. Palutikof and Co-Authors 2007 IPCC, 2007: *'Technical Summary'. Climate Change 2007: Impacts, Adaptation and Vulnerability*. Contribution of Working Group II to the *Fourth Assessment Report of the Intergovernmental Panel on Climate Change*, M.L. Parry, O.F. Canziani, J.P. Palutikotf, P.J. van der Linden and C.E. Hanson, Eds., Cambridge University Press, Cambridge, UK, 23-78.

Parsons, T. (1961). *Structure and process in modern societies.* Glencoe: The Free Press.

Paton, D. & Johnson, D. (2001) *'Disasters and Communities: Vulnerability, Resilience and Preparedness', Disaster Prevention and Management,* vol 10, no. 4, Pp. 270-277.

Patterson, O., Weil, F., & Patel, K. (2010). The role of community in disaster response: Conceptual models. *Popul Res Policy Review , 29,* 127-141.

Paul, S. (1987). Community participation in development projects: The World Bank experience. *Readings in Community Participation*. Washington: World Bank.

Pearson, C.M., & Clair, J.A. (1998). Reframing Crisis Management. *The Academy of Management Review. 23(1),* 59-76.

Perry, R.W., & Lindell, M.K. (1992). *Behavioral foundations of community emergency planning*. Washington, DC: Hemisphere Books.

Petak, W.J. (1985). Emergency management: A challenge for public administration. *Public Administration Review, 45,* 3-7.

Piers, B., Cannon, T., Davis, Ian., & Wisner, Ben (Eds.). (1994). *AT RISK: Natural hazards, people's vulnerability and disasters.* London: Routledge.

Pieterse, J.N. (2001). *Development theory: Deconstructions/ reconstructions*. London: Sage.

Planning Commission. (2002). *Tenth five year plan, 2002-2007.* Government of India, New Delhi.

Planning Commission. (2008). *Eleventh five year plan 2007-2012.* Government of India, New Delhi.

Planning Commission. (2013). *Twelfth five year plan 2012-2017.* Government of India, New Delhi

Pollitt, C. (2003). *The essential public manager*. Buckingham: McGraw-Hill.

Prabhakar, S.V.R.K., Srinivasan, A., & Shaw, R. (2008). Climate change and local level disaster risk reduction planning: Need, opportunities and challenges. *Journal of Mitigation and Adaptation Strategies for Global Change, 14,* 7-33.

Practical Action. (2009). *South Asia disaster report, special Copenhagen issue*: Retrieved from http://practicalaction.org/docs/region_south_asia/south-asia-disaster-report-copenhagen.pdf

Prasad, K., & Singh, B.D. (1994). *Drought disaster and development.* New Delhi: Mittal Publication.

Prashanta B. (2013) in case study "*Reducing Loss and Damage of Poor*" on cyclone phailin published in special issue No 99 , 2013 of southasiadisasters.net.

Prater, C., Peacock, W.G., Arlikatti, S. & Grover, H. (2006). *Social capacity in Nagapattinam, Tamil Nadu after the December 2004 great Sumatra earthquake and tsunami, 22(S3),* 715-729.

Praxis. (2008). *Accountability overdue, learnings from participatory engagement with the tsunami affected.* New Delhi: Praxis Publications.

Pushpanath, K. (1994). Disaster without memory: Oxfam's drought programme in Zambia. *Development in Practice, 4(2),* 81-91.

Putnam, R.D. (2001). Social capital: Measurement and consequences. *Isuma: Canadian Journal of Policy Research, 2,* 41-51.

Quarantelli, E. L. (1993). The environmental disasters of the future will be more and worse but the prospect is not hopeless. *Disaster Prevention and Management, 2,* 11- 25.

Quarantelli, E. L. (Ed.). (1998). *What is a disaster? Perspectives on the question.* London: Routledge.

Quarantelli, E.L. (1988). *Local emergency management agencies: Research findings on their progress and problems in the last two decades.* Preliminary Paper 126. Newark, DE: University of Delaware Disaster Research Center.

Quarantelli, E.L. (1995). *Disaster planning, emergency management and civil protection: The historical development and current characteristics of organized efforts to prevent and respond to disasters.* Preliminary Paper 227.Newark, DE: University of Delaware Disaster Research Center.

Quarantelli, E.L. (1999). Implications for programmes and policies from future disaster trends. *Risk Management, 1(1),* 9-19.

Rai, N., & Singh, A.K (Eds.). (2007). "*Disaster management in INDIA: Perspective, issues and strategies.* Lucknow: New Royal Book Company.

Ramchandran, H. (1990). *Environmental issues in agriculture.* New Delhi: Concept Publishing.

Rao, H. K., & Rao, P.S.S (Eds.). (2008). *Disaster management.* New Delhi: Serials Publications.

Rautela, P., & Pande, R.K. (2005). "Traditional inputs in disaster management: The case of Amparav North India. *International Journal of Environmental Studies, 62(5),* 505 – 515.

Ray, C.N. (2005). A note on the Disaster Management Bill, 2005. *Economic and Political Weekly, 40(47),* 4877-4881.

Rettger, M. J., & Boisver, R.N. (1979). *Flood insurance or disaster loans: An economic evaluation.* American Journal of Agricultural Economics, 61(3), 496-505.

Richards, D., & Smith, M. (2004). Interpreting the world of political elites. *Public Administration Review , 82 (4),* 777-800.

Ritchie, G. N. (1979). *Disaster and the third world*: *Third World Quarterly, 1(2),* 109-116.

Riyer, R. (2008). Floods, himalayan rivers, Nepal: Some heresies. *Economic and Political Weekly, 43(46),* 37-40.

Rogers, D. & Tsirkunov. V. (2011) *Costs and Benefits of Early Warning Systems* (Paper for UNISDR, 2011)

Rorabacher, Albert J., (2008), "Gerrymandering, Poverty and Flooding: A Perennial Story of Bihar", *Economic and Political Weekly,* February 16, Mumbai.

Roth, R. (1970). Cross cultural perspectives on disaster response. *American Behavioral Scientist, 13,* 440-451.

Roy, A. (2004). *Natural disaster management: A case study of Tornado.* New Delhi: Abhijeet Publications.

Roy, T. (2012). *Natural disasters and Indian history.* New Delhi: Oxford University Press.

Rubin, C.B., & Barbee, D. (1985). Disaster recovery and hazard mitigation: Bridging the intergovernmental gap. *Public Administration Review. 45,* 57-71.

Rubin, C.B., Drabek, T.E., & Hoetmer, G (Eds.). (1991). *Emergency management: principles and practice for local government.* Washington, D.C.: International City Management Association.

Sahni, P. (2001). *Views expressed in a faculty seminar.* New Delhi.

Sahni, P. (2002). *A report on initiatives on disaster risk management in India, (Un Published work).*

Sahni, P. (2007). Disaster preparedness: Role of education. In a workshop on *Enabling Sustainable Development through ODL, TechMODE* at UNESCO and Government of India. Ahmedabad: 4th International Conference in Environmental Education, Centre for Environment Education.

Sahni, P. (2008). Disaster research in South Asia: Major issues. In Roundtable Conference on *Researching Disasters: Prospects and Dilemmas.* Mumbai: Jamsetji Tata Centre for Disaster Management.

Sahni, P., Dhameja, A., & Medury, U (Eds.). (2001). *Disaster mitigation: Experiences and reflections.* New Delhi: Prentice-Hall of India.

Sahni, P., & Ariyabandu, M.M. (2003). *Disaster risk reduction in south asia.* New Delhi: Prentice-Hall India.

Sakurai, M. & Watson R. T. (2015). Paper on "*Securing Communication Channels in Severe Disaster Situations – Lessons from a Japanese Earthquake*" , University of Georgia. Retrieved from http:// iscram2015.uia.no/ wp-content /uploads / 2015/05/11-b.pdf

Sanchez, J. I., Korbin, W.P., & Viscarra, D.M. (1995). Corporate support in the aftermath of a natural disaster: Effects on employee strains. *The Academy of Management Journal,* 38(2), 504-521.

Saran, P. (1976). *The provincial government of the mughal, 1526-1658.* Lahore: Faran Academy.

Sarkar, S., & Sharma, A. (2006). Disaster Management Act: A disaster in waiting? *Economic and Political Weekly,* 41(35).

Sarun, S. (2011). "*Disaster risk communication over early warning technologies*" , A case study of coastal Kerala. Paper presented at the Disaster, Risk and Vulnerability Conference, Trivandrum, India. Retrieved from http:// drm. cenn.org/ pdf/

Satendra., & Sharma.V.K. (2004). *Sustainable rural development for disaster mitigation.* New Delhi: Concept Publishing Company.

Schmidt, V. (2008). Discursive institutionalism: The explanatory power of ideas and discourse. *Annual Review of Political Science, 11,* pp. 303-326.

Schwab, J.C., Gori, P.L., & Jeer, S (Eds.). (2005*).* Landslide hazards and planning, *planners.* Chicago: Book Service, American Planning Association.

Scott, J. (2000). Rational choice theory. In G. Browning, A. Halcli, & F. Webster (Eds.), *Understanding contemporary society: Theories of the present* (pp.126-139). London: Sage Publications.

SDMC. (2013). *South Asia disaster report 2011.* New Delhi: SDMC.

Second Administrative Reforms Commission. (2006). *Crisis management. From despair to hope.* Government of India. New Delhi.

SEEDS. (2002). Corporate social responsibility and disaster reduction: An Indian overview. Retrieved from http://r4d.dfid.gov.uk/PDF/Outputs/Mis_SPC/R7893CSRIndia.pdf

Settle, A.K. (1985). Financing disaster mitigation, preparedness, response, and recovery. *Public Administration Review, 45,* 101-106.

Shaluf, I.M. (2007). *An overview on disasters. Disaster Prevention and Management, 16(5),* 687-703.

Sharma, V.K. (2001). *Disaster management.* New Delhi: Indian Institute of Public Administration.

Shaw, R. (2006). Recovery from the Indian ocean tsunami disaster. *Disaster Prevention and Management,* 15(1)

Sheets, H., & Morris, R. (1974). Disaster in the desert. *A Journal of Opinion, 4(1),* 24-43.

Simon, H.A. (1957). *Models of man: Social and rational.* New York: John Wiley and Sons, Inc.

Singh, R.B (Ed.). (2006). *Natural hazards and disaster management: Vulnerability and mitigation.* New Delhi: Rawat Publications

Singh, T. (1995). *Drought disaster and agricultural development in India.* New Delhi: People's Publishing House.

Sinha, R. (2008). Kosi: Rising waters, dynamic channels and human disasters. *Economic and Politickal Weekly, 43(46),*42-46.

SIPA (2013) "*Communicating During Disasters Examining the Relationship between Humanitarian Organizations and Local Media*", Columbia University's School of International and Public Affairs . Retrieved from www.internews.org/ sites/ default/ files/ resources/ Internews_SIPA_communicating_disastes_2013-09. pdf

Smith, K. (1996). *Environmental hazards.* London: Routledge.

Smith, W., & Dowell, J. (2000). A case study of coordinative decision-making in disaster management. *Ergonomics, 43(8),* 1153-1166.

Songer, T. (2003). *Epidemiology of disasters.* Retrieved from http://www.pitt. edu/~epi2170/lecture15/sld001.htm

Stallings, R. A., & Quarantelli, E. L. (1985). Emergent citizen groups and emergency management. *Public Administration Review , 45,* 93-100.

Statistical Hand Book of Tamil Nadu. (2012). Retrieved from http://www.tn.gov. in/deptst/

Stephen, A. (2012). Natural disasters in India with special reference to Tamil Nadu. *Journal of Academia and Industrial Research, 1(2),* 59-67.

Stephens, L.H., & Green, S.J (Eds.). (1979). *Disaster assistance, appraisal, reform and new approaches.* New York: MacMillon Press.

Strydom, P. (2002). *Risk, environment and society.* Buckingham: Open University Press.

Suarez, P., Givah, P., Storey, K., and Lotsch, A. (2008). *HIV/AIDS, climate change and disaster management: Challenges for institutions in Malawi.* Policy Research Working Paper 4634. Washington: The World Bank Development Research Group Sustainable Rural and Urban Development Team.

Susana A. Barrantes (2009) (Eds.) *Information management and communication in emergencies and disasters: Manual For Disaster Response Teams Social Communicator*. Washington, D.C. At http:// reliefweb.int/ sites/ reliefweb. int/ files/resources

Tampere 1998, The Intergovernmental Conference on Emergency Telecommunications (ICET- 98) held on 18 June, 1998 , Finland.

Tang A.P. (2007) paper on "*New trends of seismic disaster management in China*" (Geophysical Research Abstracts, Vol. 9, 11225, 2007 SRef-ID: 1607-7962/ gra/EGU2007-A-11225 © European Geosciences Union 2007, Harbin, China.

Task Force (2013). *Report of the Task Force on Review of the Disaster Management Act, 2005*, Ministry of Home Affairs, Government of India. New Delhi

Telecommunications Reform 2003, Geneva , ITU.

The Hindu. (2013, July 19). *Supreme court issues notice to centre, states on disaster management.* Chennai.

Times of India. (2013, May 15). *India: CAG hauls up Tamil Nadu for poor disaster management systems.* Chennai.

TISS. (2005). *The state and civil society in disaster response: An analysis of the Tamil Nadu tsunami experience.* Mumbai: TISS.

Tobin, G.A. (1999) 'Sustainability and Community Resilience: the Holy Grail of Hazards Planning?' *Environmental Hazards*, vol 1, pp. 13-25.

Tocqueville, A. (2000). Democracy in America. Chicago: University of Chicago Press.

TRAI (2012), Telecom Regulatory Authority Of India Consultation Paper On " *Telecom Network Failures During Emergencies / Disasters Priority Call Routing Of Calls Of Persons Engaged In Response And Recovery".* New Delhi, Government of India.

TRAI (2013) Telecom Regulatory Authority Of India Consultation Paper On *"Universal Single Number Based Integrated Emergency Communication and Response System".* New Delhi, Government of India.

TRAI(2010), "*Recommendation on National Broadband plan*" New Delhi, Government of India.

Tsunami, Disaster Management Task Force Plan, Nagapattinam district. (n.d.). Retrieved from http:// www.tn.gov.in/tsunami/ digitallibrary/ebooksweb

Twigg J. (2003). ' *Lessons from disaster preparedness*' Notes for presentation to workshop 3: 'It pays to prepare' at international conference on climate change and disaster preparedness. Hague Netherlands.

Twigg, J. (2001). *Corporate social responsibility and disaster reduction: A global overview.* Londong: Benfield Greg Hazard Research Centre, University College.

UN Country Team for India. (2005). '*Recovery framework in support of Government of India for post-tsunami rehabilitation and reconstruction programme*'. Retrieved from http://www.undp.org/content/dam/undp/documents/projects/IND/00037989/UNCT%20recovery%20frame%20work.pdf

UN. (1987). *Report of the world commission on environment and development, Our common future.* Retrieved from http://conspect.nl/pdf/Our_Common_Future-Brundtland_Report_1987.pdf

UN. (1989).*85th Plenary Meeting*. Retrieved from http: //www.un.org /documents /ga/res/44/a44r236.htm

UN. (1999). *UN Press release, SM/SM/7060*. Retrieved from http: //www.un.org / News /Press/docs/1999/19990706.SGSM7060.html

UNDP (2005), "*Human Development Report*", at hdr.undp.org/en/media/hdr_2005

UNDP (2007) "*Human Development Report*", at hdr.undp.org/en/media/hdr_2007

UNDP. (1992). *Human development report.* New York: Oxford University Press.

UNDP. (2004). *Human development report 2004: Cultural liberty in today's diverse world.* Retrieved from http://hdr.undp.org/en/media/hdr04_complete.pdf

UNDP. (2005). *CRZ notification and post-tsunami rehabilitation and reconstruction-A zone of contention.* New Delhi.

UNDP. (2005). *Evolving strategies for long-term rehabilitation on shelter and habitat development in the tsunami affected areas of Tamil Nadu*. New Delhi.

UNDP. (2005, January 19). *Tsunami sets back development by 20 years in Maldives.* United Nations Press Release.

UNDRO. (1982). *Natural Disasters and Vulnerability Analysis.* Geneva: Office of United Nations Disaster Relief Coordinator.

UNISDR Handbook (2010) on Early warning systems: "*Early Warning Practices can Save Lives: Selected Examples Good Practices and Lessons Learned Bonn, August 2010.* Retrieved from: http://www.unisdr.org/

UNISDR. (1994). *Yokohoma strategy and plan of action for a safer world: Guidelines for natural disaster prevention, preparedness and mitigation*. Retrieved from http://www.unisdr.org/files/8241_doc6841contenido1.pdf

UNISDR. (2004). *Living with risk: A global review of disaster reduction initiatives.* Retrieved from http://www.unisdr.org/files/657_lwr21.pdf

UNISDR. (2005). *Hyogo framework for action 2005-2015: Building the resilience of nations and communities to disasters*. Retrieved from http://www.unisdr.org/files/1037_hyogoframeworkforactionenglish.pdf

UNISDR. (2009). *Global assessment report on disaster risk reduction*. Retrieved from www.preventionweb.net/english/hyogo/gar/report/index.php?id=9413&pid:34&pif:3

UNISDR. (2009). *UNISDR terminology for disaster risk reduction*. Retrieved from: http://www.unisdr.org/we/inform/publications/7817

UNISDR. (2012). *Analysis of National Legislation, agenda, and development plans from Asia and the Pacific: A background paper for the Asia-Pacific Disasters Report 2012*. Bangkok, Thailand: Berse.

University of Roorkee. (2000). *A report on Chamoli earthquake of March 29,1999*. Roorkee: Department of Earthquake and Engineering.

Urban Risk Forum. (2000). *Disaster briefing notes*. Oxford, UK: Oxford Center for Disaster Studies.

Ursano, R.J., & Fullerton, C.S. (2003). *Terrorism and disaster: Individual and community mental health interventions*. Cambridge: Cambridge University Press.

Ursano, R.J., & Norwood, A.E. (2003). *Trauma and disaster: Response and management*. Virgina: American Psychiatric Publishing, Inc.

Vale, L.J., and Campanella, T.J (Eds.). (2005). The resilient city: How modern cities recover from disaster. London: Oxford University Press.

Vanderford M., Nestoff T.,Telfer J.L. and Bonzo S,E, (2007) Emergency Communication Challenges in response to Hurricane Katerina: Lessons for the Centre For Disease Control and Prevention. Journal on applied communication research.

Verchick, R.R.M. (2007). *Risk, fairness and the geography of disaster*. International Law Forum of the Hebrew University of Jerusalem Law Faculty Research Paper Series No. 01-07. Retrieved from http://papers.ssrn.com/sol3/papers.cfm?abstract_id=959247

Veronica F. G. & Singh .A. (2012) Early warning Systems : State -of -Art Analysis and Future Directions , United Nations Environment programme (UNEP) available http://www.unep.org

Wallace, W. A., & Balogh, F.D. (1985). Decision support systems for disaster management. *Public Administration Review, 45,* 134-146.

Wattegama Chanuka (2007) "ICT for Disaster Management" for UNDP-APDIP and Asian and Pacific Training Centre for Information and Communication Technology for Development (APCICT) – Website: http://www.apdip.net

Watts, M. (1983). On the poverty of theory: Natural hazards research in context. In K. Hewitt (Ed.), *Interpretations of Calamity*. Boston: Allen and Unwin.

Waugh Jr, W. L., & Streib, G. (2006). Collaboration and leadership for effective emergency management. *Public Administration Review , 66(s1),*131-140.

Waugh, W. L. (2006). The Political costs of failure in the response to hurricanes Katrina and Rita. *Annals of the Academy of Political and Social Science 60(4),* 10-25.

WCNIS (2010) : Wireless Communications Networking and Information Security; an international conference of Institute of Electrical and Electronics Engineers (IEEE)

WDR (World Disaster Reports) 1997, 1998, 1999, 2002, 2003, 2004, 2005, 2007, 2009, 2012 , 2013 International Red Cross and Red Crescent Society, Geneva. Retrieved from http://www.ifrc.org/Global/Publications/disasters/WDR/69001-WDR2005-english-LR.pdf , http://www.ifrc.org/PageFiles/99876/WDR2007-English.pdf

Whitmore, G.A. (2008). *Using principles of risk management and engineering reliability to reduce disaster risks.* Retrieved from http://papers.ssrn.com/sol3/papers.cfm?abstract_id=1032157

WHO. (1989). Coping with natural disasters: The role of local health personnel and the community. Geneva: League of Red Cross and Red Crescent Societies.

Wisner, B., Blaikie, P., Cannon, T. & Davis, I. (2003) *At Risk. Natural hazards, People's Vulnerability and Disasters*, Routledge, London.

Wolfenstein, M. (1957). *Disaster: A psychological essay*, New York, USA: The Free Press.

World Bank Handbook (2010*) "Safer Homes, Stronger Communities: A Handbook for Reconstructing after Natural Disasters"*, published in January 2010. https://openknowledge.worldbank.org

World Bank Report on Disaster (2009) Available at http://www.worldbank.org/

Yin, R. (2002). Case study research. Thousand Oaks, California: Sage Publications

Zeitschrift für Soziologie (2012) "*Communicating Disaster – A Case for Qualitative Approaches to Disaster Research*" Report of a Research Group at the Center for Interdisciplinary Research (ZiF), Bielefeld University. Retrieved from http://geo.aau.at/ images/ stories/ pdf/

Websites

http://www.adpc.at.in

http://www. bmtpc.org

https://books.google.co.in

http://www.dfid.gov.uk

http://www.disaster-info.net

http://www.disasterlesskerala.org

http://www.dmc.engr.wisc.edu

http://economictimes.indiatimes.com

http://education-portal.com

http://www.emdat.be

http://www.en.wikipedia.org

http:// www. eqs. metu.edu.tr

http://www.fao.org/sd

http://www.fema.gov

http://www.gdrc.org

http://www.icrc.org

http://www.ifrc.org

http://www.imd.gov.in

http://www.infochangeindia.org

http://www.iom.in

http://www.isdr.org

http://www.isuma.net

http://www.itu.int

http://www.mapsofindia.org

http://www.mindtools.com

http://www.mospi.nic.in

http://www.ncrc.in

http://www.ncmrwf.gov.in

http://www.ndma.gov.in

http://www.ndmindia.nic.in

http://www.nidm.net

http://www.nio.org/jsp

http://www.paho.org

http://www.panda.org

http://www.primidi.com

http://www.ready.gov

http://www.resource-alliance.org

http://www.sahana.lk

http://www.static.teriin.org

http://www.tnmaps.tn.nic.in

http://www.trai.gov.in

http://www.training.fema.gov

http://www.un.org

http://www.undp.org

http://www.unep.org

http:// www.unfccc.int

http://www.unisdr.org

http://unpan1.un.org/

http://www.usaid.gov

http://www.us.oneworld.net

http://www.usgs.gov

http://wiki.sahanafoundation.org

https://www.wmo.int

https://www.webopedia.com

INDEX

F

G

H

I

AUTHORS PROFILE

Dr Preeti Banzal is Director in Telecommunication Engineering Centre under the Department of Telecommunications, Ministry of Communications, Government of India, New Delhi. She is B.E. (Hons) (Electronics and Telecommunications) , MBA (HR) and PhD. She has got her Ph.D. for her research on "Disaster Communication and Early Warning Systems in India: A Prognostic Analysis" from the Indira Gandhi National Open University. She has about over 25 years of professional experience in Telecom Technology Operations, Equipment Installations, Project Planning, Implementation and Management, Administration, Corporate Restructuring, Capacity Building and Trainings Management. While she worked at the Advanced Level Telecom Training Centre, she coordinated many capacity building and training programmes with reputed international telecom organisations. After her graduation she had taught as a lecturer in Government College of Engineering and technology (NIT Raipur, CG, India). She has also worked at the National Disaster Management Authority as Director and Joint Advisor (Finance, Capacity Building and Training). At NDMA, she implemented capacity building projects with IGNOU, Ministry of Earth sciences, IITs, CSIR , INCOIS, LBSNAA Mussoorie, AIIMS Trauma centre New Delhi, UNDMT, UNFPA etc. Dr. Preeti Banzal has written many papers and articles on diverse topics related to communication and management.

Dr Anil K Gupta is Head of Policy Planning Division Head of Training and Capacity Building at Government of India's National Institute of Disaster Management, under the Ministry of Home Affairs at New Delhi, as Associate Professor of Policy Instruments, Environment & Climate Change. He is Nodal Officer from NIDM for the Ministries - Environment Forests & Climate Change, Science & Technology, Agriculture and Farmers Welfare, in Government of India. He is Core Group Member of Council of Ecosystem Management at IUCN South Asia, and Fellow, Society of Earth Scientists. He was Reader & Director of Institute of Environment and Development Studies at Bundelkhand University under Government of Uttar Pradesh. He also worked at Government of Madhya Pradesh (Disaster Management Institute), PSU (National Mineral Development Corporation), CSIR-National Environmental Engineering Research Institute Nagpur and Lucknow based BB Ambedkar Central University at academic and administrative positions. He steered international projects – ekDRM (GIZ Germany), Climate Resilient DRR (CDKN UK), UNDP, UNEP, ESCAP, and various Ministries. W of CCA-DRR, and Ecosystem Approach (ecoDRR), and known for the National Disaster Management Plan, National Human Resource Plan, and guiding training planning, need analysis, disaster and climate action planning and implementation at national and sub-national levels. He has more than 170 publications to his credit including 22 books and 47 papers. He has been recipient of Young Scientist Award 1996 in Environment & Forestry from Government of Madhya Pradesh and Excellence Award of the Indian Society of Environmental & Occupational Health.

www.ingramcontent.com/pod-product-compliance
Lightning Source LLC
Chambersburg PA
CBHW071239230926
51849CB00001B/158